Land Surface Remote Sensing in Continental Hydrology

Remote Sensing Observations of Continental Surfaces Set

coordinated by
André Mariotti

Land Surface Remote Sensing in Continental Hydrology

Edited by

Nicolas Baghdadi
Mehrez Zribi

ISTE PRESS

ELSEVIER

First published 2016 in Great Britain and the United States by ISTE Press Ltd and Elsevier Ltd

ISTE Press Ltd
27-37 St George's Road
London SW19 4EU
UK

www.iste.co.uk

Elsevier Ltd
The Boulevard, Langford Lane
Kidlington, Oxford, OX5 1GB
UK

www.elsevier.com

British Library Cataloguing-in-Publication Data
A CIP record for this book is available from the British Library
Library of Congress Cataloging in Publication Data
A catalog record for this book is available from the Library of Congress
ISBN 978-1-78548-104-8

Contents

Foreword

I have been entrusted by ISTE Science Publishing with the responsibility for a multidisciplinary editorial line: Earth System – Environment, and within this framework it gives me great pleasure today to present a set of books dedicated to the topic of remote sensing, compiled and edited by Nicolas Baghdadi and Mehrez Zribi.

Both the content and the organization of this collection have largely been inspired by reflections, analyses and prospective works conducted by almost 200 authors and researchers with a high level of international expertise in this discipline.

This community, which is recognized for its scientific merit, has sought to expand its research activities under the direction the two editing authors founded on a solid effort in the area of acquisition and wider dissemination of knowledge within this field.

This represents a community characterized by the firm commitment to adopting a holistic or even an ecosystem approach within the context of an interdisciplinary science of the Earth system. In this scientific context where the complexity of natural systems is compounded with the complexity of societies, the authors have given careful consideration to depicting a finalizable and "public" type of discipline, open to decision makers, managers and all those beyond the scientific community who are interested in the future of our planet.

Two main tools are necessary in order to satisfy the requirements in terms of understanding and characterizing our environment and its evolution: process modeling and observation.

Remote sensing observations in conjunction with measurements and modeling constitute a discipline that makes it possible to understand the functional properties of the observed system and their dependence on its structural properties. This is one of the key disciplines that allow the analysis and provide access to the understanding of the functioning of our environment: in general, this is dedicated to aspects such as the analysis of climate change, the effects of anthropogenic and demographic pressure, natural disasters, the increasing decline in resources (water, etc.), the degradation of biodiversity across all environments, desertification, the need to nourish the planet (for example, mapping of crops and yield prediction), etc.

Analyzing and understanding these different types of problems is rendered possible:

– by analyzing the detected, structural and functional objects (soils, hydrosystems, vegetation, etc.);

– by understanding the main basic processes, which incorporate these main elements: water flow covering all scales and compartments, erosion, meteorology, crop development, soil pollution, etc;

– by developing indicators in order to evaluate the short-, medium- and long-term evolution of all environmental compartments and variables.

The importance of these scientific questions has led to a general mobilization of the international organizations by means of various international conventions and agreements to protect the environment and meet the specific requirements in terms of observation. Various international networks have been developed over the past few years with the purpose of conducting continuous measurements. However, these punctual measurements could not provide sufficient spatiotemporal monitoring, in particular in difficult-to-access regions. Within this context, spatial observation could be implemented to its full potential, both by means of considerable progress in terms of instrumentation and by means of the development of effective data processing and analysis methods, data whose provision becomes increasingly free of rights.

Under the initiative of numerous space agencies (in particular European, North-American, Japanese, etc.), important space missions were launched for the purpose of conducting Earth observations, among which the following may be mentioned:

– Sentinel, within the framework of the Copernius program (formerly referred to as GMES for Global Monitoring for Environment and Security) implemented in numerous areas such as land and marine environment monitoring, emergency management (for example, natural disasters) and climate change monitoring (radar and optical imaging);

– Landsat;

– ALOS, launched by the Japan Aerospace Exploration Agency, in particular for deforestation monitoring;

– SMOS and SMAP, in particular for the global mapping of soil moisture, etc.

Although remote sensing represents a field in which specialist knowledge is required in order to conduct a better analysis and interpretation of data, this programmatic development is undoubtedly associated with a significant progress with respect to the implementation of space-based Earth observations at the level of an increasing number of laboratories across both developed and emerging nations. This development is likewise associated with new disciplines and thematic backgrounds, among which numerous areas of the humanities and social sciences which enrich and extend the primarily physical foundations of remote sensing may be mentioned.

It would be superfluous to list all remote sensing applications along with the disciplines and scientific questions which adopted this concept, as this would also inevitably result in regrettable omissions: nonetheless, the extensive implementation of spatial observation grants the latter a strong interdisciplinary status.

The launch of new large-scale space missions, the higher degree of convenience, including financial convenience, as well as the access to data will facilitate an intensification and generalization of the use of spatial observation data and products: new scientific subjects, new users (managers, decision makers, etc).

The high demand for educational material containing updated information on the various remote sensing concepts and methods and the main applications thereof, in particular at the level of continental surfaces, are derived therefrom.

It is within this framework that this collection of books is proposed, which aims to provide researchers, students in masters, engineer and PhD programs, as well as decision makers, engineers specialized in management services on a territorial, departmental, regional or national scale and players in the decision-making authorities with a tool which incorporates both the foundations of the physical principles underlying various spatial applications and the implementation methods and exemplification at the level of various applications based on spatial observation.

In these six volumes, Nicolas Baghdadi and Mehrez Zribi have mobilized almost 200 internationally recognized researchers to propose a comprehensive "toolkit", describing the latest scientific methods and actions in terms of the implementation of spatial observation.

The first two volumes describe the physical principles underlying various techniques which cover the frequency spectrum ranging from visible to microwaves. The third volume illustrates the agricultural and forestry applications of spatial observation. The fourth volume presents the applications of spatial observation in the field of continental hydrology. The fifth volume is dedicated to the observation of urban and coastal areas, whereas the final volume presents the implementation of spatial observation within the context of risk assessment and understanding.

Thanks are due to Nicolas Baghdadi and Mehrez Zribi for taking the time to draft, harmonize and partially edit these volumes and committing to this effort in terms of putting this modern and high-quality knowledge across and making it accessible to a diverse and vast scientific audience.

I wish to thank them both for their altruism, perseverance and devotion in service of the success of this endeavor.

André MARIOTTI

June 2016

Acronyms

2D	Two dimensions
3D	Three dimensions
4AOP	Automatized Atmospheric Absorption Atlas Operational
6S	Second Simulation of Satellite Signal in the Solar Spectrum
AATSR	Advanced Along-Track Scanning Radiometer
ACORN	Atmospheric Correction Now
ADC	Analog-to-digital converter
ADCP	Acoustic Doppler Current Profiler
ADEME	French Environment and Energy Management Agency
ADEOS	Advanced Earth Observing Satellite
AERONET	Aerosol Robotic Network
AET	Actual Evapotranspiration
AFRITRON	African Tropical Rainforest Observation Network
AGB	Above-Ground Biomass
AGNES	Agglomerative Nesting
AHS	Airborne Hyperspectral Scanner
AHT	Astronomical High Tide
AIEM	Advanced Integral Equation Model
AirSAR	Airborne Synthetic Aperture Radar
ALB	Airborne LiDAR Bathymeter
ALEXI	Atmosphere-Land Exchange Inverse
ALS	Airborne Laser Scanning

AltBOC	Alternate Binary Offset Carrier
AMARTIS	Advanced Modeling of the Atmospheric Radiative Transfer for Inhomogeneous Surfaces
AMMA	African Monsoon Multidisciplinary Analysis
AMSR	Advanced Microwave Scanning Radiometer
ANA	*Agência Nacionalde Aguas* (Brazilian National Water Agency)
AOL	Airborne Oceanographic LiDAR
APD	Avalanche Photodiode
API	Antecedent Precipitation Index
APOM	Aerosol Plume Optical Model
ARVI	Atmospherically Resistant Vegetation Index
ASAR	Advanced Synthetic Aperture Radar
ASCAT	Advanced Scatterometer
ASDF	Averaged Square Difference Function
ASI	*Agenzia Spaziale Italiana* (Italian Space Agency)
ASI	Agriculture Stress Index
ASTER	Advanced Spaceborne Thermal Emission and Reflection Radiometer
ATCOR	Atmospheric and Topographic Correction
ATLAS	Advanced Topographic Laser Altimeter System
ATM	Airborne Topographic Mapper
ATREM	Atmospheric Removal (atmospheric correction method)
ATSR	Along Track Scanning Radiometers
AUC	Area Under Curve
AVHRR	Advanced Very High Resolution Radiometer
AVIRIS	Airborne Visible/Infrared Imaging Spectrometer
AWiFS	Advanced Wide Field Sensor
BEIDOU/ COMPASS	Chinese Navigation Satellite System
BELA	BepiColombo Laser Altimeter
BGB	Below-Ground Biomass
BIL	Band Interleaved by Line
BIOMOD	Biomolecular Design (Computer platform implemented in R)
BIP	Band Interleaved by Pixel

BLUE	Best Linear Unbiased Estimator
BOC	Binary Offset Carrier
BPSK	Binary Phase Shift Keying
BPT	Binary Partition Tree
BRDF	Bidirectional Reflectance Distribution Function
BRGM	*Bureau de Recherches Géologiques et Minières* (French Geological Survey)
BSA	Back Scatter Alignment
BSQ	Band Sequential
BUI	Build-up Index
BWI	Bassist Wetness Index
C/A	Coarse Acquisition
CaCO3	Calcium Carbonate
CAIC	Consistent Akaike's Information Criterion
CALIOP	Cloud-Aerosol LiDAR with Orthogonal Polarization
CALIPSO	Cloud-Aerosol LiDAR Pathfinder Satellite Observation
CanEx-SM	Canadian Experiment for Soil Moisture
CART	Classification and Regression Trees
CASI	Compact Airborne Spectrographic Imager
CBERS	China–Brazil Earth Resources Satellite
CBOC	Composite Binary Offset Carrier
CCD	Charged Coupled Devices
CCDAS	Carbon Cycle Data Assimilation System
CCI	Climate Change Initiative
CCRS	Canada Center for Remote Sensing
CDF	Cumulative Density Function
CDMA	Code Division Multiple Access
CDOM	Colored Dissolved Organic Matter
CEC	Cation-Exchange Capacity
CEM	Constrained Energy Minimization
CEOS	Committee on Earth Observation Satellites
CERES	Crop Environment Resource Synthesis

CESBIO	*Centre d'Études Spatiales de la Biosphère* (Center for the Study of the Biosphere from Space)
CFC	Chlorofluorocarbons
CFD	Constant Fraction Discriminator
CFFDRS	Canadian Forest Fire Danger Rating System
cGNSS	Conventional Global Navigation Satellite Systems
CHAMP	Challenging Minisatellite Payload (German geosciences satellite)
CHL	Chlorophyll Content
CHM	Canopy Height Model
CHRIS	Compact High Resolution Imaging Spectrometer
CIRAD	Agricultural research for development (France)
CLASlite	Carnegie Landsat Analyse System Lite
CLM	Community Land Model
CLMGW	Community Land Model with a Ground Water Parameterization
CLPX	Cold Land Processes Field Experiment
CLS	Collecte Localisation Satellite
CLSM	Catchment Land Surface Model
CM	Code Moderate (for GNSS)
CMC	Canadian Meteorological Centre
CMEM	Community Microwave Emission Model
CNES	*Centre National d'Études Spatiales* (French Space Agency)
CNRS	*Centre National de Recherche Scientifique* (National French Center for Scientific Research)
COD	Controlled Origin Designation
COG	Center of Gravity
CONUS	Contiguous United States
COST	Cosine Estimation of Atmospheric Transmittance
CR	Continuum Removed
CRI	Carotenoid Reflectance Index
Cryosat	Satellite Radar Altimeter
CSA	Canadian Space Agency
CSR	Centre for Space Research
CTA	Classification Tree Analysis

CTFC	Center for Tropical Forest Science (Cameroon)
CTFS	Center for Tropical Forest Science
CTMF	Cluster Tuned Matched Filter
CW	Continuous Waves
CWFIS	Canadian Wildland Fire Information System
CyGNSS	NASA's Cyclone Global Navigation Satellite System
CZCS	Coastal Zone Color Scanner
CZMIL	Coastal Zone Mapping and Imaging LiDAR
DAM	Dry Aerial Mass
DART	Discrete Anisotropic Radiative Transfer
DATAR	Interministerial Delegation for Territorial Planning and Regional Attractiveness
DBH	Diameter at Breast Height
DC	Drought Code
DCA	Dual Channel Approach
DCM	Digital Canopy Model
DDM	Delay Doppler Map
DEGRAD	Forest Degradation program
DEIMOS	Deep Imaging Multi-Object Spectrograph
DEM	Digital Elevation Model
DEOS	Delft Institute of Earth Observations and Space Systems
DERD	Double bounce Eigenvalue Relative Difference
DETER	*Detecçao de Desmatamento em Tempo Real* (Near real-time deforestation detection system)
DGPS	Differential Global Positioning System
DHSVM	Distributed Hydrology Soil-Vegetation Model
DIACT	Inter-ministerial Agency for Spatial Planning and Competitiveness
DIANA	Divise Analysis
DIC	Digital Image Correlation
DIMAP	Digital Image MAP
DInSAR	Differential SAR Interferometry
DisALEXI	Disaggregated Atmospheric Land Exchange Inverse
DLIS	Desert Locust Information Service

DLR	*Deutsches Zentrum für Luft und Raumfahrt* (German Space Agency)
DM	Dry Matter
DMC	Disaster Monitoring Constellation or Duff Moisture Code, depending on the application
DMRT	Dense Media Radiative Transfer
DMSP	U.S. Air Force Defense Meteorological Satellite Program
DMU	De Monfort University
DOAS	Differential Optical Absorption Spectroscopy
DOS	Dark Object Substraction
DOY	Day-of-Year
DPSS	Diode-Pumped Solid-State
DS	Diffuse Scatterer
DSM	Digital Soil Mapping
DSM	Digital Surface Model
DTC	Dry Troposphere Correction
DTED	Digital Terrain Elevation Data
DTM	Digital Terrain Model
DVI	Difference Vegetation Index
DW	Dry Weight
DWBA	Distorted Wave Born Approximation
EAARL	Experimental Advanced Airborne Research LiDAR
ECDC	European Centre for Disease Prevention and Control
ECMWFMMT	European Centre for Medium-Range Weather Forecasts Mobile Mapping Technology
ECV	Essential Climate Variables
EDF	*Électricité de France* (French electric company)
EEA	European Environment Agency
EFA	Effective Field Approximation
EFFIS	European Forest Fire Information System
EGNOS	European Geostationary Navigation Overlay Service
EID-Méditerranée	Interdepartmental Agreement for Mosquito Control on the Mediterranean Coast
EKF	Extended Kalman Filter

ELBARA	ETH L-Band Radiometer
ELUE	Effective Light Use Efficiency
EM	Electromagnetic
ENEA	*Energia Nucleare ed Energie Alternative* (Italian National Agency for New Technologies, Energy and Sustainable Economic Development)
EnKF	Ensemble Kalman Filter
ENMAP	Environmental Monitoring and Analysis Program
ENSO	El Niño Southern Oscillation
ENVEO	Environmental Earth Observation Information
ENVISAT	Environmental Satellite
EOS	Earth Observing System
EPICA	European Project for Ice Coring in Antarctica
EPS	European Polar System
EQeau	Model developed by INRS-Ete with the objective to extract the soil water content from SAR images
ERM	Exact Repeat Missions
ERS	European Remote-sensing Satellite
ESA	European Space Agency
ESCAT	ERS Scatterometer
ESSA	Environmental Science Services Administration Satellite
ET	Evapotranspiration
ETM	Enhanced Thematic Mapper
EUFAR	European Facility for Airborne Research
EUMETSAT	European Organization for the Exploitation of Meteorological Satellites
EVASPA	Evapotranspiration Assessment from Space
EVI	Enhanced Vegetation Index
EWT	Equivalent Water Thickness
EZW	Embedded Zerotrees of Wavelet transforms
FAA	Federal Aviation Administration (USA)
FAI	Floating Algae Index
FAO	Food and Agriculture Organization of the United Nations
FAPAR	Fraction of Absorbed Photosynthetically Active Radiation

FBD	Fine-Beam Double Polarization
FBP	Fire Behavior Prediction
Fcover	Fraction of Vegetation Cover
FDTD	Finite Difference Time Domain
FFMC	Fine Fuel Moisture Code
FFT	Fast Fourier Transform
FI	Fine Particles Index
FIPAR	Fraction of Intercepted Photosynthetically Active Radiation
FLAASH	Fast Line-of-sight Atmospheric Analysis of Spectral Hypercubes (Atmospheric correction)
FLOE	Fish LiDAR Oceanographic Experimental
FMA	Fore-Mid-Aft
FMCW	Frequency Modulated Continuous Waves
FOTO	Fourier-based Textural Ordination
FOV	Field of View
FSA	Forward Scatter Alignment
FST	Spectro-Transfert functions
FT	Fourier Transform
FW	Full Waveform or Fresh Weight, depending on the application
FWHM	Full Width at Height Maximum
FWI	Fire Weather Index
FWS	Fraction of Water Surface
GAGAN	GPS Aided GEO Augmented Navigation system
GAI	Green Area Index
GALILEO	European Global navigation satellite system
GCOS	Global Climate Observing System
GCP	Ground Control Points
GDAL	Geospatial Data Abstraction Library
GDR	Geophysical Data Record
GEO	Group on Earth Observations
GEOCAPI	Geostationary Ocean Color Advanced Permanent Imager
GEOGLAM	Group on Earth Observations Global Agricultural Monitoring
GEOSS	Global Earth Observing System of Systems

GEOSUD	Geoinformation for Sustainable Development (France)
GEOTIFF	Geostationary Earth Orbit Tagged Image File Format
GEROS-ISS	GNSS Reflectometry, Radio Occultation and Scatterometry Onboard International Space Station
GF	Green Fraction
GFO	Geosat Follow-On (radar altimeter satellite)
GFZ	*GeoForschungs Zentrum* (German research centre for Geosciences)
GHG	Green House Gase
GIAM	Global Irrigated Area Map
GIEWS	Global Information and Early Warning System on Food and Agriculture
GIM	Global Ionospheric Model
GIMMS	Global Inventory Modeling and Mapping Studies
GIS	Geographic Information System
GLAI	Green Leaf Area Index
GLAS	Geoscience Laser Altimeter System
GLCM	Grey Level Co-occurrence Matrix
GLDAS	Global Land Data Assimilation System
GlobCover	Global Land-Cover Map
GLONASS	Global Navigation Satellite System (Russia)
GLP	Global Land Project
GLRT	Generalized Likelihood Ratio Test
GLS	Global Land Survey (Landsat)
GM	Geodetic Mission
GMES	Global Monitoring for Environment and Security
GMFS	Global Monitoring of Food Security
GMM	Gaussian Mixture Model
GNSS	Global Navigation Satellite System
GNSS-R	Global Navigation Satellite System – Reflectometry
Go	Giga-Octet (1,000,000,000 octets)
GOCE	Gravity field and Steady-state Ocean Circulation Explorer (ESA)
GOES	Geostationary Operational Environmental Satellite
GOFC Fire IT	Global Observation of Forest Cover Fire Implementation Team

GOM	Geometrical Optics Model
GORS	GNSS Occultation Reflectometry Scatterometry
GPCC	Global Precipitation Climatology Centre
GPP	Gross Primary Production
GPR	Gaussian Process Regression
GPS	Global Positioning System
GRACE	Gravity Recovery and Climate Experiment (satellite)
GRDC	Global Runoff Data Center
GRGS	Space Geodesy Research Group
GtC	Gigatonnes of Carbon (10^9 tons)
GUS	Ground Uplink Stations
GVI	Difference Vegetation Index
GWIS	Global Wildfire Information System
Ha	Hectare (= 10,000 m^2)
HCFC	Hydro-chlorofluorocarbons
HEC-RAS	Hydrologic Engineering Center River Analysis System (hydraulic model)
HgCdTe	Mercury Cadmium Telluride
HiRI	High Resolution Optical Imager
HITRAN	High-Resolution Transmission Database
HOG	Histogram of Oriented Gradient
HPC	High Performance Computing
HRG	High Resolution Geometry
HSC	Height-Scaled Crown
HSCOI	Height-Scaled Crown Openness Index
HSR	High Spatial Resolution
HYMAP	Hyperspectral Mapper (airborne hyperspectral sensor)
HYPXIM	Hyperspectral-X Imagery
HYSPEX	Hyperspectral Imaging System
HyspIRI	NASA's Hyperspectral Infrared Imager
IASI	Infrared Atmospheric Sounding Interferometer
ICA	Independent Component Analysis
ICARE	Inversion Code for urban Areas Reflectance Extraction

ICESat	Ice, Cloud and Land Elevation Satellite
ICF	Interferometric Complex Field
ICP	Iterative Closest Point technique
IDAN	Intensity Driven Adaptative Neighborhood
IEM	Integral Equation Model
IFN	French National Forest Inventory
IFOV	Instantaneous Field of View
IGN	French National Geographic Institute and Forest Information
iGNSS	Interferometric GNSS
IHS	Intensity, Hue, Saturation
IMU	Inertial Measurement Unit
INERIS	French National Institute for Environmental Protection and Industrial Risks
InGaAs	Indium Gallium Arsenide
INPE	*Instituto Nacional de Pesquisas Espaciais* (Brazilian Institute of Space Research)
INRA	French National Institute for Agricultural Research
INSAR	Interferometric Synthetic Aperture Radar
InSb	Indium antimonide
INSEE	French National Institute for Statistics and Economic Studies
IOD	Indian Ocean Dipole
IPCC	Intergovernmental Panel on Climate Change
IPT	Interference Pattern Technique
IRD	French Research Institute for Development
IRNSS	Indian Regional Navigational Satellite System
IRSTEA	French National Research Institute of Science and Technology for Environment and Agriculture
ISBA	Interactions Soil-Biosphere-Atmosphere (model)
ISDC	Integrated System Data Center
ISI	Initial Spread Index
ISODATA	Iterative Self-Organizing Data Analysis Technique
ISRO	Indian Space Research Organisation
ITC	Individual Tree Crown
ITCZ	Intertropical Convergence Zone

ITG	Institute of Theoretical Geodesy
IWPB	Institute of Water Problem of Bishkek, Kyrgyzstan
Jason	Radar altimeter
JAXA	Japan Aerospace Exploration Agency
JECAM	Joint Experiment for Crop Assessment and Monitoring
JPEG	Joint Photographic Experts Group (image format)
JPL	Jet Propulsion Laboratory
JRC	European Commission's Joint Research Centre
KBR	K-Band Microwave Ranging
KLT	Karhunen–Loeve transform
LaDAR	Laser Detection And Ranging
LADS	Laser Airborne Depth Sounder
LAGEOS	Laser Geodynamics Satellite
LAI	Leaf Area Index
LANDSAT	LAND + Satellite
LAUVA	Airborne Ultraviolet Aerosol LiDAR
LAX	Maximum LAI
LBAS	Local Based Augmentation System
LCCS	Land Cover Classification System
LCLU	Land Cover / Land Use
LDAS	Land Data Assimilation System
LEGOS	Laboratory for Studies in Geophysics and Spatial Oceanography (France)
LEnKS	Local EnKF Smoother
LEO	Low Earth Orbit
LEP	Leading Edge Position
LEWIS	L-band for Estimating Water In Soils
LFMC	Live Fuel Moisture Content (%)
LHCP	Left Hand Circular Polarization
LiDAR	Light Detection and Ranging
LISAH	Laboratory for Soil, Agrosystems and Water Systems (France)
LISFLOOD-FP	Two-dimensional Hydrodynamic Model
LMA	Leaf Mass per Area

LMM	Linear Mixed Model
LOADDT	Spatial planning and territorial development
LOLA	Lunar Orbiter Laser Altimeter
LOV	Villefranche Oceanography Laboratory (France)
LPCA	Laboratory for Physico-Chemistry of the Atmosphere
LPRM	Land Parameter Retrieval Model
LRM	Low Resolution Mode
LSCE	Climate and Environment Sciences Laboratory (France)
LSM	Land Surface Model
LSSM	Least Squares 3D Surface Matching
LST	Land Surface Temperature
LULCC	Land Use and Land Cover Change
LUT	Look-Up Table
LWIR	Long-Wave Infrared
LZW	*Lempel-Ziv-Welch* (compression algorithm)
MACCS	Multisensor Atmospheric Correction and Cloud Screening processor
MARS	Monitoring of Agriculture with Remote Sensing
MATISSE	Advenced Earth Modeling for Imaging and Scene Simulation
MaxEnt	Maximum Entropy Method
MBOC	Multiplexed Binary Offset Carrier
MCT	Mercury Cadmium Telluride
MEB	Microwave Emission of the Biosphere
MERIS	Medium Resolution Imaging Spectrometer
MESA	Monitoring of Environment and Security in Africa
METEOSAT	METEO + Satellite
METOP	Meteorological Operational Satellite Programme/Advanced Scatterometer
METRIC	Mapping Evapotranspiration at High Resolution with Internalized Calibration
MGVI	MERIS Global Vegetation Index
MHz	Mega-Hertz (= 1,000,000 Hz)
MIMR	Multichannel Microwave Imaging Radiometer

MIPERS	Multistatic Interferometric Polarimetric Electromagnetic model for Remote Sensing
MIR	Middle Infrared
MISDc	*Modello Idrologico Semi Distribuito in Continuo* (Continuous rainfall-runoff model)
MISR	Multi-angle Imaging Spectro Radiometer
MISTIGRI	Microsatellite for Thermal Infrared Ground Surface Imaging (CNES, France)
MLR	Multiple Linear Regression
MLS	Mobile LiDAR Scanner
MMD	Minimum–Maximum Difference
MMS	Mobile Mapping Systems
MMU	Minimum Mapping Units
MMV	Mobile Mapping Vehicle
MNDWI	Modified Normalized Difference Water index
MNF	Maximum Noise Fraction
Mo	Mega-octet (1,000,000 octets)
MODCOU	Hydrogeological model
MODIS	Moderate Resolution Imaging Spectroradiometer
MODTRAN	Moderate Resolution Atmospheric Transmission
MOLA	Mars Orbiter Laser Altimeter
MPE	Maximum Permissible Exposure
MSAS	Multi-functional Satellite-based Augmentation System
MSG	Meteosat Second Generation
MSI	Moisture Stress Index
MSI	Multispectral Instrument (Sentinel-2)
MTSAT	Multi-functional Transport Satellites
MVSA	Minimum Volume Simplex Analysis
MWIR	Mid-Wavelength Infrared
NAOMI	New AstroSat Optical Modular Instrument
NASA	National Aeronautics and Space Administration (USA)
NBR	Normalized Burn Ratio
NCC	Normalized Cross-correlation
NCEP	National Centers for Environmental Prediction

Nd:YAG	Neodymium-doped Yttrium Aluminium Garnet
NDSI	Normalized Difference Snow Index
NDVI	Normalized Difference Vegetation Index
NDVITM	Normalized Difference Vegetation Index Threshold Method
NDWI	Normalized Difference Water Index
NEBN	Noise Equivalent Beta Naught
NEDT	Noise Equivalent Delta Temperature
NEE	Net Ecosystem Exchange
NEF	Noise Equivalent Flux
NEM	Normalized Emissivity Method
NEP	Net Ecosystem Productivity
NEP	Noise Equivalent Power
NIR	Near Infrared
NLES	Navigation Land Earth Station
NLRI	Near Laser Ranging Investigation
NMC	National Meteorological Center (USA)
NMF	Non-negative Matrix Factorization
NOAA	National Oceanic and Atmospheric Administration
NOHD	Nominal Ocular Hazard Distance
NORUT	Norut Northern Research Institute (Norway)
NPV	Non-photosynthetic Vegetation
NPW	Numerical Weather Prediction models
NSC	NarynSyrdarya Cascade
NSCAT	NASA Scatterometer
NSIDC	National Snow and Ice Data Center (USA)
NWP	Numerical Weather Prediction
OA	Overall Accuracy
OBIA	Object Based Image Analysis
OLCI	Ocean and Land Colour Instrument
OLI	Operational Land Imager
OLS	Operational Linescan System
OM	Organic Matter
OMI	Ozone Monitoring Instrument

ONERA	French Aerospace Research Agency
ONF	French National Forest Office
OS	Open Service
OSCAT	OceanSat-2 Scatterometer
OTB	OrfeoToolBox
PA	Producer's Accuracy
PACE	Pre-Aerosol, Clouds and Ocean Ecosystems
PAH	Polycyclic Aromatic Hydrocarbons
PAI	Plant Area Index
PARIS-IoD	Passive Reflectometry and Interferometry System In Orbit Demonstrator
PCA	Principal Components Analysis
PERSIANN	Precipitation Estimation from Remotely Sensed Information using Artificial Neural Networks
PET	Potential Evapotranspiration
PFT	Phytoplankton Functional Types
PGN	Permanent GNSS Network
PHR	Panchromatic High Resolution
PIF	Pseudo-Invariant Features
piGNSS-R	Partial Interferometric GNSS-R
PLF	Polarization Loss Factor
PLOF	Local Land Tenure Plan (France)
PLSR	Partial Least Squares Regression
PMT	Photomultiplier Tubesv
POA	Polarization Orientation Angle
PODAAC	Physical Oceanography Distributive Active Data Center
POLDER	Polarization and Directionality of the Earth's Reflectances
PolInSAR	Polarimetry-interferometry Synthetic Aperture Radar
PolSAR	Polarimetry SAR
POLYMER	Polynomial Based Algorithm Applied to Meris
POM	Physical Optics Model
PPCDAM	Plan for Preventing and Controlling Deforestation in Amazônia Legal
PRESS	Prediction Sum of Squares

PRF	Pulse Repeat Frequency
PRISMA	*Precursore Iperspettrale della Missione Applicativa* (Italian hyperspecral mission)
PRN	Pseudo Random Noise
PROBA	Project for On-board Autonomy
PRODES	*Programa de Cálculo do Desflorestamento da Amazônia* (Brazilian Amazon Deforestation Monitoring Program)
PROSPECT	Leaf Optical Properties Spectra (radiative transfer model)
PRS	Public Regulated Service
PS	Persistent Scatterers
PSI	Persistent Scatterers Interferometry
PSR	Penalized-Spline Regression
PV	Photosynthetic Vegetation
PVI	Perpendicular Vegetation Index
PWC	Plant Water Content
QBO	Quasi-Biennal Oscillation
QCA	Quasi-Crystalline Approximation
Qgis	GIS software (open source)
QPSK	Quadrature Phase Shift Keying
QUIKSCAT	Quick Scatterometer (NASA)
QZSS	Quasi-Zenith Satellite System
RADAR	Radiodetection and Ranging
RAF	French Altimetric System
RAINFOR	Amazon Forest Inventory Network
RAMSES	ONERA Airborne Multi-frequency SAR Imaging System (France)
RAN	Royal Australian Navy
RANSAC	Random Sample Consensus
RCA	Radio Corporation of America
RCM	RADARSAT Constellation Mission
RCS	Radiometric Control Sets
REDD	Reduction of Emissions from Deforestation and Forest Degradation

REDDAF	Reducing Emissions from Deforestation and Degradation in Africa (European Project)
RENAG	French National GNSS Permanent Networks
RESIF	French Seismological and Geological Network
RF	Random Forests (classifier)
RFI	Radio Frequency Interference
RG	Relative Greenness
RGB	Red Green Blue
rGNSS-R	Reconstructed GNSS-R
RHCP	Right Hand Circular Polarization
RMSD	Root Mean Square Difference
RMSE	Root Mean Square Error
RPAS	Remotely Piloted Aircraft System
RPC	Rational Polynomial Coefficients
RPD	Ratio of Performance to Deviation
RPIQ	Ratio of Performance to Inter-Quartile range
RTK	Real Time Kinematic
RTM	Radiative Transfer Model
RUE	Rain Use Efficiency
RVI	Radar Vegetation Index
RVoG	Random Volume over Ground
RXD	Reed-Xiaoli Detector
SAFY	Simple Algorithm For Yield estimate
SAM	Spectral Angle Mapper
SAMIR	Satellite Monitoring of Irrigation (model)
SAR	Synthetic Aperture Radar
SAR in	SAR interferometric
Saral/Altika	Radar Altimeter (French–Indian altimetry mission)
SASS	SEASAT Advanced Scatterometer System
SAVI	Soil-Adjusted Vegetation Index
SBAS	Satellite Local-based Augmentation System
SBAS	Small Baselines
SCA	Snow Cover Area

SCARAB	Scanning Radiometer for Radiation Balance
SCF	Snow Cover Fraction
SDC	Snow Depletion Curve
SDS	Science Data System (USA)
SEAS	Survey of the Environment Assisted by Satellite
SeaWiFS	Sea-viewing Wide Field-of-view Sensor (satellite)
SEBAL	Surface Energy Balance Algorithm for Land
SEBS	Surface Energy Balance System
SEC	Standard Error of Calibration
SEKF	Self-extended Kalman Filter
SEP	Standards Error of Prediction
SER	Section Efficace Radar
SERD	Single bounce Eigenvalue Relative Difference
SEVIRI	Spinning Enhanced Visible and Infrared Imager
SFCW	Stepped Frequency Continuous Waves
SfM	Surface-from-Motion
SFT	Strong Fluctuation Theory
SGBM	Semi-Global Block Matching algorithm
SHALOM	Spaceborne Hyperspectral Applicative Land and Ocean Mission
SHI	State Hydrological Institute (St. Petersburg, Russia)
SHOALS	Scanning Hydrographic Operational Airborne LiDAR Survey
SHOM	French Navy's Hydrographic and Oceanographic Service
SID	Spectral Information Divergence
SIERRA	Spectral Reflectance Image Extraction from Radiance with Relief and Atmospheric Correction
SIFT	Scale Invariant Feature Transform
SIGMA	Simulation Innovation for Global Monitoring of Agriculture
SLA	*Scanner LiDAR aérien* (aerial LiDAR scanner)
SLC	Single Look Complex
SLR	Single-Lens Reflex
SM	Soil Moisture
SMA	Spectral Mixture Analysis
SMAC	Simplified Method for Atmospheric Correction

SMAP	Soil Moisture Active and Passive mission (Radiometer)
SMAPVEX	SMAP Validation Experiment
SMEX02	Soil Moisture Experiment 2002
SMF	Spectral Matched Filter
SMLR	Stepwise Multiple Linear Regression
SMMR	Scanning Multichannel Microwave Radiometer
SMOS	Soil Moisture and Ocean Salinity mission (satellite)
SNAS	Chinese Satellite Navigation Augmentation System
SNR	Signal to Noise Ratio
SNSB	Swedish National Space Board
SNV	Standard Normal Variate
SPAD	Single-Photon Avalanche Diode
SPM	Small Perturbation Model
SPM	Suspended Particulate Matter
SPOT	Satellites for Earth Observation
SRM	Snowmelt-Runoff Model
SRTM	Shuttle Radar Topography Mission
SSA	Surface Specific Area
SSC	Soil Surface Characteristics
S-SEBI	Simplified Surface Energy Balance Index
SSM	Soil Surface Moisture
SSMI	Special Sensor Microwave Imager (satellite)
SST	Sea Surface Temperatures
STF	Spectrotransfer Functions
STICA	Socio-technical Information and Communication Arrangements
STICS	Crop model
SUCROS	Simple and Universal Crop Growth Simulator
SUHI	Surface Urban Heat Islands
Suomi-NPP	Suomi National Polar-Orbiting Partnership
SURFEX	Surface model platform (Météo France)
SVAT	Soil–Vegetation–Atmospheric Transfer
SVM	Support Vector Machine
SVMR	Support Vector Machine Regression

SVR	Support Vector Regression
SWAT	Soil and Water Assessment Tool
SWE	Snow Water Equivalent
SWI	Soil Wetness Index
SWIR	Short-wave Infrared
SWOT	Surface Water Ocean Topography (satellite)
SYSIPHE	Airborne hyperspectral imaging system
T/P	Topex/Poséidon (Franco-American altimeter)
TDR	Time Domaine Reflectometry
TEC	Total Electron Content
TES	Emissivity Separation algorithm
THEIA	French Land Data Centre
THIRSTY	Thermal Infrared Spatial System (satellite project by CNES and NASA)
TIFF	Tag Image File Format
TIN	Triangular Irregular Network
TiO_2	Titanium dioxide
TIR	Thermal Infrared
TISI	Temperature Independant Spectral Indices
TLS	Terrestrial Laser Scanning
TM	Thematic Mapper
TMBOC	Time Multiplexed Binary Offset Carrier
TNT2	Topography based Nitrogen Transfer and Transformation
To	Tera-octet (1,000,000,000,000 octets)
TOA	Top of Atmosphere
TOC	Top of Canopy
TomoSAR	Tomography SAR
Topex/ Poseidon	Radar altimeter
TOPLATS	Topographic Land Atmosphere Transfer Scheme
TRIP	Total Runoff Integrating Pathways
TRMM	Tropical Rainfall Measuring Mission (satellite)
TSAVI	Transformed Soil Adjusted Vegetation Index

TSEB	Two-Source Energy Balance (model)
TTL	Transistor-Transistor Logic
TWAP	Transboundary Water Assessment Program
TWS	Terrestrial Water Storage
UA	User's Accuracy
UAA	Utilized Agricultural Area
UAV	Unmanned Aerial Vehicles
UHI	Urban Heat Islands
ULICE	Ultraviolet LiDAR for Canopy Experiment
UNEP	United Nations Environment Program
USDA	United States Department of Agriculture
USGS	United States Geological Survey
USO	Ultra-Stable Oscillator
UTC	Coordinated Universal Time
UTM	Universal Transverse Mercator
UV	Ultra-violet
VARI	Visible Atmospherically Resistant Index
VCA	Vertex Component Analysis
VD	Virtual dimensionality
VHF	Very high frequency
VHI	Vegetation Health Index
VHSR	Very High Spatial Resolution
VIC	Variable Infiltration Capacity
VIIRS	Visible Infrared Imager Radiometer Suite
VIS	Visible
VISAT	Video, Inertial, and Satellite GPS
Vis-NIR	Visible and Near Infrared
VITO	Flemish Institute for Technological Research (Belgium)
VLA	Very Large Array
VLBI	Very Large Baseline Interferometry
VOS	Volatile Organic Compound
VPD	Vapor Pressure Deficit
VSDI	Visible and Shortwave Infrared Drought Index

VWC	Vegetation Water Content
WALID	Water LiDAR Simulation Model
WASS	Wide Area Augmentation System
WDI	Water Deficit Index
WGHM	Water GAP Global Hydrology Model
WGS	World Geodetic System
WMA	Winter Metric Anomaly
WMO	World Meteorological Organization
WMS	Wide-area Master station
WRS	Wide-area Reference Stations
WSI	Water stress index
WTC	Wet Troposphere Correction
XML	Extensible Markup Language
ZSSD	Zero-mean Sum of Squared Difference

Introduction

Continental hydrological reservoirs represent a very small fraction of the total water on Earth (about 0.025%). Despite this, they play a key role for life on Earth and climate dynamics, because of their contribution to the interface of the continents and the atmosphere. In addition to the polar caps, fresh water is stored in the different reservoirs such as snow packs, glaciers, aquifers, the root zone that is within the first few meters of the soil, and surface waters which include streams and rivers, lakes, reservoirs due to human activity and wetlands. Despite this, the continental hydrological cycle remains one of the least well understood of the climate system components. The understanding of the different processes involved and the prediction of their evolution is an important issue in hydrology and meteorology. This explains the important efforts made by the scientific community in this field.

Understanding the continental hydrological cycle requires both consistent observation of essential variables and the development of models representing the different processes involved. The accuracy of the models is generally limited by our imperfect knowledge of physical phenomena, initial conditions and the limit conditions of the modeled system. Observations taking into account the spatial and temporal variabilities are then needed to calibrate the models and control their forecasts. Until recently, the only observations used in modeling hydrological processes were punctual and often unrepresentative of the modeled spatial scales.

Remote sensing now provides access to useful parameters in land surface monitoring. The assimilation of satellite measurements and products in the

---

Introduction written by Nicolas BAGHDADI and Mehrez ZRIBI.

models describing the functioning of hydrological processes and water management procedures facilities an improvement in the understanding of the continental water cycle.

This book, part of the *Remote Sensing Observations of Continental Surfaces* Set, focuses on the use of remote sensing in hydrology. It is written by world-renowned scientists in their field. It will allow for the actualization of new knowledge and description of the challenges in research and development for years to come. It is designed for remote sensing or hydrology research teams and students in 2nd (engineering schools, Master's) and 3rd (PhD) university cycles.

The first part of this book addresses the use of remote sensing to characterize continental soil surface properties. These soil surface properties play an essential role in understanding and modeling different processes (infiltration evapotranspiration, runoff, etc.). Chapter 1 provides a detailed analysis of the potential of the high resolution high resolution SAR (synthetic aperture radar) remote sensing in the description of the surface soil properties (hydric conditions, roughness, salinity, texture). Chapters 2 and 3 analyze the same question, with microwave techniques (active and passive), but with low resolution sensors adapted to regional or global uses. Chapters 4 and 5 present the contribution of optical and radar remote sensing data in monitoring snow, which fulfills a key function as a temporary storage of winter precipitation.

The second part presents the use of space observation in monitoring underground and surface water. Changes affecting freshwater supplies (lakes, ponds, wetlands) and changes in the main river flow are crucial to the functioning of the continental water cycle. Chapter 6 analyzes the potential of satellite altimetry to meet this need. Chapter 7 discusses the use of the same technique for for monitoring Antarctica ice sheet. Chapter 8 is dedicated to methods based on spatial gravimetry techniques for remote monitoring groundwater reserves, especially for the most threatened areas in the globe by the lack of water and overexploitation of aquifers. Chapter 9 discusses the potential of new GNSS-R (Global Navigation Satellite System Reflectometry) technique meeting the same objectives.

The final part discusses the use and assimilation of remote sensing measurements and products in various hydrological process models. Chapter 10 discusses surface–atmosphere exchanges, particularly evapotranspiration.

Chapter 11 analyzes the assimilation of space observations in hydrological models developed on a watershed. Finally, Chapter 12 analyzes in a larger, regional or global scale, the contribution of spatial data in the modeling of water and carbon cycle.

Finally, we would like to thank the people who contributed to the development of this volume. First, the scientists, the authors of the chapters and also the experts of the Scientific Committee for their review of the chapters. This project was conducted with support from the IRSTEA (French Institute for Research in Science and Technology for Environment and Agriculture), CNRS (French National Center for Scientific Research) and CNES (French National Center of Space Studies).

We also thank our families for their support in making this project happen and Dr. André Mariotti (Emeritus Professor, Pierre and Marie Curie University) and Dr. Pierrick Givone (Scientific Director, IRSTEA) for their encouragement.

1

Characterization of Soil Surface Properties Using Radar Remote Sensing

1.1. Thematic introduction

Soil surface characteristics (SSC) play a key role in the understanding of different processes taking place at the soil–vegetation–atmosphere interface (runoff, infiltration, soil erosion, exchange of water and energy streams). Until the 1990s, the only observations used for the modeling of this interface were limited and often unrepresentative of the spatial scales modeled [LOU 91]. Radar remote sensing now allows spatial parameters to be accessed for the monitoring of the soil surface and the modeling of its functioning. In fact, signals acquired by radar are strongly correlated to some physical variables that are linked to soil surface conditions, such as soil moisture and surface roughness. The assimilation of these data in functional models (hydrologic, erosion, SVAT (Soil–Vegetation–Atmosphere Transfer) etc.) has shown a clear improvement in the simulation of physical processes (see Chapters 11 and 12).

Active microwave remote sensing (radar) is particularly well adapted to the characterization of soil surface conditions in agricultural fields. Contrary to optical remote sensing techniques, Synthetic Aperture Radar (SAR) allows all-weather measurements, independently of meteorological and lighting conditions (cloud cover, day/night, etc.). The disadvantage of optical

Chapter written by Nicolas BAGHDADI and Mehrez ZRIBI.

techniques based on thermal infrared, connecting soil moisture to the surface temperature, is their dependence on ambient conditions. Radar uses microwave frequencies (wavelengths between 1 mm to 1 m) that are very sensitive to the geometric and dielectric properties of the measured medium, which are themselves dependent on surface parameters (roughness, soil moisture, soil composition, vegetation cover). A SAR signal also depends on different instrumental parameters, polarization, incidence angle and radar wavelength. In the presence of vegetation, the scattered radar signal is a combination of soil and vegetation contributions. The soil contribution decreases when the radar wavelength decreases.

The first studies using radar remote sensing started at the end of the 1970s with *in situ* or airborne scatterometers [ULA 78]. Important scientific developments started in the 1990s with satellite and airborne SAR (ERS-1/2, JERS, SIR-C, RADARSAT-1/2, PALSAR-1/2, ASAR, TerraSAR-X, COSMO-SkyMed, etc.). Most studies were carried out in the L-band (wavelength ~22 cm), C-band (wavelength ~6 cm), and more recently, X-band (wavelength ~3 cm). The first satellite SAR sensors accessible to the scientific community had an instrumental configuration of mono-polarization and a single incidence angle (ERS-1/2, JERS). The second generation of radar sensors with new instrumental configurations (RADARSAT, ASAR/ENVISAT, PALSAR/ALOS, TerraSAR-X, COSMO-SkyMed, Sentinel-1) allowed the scientific community to gather images in multi-polarization and sometimes polarimetric mode (scattering matrix) with frequencies ranging from the L band to the X band. Additionally, the new SAR sensors have a revisit time and spatial resolution allowing temporal acquisitions adapted to hydrological and agronomic applications on local or regional scales, for which fine spatial and temporal resolutions are sometimes necessary. In fact, these new SAR sensors provide images in high spatial resolution (around 1 m for TerraSAR-X and COSMO-SkyMed) and with a high revisit time (more than one image per week for Sentinel-1). These new metric sensors have allowed a fine analysis at the intra-plot scale. Low-resolution spatial microwave sensors (several km) also exist, but they are better adapted to the needs of meteorological and climatic applications on a global scale, like AMSR-E, AMSR2, SMOS (microwave radiometers) and ASCAT/METOP (C-band scatterometer), which provide users of soil moisture products with a temporal frequency in the order of a few days and a spatial resolution of around 25–40 km.

This chapter describes the influence of different instrumental parameters on radar backscattering in the case of bare or scarcely covered soils. Section 1.2 describes the soil parameters and the *in-situ* methods for characterizing them, in particular the roughness and soil moisture. The sensitivity of the radar signal to these soil parameters is presented in section 1.3. Section 1.4 presents studies of radar signal modeling. Section 1.5 describes inversion approaches for the estimation of soil parameters. Finally, section 1.6 presents development prospects for the years to come.

1.2. Description of soil parameters

1.2.1. *Soil roughness*

Different approaches have been proposed for the description of soil roughness. For radar applications, the surface Z (x,y) is generally considered to be stationary and ergodic. The description of the surface is then based on the calculation of the autocorrelation function ρ (u,v), defined as:

$$\rho(u,v)=\left\langle\left[Z(x+u,y+v)-\langle Z\rangle\right]\left[Z(x,y)-\langle Z\rangle\right]\right\rangle \quad [1.1]$$

where $\langle Z\rangle$ is the average height of altitudes measured from the roughness profile Z (x,y).

Generally, two roughness parameters are used and estimated based on the autocorrelation function. The first of these is the standard deviation of the surface height (root mean square surface height, Hrms), which defines the vertical scale of the roughness and is computed as:

$$Hrms^2=\left\langle\left[Z(x,y)-\langle Z\rangle\right]^2\right\rangle \quad [1.2]$$

The second roughness parameter is the correlation length (*L*), which is usually defined as the horizontal displacement for which the autocorrelation function of the profile decreases to 1/e.

When the roughness is weak and the soil is smooth (*Hrms* lower than approximately 1 cm), the autocorrelation function has a generally exponential shape. Inversely, for higher roughness, the autocorrelation function has a shape close to a Gaussian. Zribi [ZRI 98] introduced the fractal dimension to the description of the autocorrelation function's shape for bare soils in agricultural fields. For one-dimensional roughness profiles, the autocorrelation functions are defined as follows:

$$\begin{aligned}\rho(x) &= Hrms^2\, e^{-\left(\frac{x}{L}\right)} &&: \text{exponential}\\ &= Hrms^2\, e^{-\left(\frac{x}{L}\right)^2} &&: \text{Gaussian}\\ &= Hrms^2\, e^{-\left(\frac{x}{L}\right)^{\alpha}} &&: \text{fractal}\end{aligned} \qquad [1.3]$$

with $\alpha = -2D + 4$, where D is the fractal dimension. When the fractal dimension varies, the shape of the autocorrelation function changes: it goes from an exponential function for $D = 1.5$ to a Gaussian shape for $D = 1$. The experimental measurements show a fractal dimension between 1.25 and 1.45, hence an autocorrelation function power α between 1.1 and 1.5.

In the case of agricultural surfaces with periodic structures (rows, with P periods), the autocorrelation function could be analytically described with the following form (in the case of a Gaussian shape, for example):

$$\rho(x) = Hrms^2 e^{-\left(\frac{x}{L}\right)^2} + S^2 e^{-\left(\frac{x}{L_S}\right)^2} \cos\left(\frac{2\pi x}{P}\right) \qquad [1.4]$$

The second term models the directional roughness variations as a narrowband Gaussian random process, centered on a frequency ($1/P$) and a band length of $2\pi/L_S$. A Fourier transform of this term allows the deduction of the three parameters describing the directional structure (the intensity S, the periodicity P, and the correlation length L_S).

The inversion of the radar signal to estimate all surface parameters of the soil (*Hrms*, *L*, *D*, and soil moisture) is impossible without simplifications in the description of the roughness or supplementary hypotheses [MAT 99, VER 08]. It is in this context that [ZRI 14a] proposed a new parameter *Zg*, which combines the standard deviation of heights, the correlation length, and the shape of the autocorrelation function (described by the power α):

$$Zg = Hrms\left(Hrms/L\right)^{\alpha} \qquad [1.5]$$

Figure 1.1(b) illustrates the best correlation between the radar signal and the normalized roughness parameter *kZg* (*k* is the radar wave number = $2\pi/\lambda$ with λ the radar wavelength), in comparison with the correlation between the radar signal and kHrms (Figure 1.1(a)). The backscattered radar signals have been simulated by a model based on the numerical moment method for large ranges of roughness conditions (kHrms, correlation length, correlation function power).

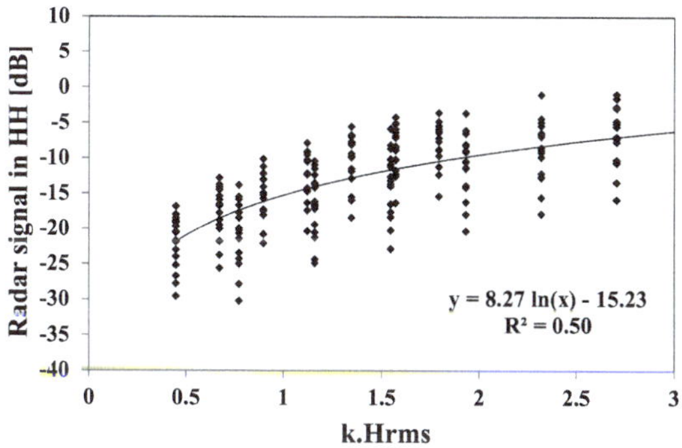

a)

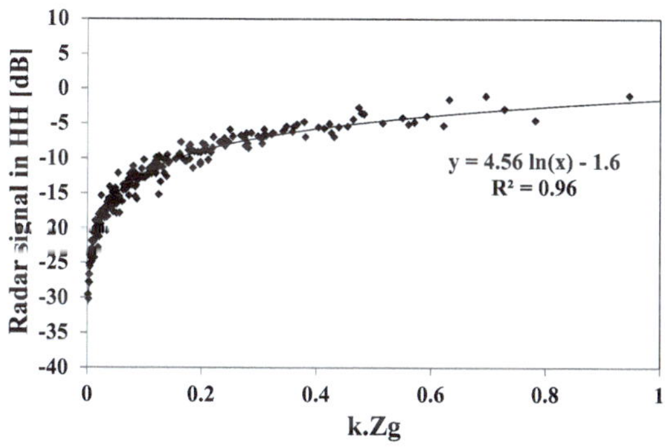

b)

Figure 1.1. *Simulated radar signal (HH-40°) as a function of the two roughness parameters: a) k Hrms, and b) k Zg*

Several techniques can be used to measure soil roughness: pin profilometer (Figure 1.2(a)), laser profilometer (Figure 1.2(b)) and 3D photogrammetry. The use of the laser or 3D photogrammetry allows for the accurate rendering of soil roughness (high spatial resolution) with a precise estimation of the roughness parameters, *Hrms* and *L*. As regards the correct characterization of soil roughness using a pin profilometer, it is necessary to have (1) long roughness profiles or several short profiles and (2) a relatively fine horizontal resolution from the profilometer (small sampling interval, Δx).

Nevertheless, the pin profilometer is very widely used, primarily for reasons of cost. Most pin profilometers are 1 or 2 m long with a sampling interval of 0.5, 1 or 2 cm. The measures are generally taken in both directions, parallel and perpendicular to the row direction, in order to consider the directional effect of soil tilling (several profiles in each direction). The autocorrelation functions, calculated using different roughness profiles of a reference plot, are averaged, and the roughness parameters *Hrms* and *L* estimated [DAV 00, MAT 99].

a)

b)

Figure 1.2. *Examples of roughness profiles: a) 1D profile from a pin profilometer (1 m long and a sampling interval of 2 cm); b) 2D profile from a laser scanner*

Various experimental and theoretical studies have demonstrated that the precision of *in-situ* measurements of the aforementioned two soil surface's roughness parameters (root mean surface height (*Hrms*) and correlation length (*L*)) is highly influenced by the length and horizontal resolution of the roughness profiles [OH 98, LIE 09]. Small profiles will lead to an underestimation of *Hrms* and *L*. Additionally, when roughness profiles are used with large sampling intervals, the estimated *Hrms* values decrease while the estimated correlation lengths increase. This effect is more pronounced in surfaces with short correlation lengths.

For agricultural soils, with around 10 profiles measuring 1 m, *Hrms* can be measured with a precision of 10%, while the precision of the estimation of *L* is around 15 to 20% (a larger error for strong *L* values). To measure *L* and *Hrms* with an error better than 5%, the sampling interval Δx must be smaller than 0.2*L* and 0.5*L* respectively [OH 98]. The *Hrms* values are generally seen between 0.3 cm (very smooth fields that have just been sown) and 4 cm (fields that have just been plowed). The correlation lengths *L* measured on the agricultural plots are predominantly between 3 and 25 cm.

1.2.2. *Soil surface moisture*

The quantity of water present in soil affects its electric properties and consequently the radar scattered signal. Soil moisture is a particularly dynamic parameter in time and space, so *in situ* soil moisture data is generally collected simultaneously with radar acquisitions. The distribution and spatial density of these measurements depend upon the level of heterogeneity of reference plots (intra-plots variations) and their size. A minimum of 20 measurements is generally taken for each reference plot (measuring at least one hectare).

There are two commonly used techniques for the measurement of soil moisture content: gravimetry and time-domain reflectometry (TDR) (or theta probe analysis).

The gravimetric method, while destructive and complicated to implement, remains a reference for the calibration of different equipment used to measure soil moisture. It consists of using a cylinder to collect soil samples, which are then placed in an oven at 105°C for 24 h. This method determines the ponderal water content (*Wp*) of a soil sample by comparing the wet weight (*Ph*) of the sample with its dry weight (*Ps*):

$$Wp\,(g.g^{-1}) = 100\left(\frac{Ph - Ps}{Ps}\right) \quad [1.6]$$

The volumetric water content *mv* (in $cm^3.cm^{-3}$or vol.%) is deduced from the ponderal water content using the soil's apparent density (*Da=Ps* / volume of a considered cylinder):

$$mv = Da\,.Wp \quad [1.7]$$

The TDR probe consists of emitting an electromagnetic microwave pulse along two (sometimes three) waveguides of a given length, inserted into the soil, and measuring the transit time of the return signal. It is recommended to proceed to a preliminary calibration of the device by comparison with measurements obtained by the gravimetric method. Finally, TDR measurements are not valid for frozen soil where a significant drop in the moistures registered by the probe can be observed.

When performing studies using radar images in the L-, C- and X-bands for the characterization of the soil surface moisture in agricultural areas, *in-situ* measurements of soil moisture are taken at a depth of between 0 and 10 cm. This measurement depth is related to the penetration depth of the radar wave (δ_p) that is generally equal to a few centimeters in the C- and X-bands. In the L-band, this depth can reach a few dozen cm for very dry soils. In fact, the thickness of this surface layer depends on the radar wavelength (λ) (more penetration with greater wavelengths) and the dielectric constant of soil (water content and soil composition) [ULA 78]:

$$\delta_p \cong \frac{\lambda\sqrt{\varepsilon'}}{2\pi\sqrt{\varepsilon''}} \quad [1.8]$$

ε' is the real part of the dielectric constant and ε' its imaginary part. [BRU 88] found that the penetration depth of the radar signal in the C-band for a clay loam soil decreases from 5 to 1 cm when the soil moisture increases from 10 to 30 vol.% (HH polarization and 15° incidence angle).

The dielectric constant is a physical quantity also known as complex permittivity. It defines the soil's electrical properties in terms of electrical losses due to the transport and absorption of energy brought by the radar wave, so it impacts the transmission capacity of reflection and absorption of microwaves. This constant is a complex number expressed in the form $\varepsilon = \varepsilon' - j\varepsilon''$. The real part ε' is more affected by the soil moisture, while the imaginary part ε'' essentially depends on the electrical conductivity of the soil solution.

The most commonly used models of the dielectric constant are those developed by [HAL 85, DOB 85] and [MIR 04]. The first is defined for a range of frequencies between 1.4 and 18 GHz. This model depends on the volumetric water content level, the size of the soil particles, the different components of the soil, the dielectric constant of water, and the radar frequency.

The empirical relationship between the volumetric water content of soil and its dielectric constant described by [TOP 80] is also widely used for its simplicity. This relation only allows the derivation of the real component of the dielectric constant:

$$mv = (-530 + 292\varepsilon - 5.5\varepsilon^2 + 0.043\,\varepsilon^3).10^{-4} \quad [1.9]$$

This relationship has been validated for numerous mineral soils and under different water content conditions.

1.3. Radar signal sensitivity to soil parameters

1.3.1. *Sensitivity of radar signal to soil roughness*

The numerous studies done in the past 30 years show that the radar signal backscattered by a bare soil increases with the rms surface height (*Hrms*) according to a logarithmic or exponential law to then become constant after a certain roughness threshold (Figure 1.3). This threshold, after which the signal becomes constant, depends on the wavelength and the radar's incidence angle. The results of different studies show a more rapid saturation of the radar signal with the soil's roughness (*Hrms*) when the wavelength and/or the incidence angle are weak. Figure 1.3 shows a saturation of the radar signal after *k.Hrms* below roughly 1. This threshold corresponds to

Hrms values of 4 cm in the L-band, 1 cm in the C-band, and around 0.5 cm in the X-band. Figure 1.3 also shows that the dynamic of the radar backscattering coefficient is weaker in the case of weak incidence angles (variation of 7 dB for 20°–25°, Figure 1.3a) than in the case of strong incidence angles (variation of 10 dB for 45°–50°, Figure 1.3(b)). Theoretical and experimental works show a slightly stronger sensitivity of the signal to the roughness of bare soil in HH polarization.

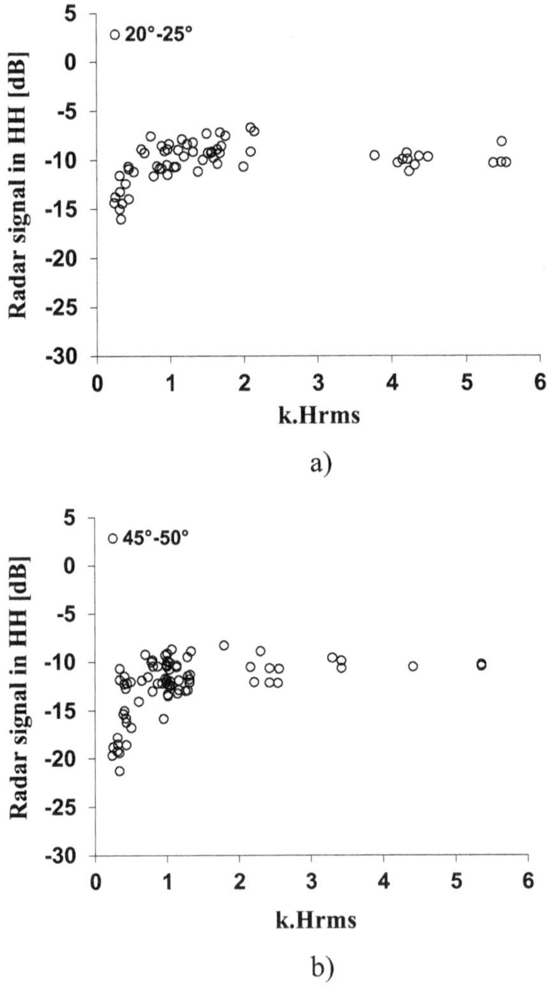

Figure 1.3. *Sensitivity of the radar signal (measurements from satellite sensors ERS, RADARSAT, ASAR and PALSAR) to soil roughness. The signal is represented as a function of k Hrms. The soil moisture values vary from 10 to 20 vol.%*

Only large wavelengths (e.g. L-band) and strong incidence angles (e.g. 45°) make it possible to map three roughness classes (smooth "sowing", rough "well plowed soil", moderately rough "slightly plowed soil") [BAG 02].

Rows (directional roughness) play an important role in the modeling of runoff and erosion on agricultural areas. The radar signal's sensitivity to these structures is proven in different theoretical and experimental studies (generally limited to airborne measurements). Observations have shown a maximum sensitivity to rows for weak incidence angles and low radar frequencies, with differences in the backscattered radar signals between measurements that are parallel and perpendicular to the direction of the rows that exceed 7 dB in the S band (~3 GHz) at 20° of incidence angle [ZRI 02].

1.3.2. *Sensitivity of the radar signal to soil moisture*

The radar signal's sensitivity to soil moisture has been analyzed in numerous studies [ZRI 13] as a function of different instrumental radar configurations (incidence angle, polarization, wavelength). These different studies have proved that the optimal configuration for obtaining the best sensitivity of the radar signal to soil moisture (weak influence of the roughness) combines HH polarization and a weak radar incidence angle [ULA 86, LET 82, BEA 90]. The results obtained from airborne or satellite radar data highlighted that radar incidence angles in the range 15°–35° are the most appropriate for the estimation of soil moisture.

Simulations made by the backscattering models illustrate an approximately logarithmic law between soil moisture and the radar signal. However, this logarithmic function is generally approximated by a linear function for soil moisture between 5 and 35 vol.%. After a soil moisture of around 35 vol.%, the radar signal stabilizes, then starts to decrease with the increase of soil moisture, so the estimation of soil moisture cannot be done without ambiguity after this threshold of around 35 vol.%. In fact, two plots having the same roughness can have the same backscattered radar signal if one has a soil moisture greater than 35 vol.% and the other a soil moisture below 35 vol.%.

Experimental studies, most numerous in the C-band, have shown a variable sensitivity of radar signal to soil moisture, approximately between

0.15 dB/vol.% and 0.3 dB/vol.%. This variability of the sensitivity is linked to different factors (temporal dynamic of the soil's roughness [ZRI 05] or heterogeneity of the vertical profile of soil moisture [ZRI 14b]). Concerning the effect of the radar wavelength, studies done by [AUB 11] have shown that the signal's sensitivity to soil moisture is twice as high in the X-band as in the C-band (around 0.35 dB/vol.% in X-band vs. 0.15 dB/vol.% in C-band) (Figure 1.4).

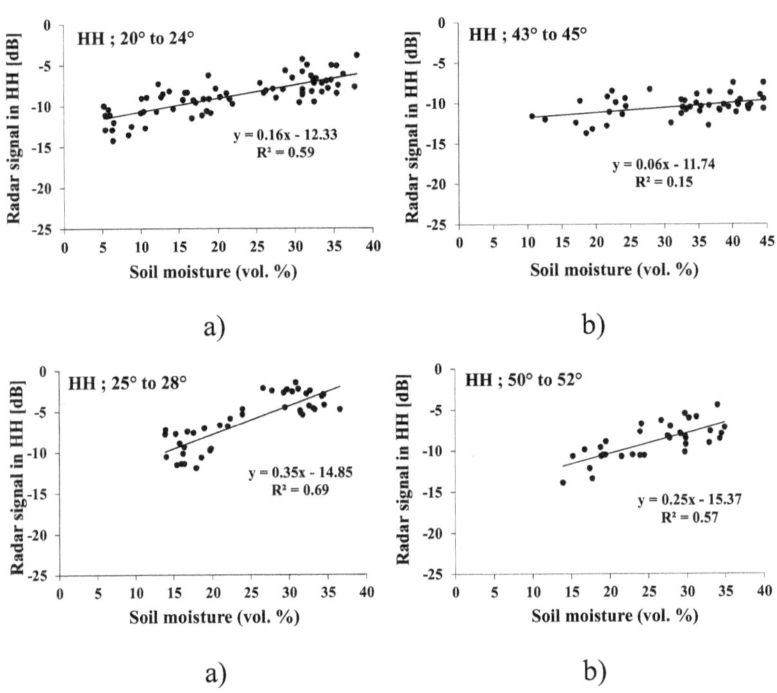

Figure 1.4. *Sensitivity of the radar signal to soil moisture in the C- and X-bands. The Hrms values vary between 0.5 and 1.5 cm: a) C-HH(20°–24°), b) C-HH(43°–45°) c) X-HH(25°–28°), d) X-HH(50°–52°)*

Observations in the L-band have approximately the same range of sensitivity to soil moisture as in the C-band [NAR 15]. This sensitivity of the signal to soil moisture decreases when the incidence angle increases (Figure 1.4). In conclusion, the X-band and allows the observation of small variations in soil moisture, undetectable by the sensors in the C- and L-bands. For a radiometric precision of the radar signal of about 1 dB, soil moisture variations in the order of 2.5 vol.% are detectable on images in the X-band versus 5 vol.% in the C- and L-bands. This strong sensitivity of the radar signal to soil moisture in the X-band coupled with the very high spatial resolution of X-band sensors (TerraSAR-X and COSMO-SkyMed) also makes it easier to detect soil moisture variations at the intra-plot scale.

1.3.3. *Soil salinity*

Salinization is a chemical process that leads to the degradation of arable soil, desertification, and biomass reduction; so, the monitoring of this process's development is necessary for planning measures to protect at-risk areas. Despite the dominant effect of the two parameters – soil moisture and roughness on the radar signal's variability, other parameters, such as salinity, have a relatively important effect. In this regard, several studies have been done to examine the relationship between the radar backscattering coefficient and soil salinity through the dielectric constant ε [LAS 07, SHA 03]. The influence of salinity is more pronounced on the imaginary part of the dielectric constant than on the real part. In the imaginary part, the influence of the soil's salinity is linked to soil moisture. This is mainly due to the solubility of salts in water leading to an increase in the soil's electrical conductivity, thus leading to a growth in the imaginary part.

Very few experimental studies have been done to estimate the salinity through SAR measurements. Radar backscattering coefficient simulations using the integral equation model (IEM) show some prospects for the rendering of this parameter with an increase of the simulated signal with salinity. The optimal configurations for estimating salinity through SAR data are wet soils, a low frequency (e.g. L-band), a strong incidence angle, and VV polarization [LAS 07]. Figure 1.5 illustrates the multi-frequency dynamic of the simulated radar signal (40° incidence, VV polarization) in relation to a variation of the salinity.

Simulations of the radar backscattering coefficient by the IEM model in the L-, C- and X-bands show a high potential of estimating this parameter only in the L-band (Figure 1.5). When changing from no salinity to saline soil (S = 100‰), there is a variation of around 2 dB.

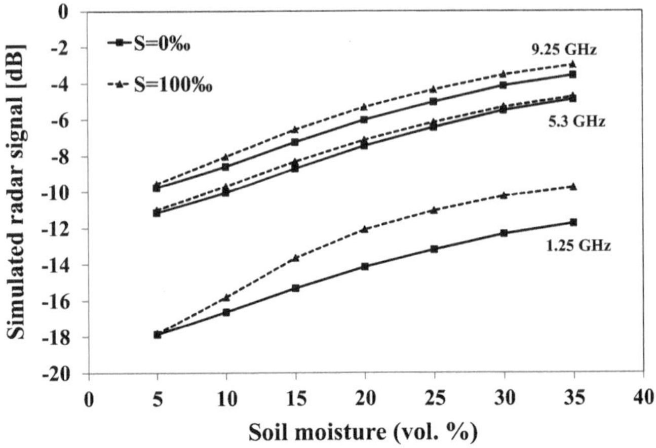

Figure 1.5. *Simulations of the backscattering coefficient (VV-40°) from IEM model as a function of soil moisture, for two salinity conditions (S=0 and S=100‰) with three frequency bands: L, C and X*

1.3.4. *Soil texture composition*

The soil's texture is a key parameter for establishing pedotransfer functions in order to derive the soil's hydrodynamic parameters (wilting point, field capacity, saturated hydraulic conductivity). Clay content is the most important parameter to be estimated for analyzing the water storage capacities for the soil and the plant. Research done to analyze the radar technique's potential is still limited, because the effect of the soil's texture is relatively weak in the calculation of the dielectric constant. In contrast, the strong link between the soil's texture (particularly the clay content) and the temporal dynamic of its water content is becoming increasingly discussed. [AUB 11] showed the ability of X-band radar measurements (TerraSAR-X) to observe slaking crusts. This potential for the detection of slaking crusts by means of radar signals is indirect and tied to the fact that soils presenting a slaking crust can have higher water content levels compared to soils without a crust, and therefore have higher signal values. [ZRI 12] quantified the clay content of these slaking crusts using TerraSAR-X measurements through the analysis of the drying rates of different types of soil (clay, sandy).

1.4. Modeling of radar backscattering on bare soil

Numerous radar backscattering models have been developed to analyze the radar signal's sensitivity to the soil's physical parameters (roughness and water content in particular). Furthermore, the effect of instrumental parameters (wavelength, incidence angle, and polarization) is also studied by these models in order to define the best radar configuration for estimating soil parameters: wavelength choice, pertinence of multi-polarization and polarimetric modes as compared to a mono-polarization mode.

Different physical analytical models have been developed to simulate radar backscattering of the soil's surface [RIC 51, BEC 63, ULA 86, FUN 92, CHE 03]. These models are all limited to domains of validity due to the considered physical approximations (Figure 1.6). For example:

– the small perturbation model (SPM) is better adapted to surfaces with a low roughness:

$$k.Hrms < 0.3 \tag{1.10}$$

$$\sqrt{2}Hrms / L < 0.3 \tag{1.11}$$

– models based on the Kirchhoff approximation (stationary phase or scalar approximations) are better adapted to rough surfaces:

$$kL > 6 \tag{1.12}$$

$$0.06(kL)^2 > kHrms \tag{1.13}$$

– the IEM developed by [FUN 92] corresponds to the largest domain of validity. Its calculation is based on the solution of the Stratton–Chu equation. In comparison with the Kirchhoff theory, a second component is added in the backscattered field calculation:

$$kHrms < 3 \tag{1.14}$$

$$\left((k\,Hrms\cos\theta)^2/\sqrt{0.46k\,Hrms}\right)\exp\left\{-\sqrt{0.92\,k\,L(1\quad\sin\theta)}\right\} < 0.25 \tag{1.15}$$

– the advanced integral equation model (AIEM) developed by [CHE 03] is an extension of the IEM. This modification of the IEM is expressed by a simplification of the terms of multiple scattering, those of the single scattering remaining the same as for the original version of the IEM. This modification allows a more precise calculation of the simple scattering for a surface with a large standard deviation of heights.

The most commonly used semi-empirical models are those of Dubois and Oh [DUB 95, OH 04]. Unlike in physical models, soil roughness is only defined by the standard deviation of heights (*Hrms*) in most semi-empirical models. Furthermore, these models cover much more important ranges of validity than those of physical models and they are better adapted to an operational use for the inversion of radar data.

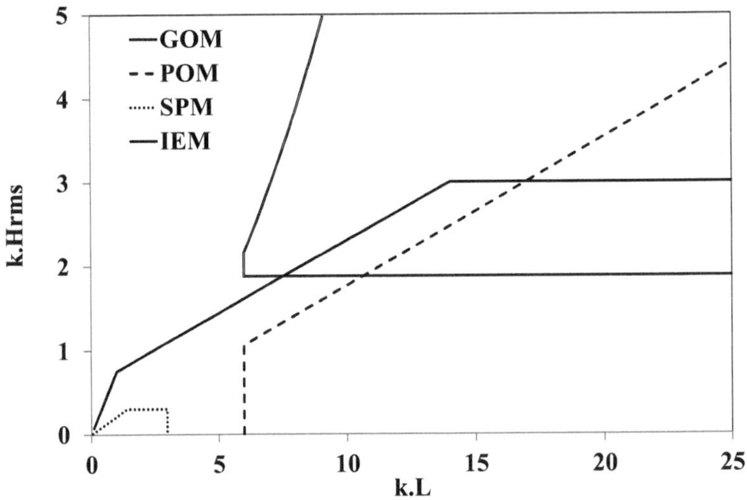

Figure 1.6. *Validity domains of the physical models: small perturbation model (SPM), physical optics model (POM), geometrical optics model (GOM) and integral equation model (IEM)*

1.4.1. *Description of the IEM*

In the IEM, the soil's physical characteristics are taken into account using input parameters like the dielectric constant (ε_r), the standard deviation of surface heights (*Hrms*), the shape of the autocorrelation function (ρ), and the correlation length (*L*). The model also considers parameters of the sensor, such as the incidence angle, polarization, and radar frequency. For a bare agricultural surface, the backscattering coefficient of the IEM in HH and VV is expressed by (surface contribution):

$$\begin{aligned}\sigma^0_{pp} &= \frac{k^2}{2}\left|f_{pp}\right|^2 e^{-4Hrms^2k^2\cos^2\theta}\sum_{n=1}^{+\infty}\frac{(4Hrms^2k^2\cos^2\theta)^n}{n!}W^{(n)}(2k\sin\theta,0)\\ &+\frac{k^2}{2}\mathrm{Re}(f^*_{pp}F_{pp})e^{-3Hrms^2k^2\cos^2\theta}\sum_{n=1}^{+\infty}\frac{(4Hrms^2k^2\cos^2\theta)^n}{n!}W^{(n)}(2k\sin\theta,0)\\ &+\frac{k^2}{8}\left|F_{pp}\right|^2 e^{-2Hrms^2k^2\cos^2\theta}\sum_{n=1}^{+\infty}\frac{(Hrms^2k^2\cos^2\theta)^n}{n!}W^{(n)}(2k\sin\theta,0)\end{aligned} \quad [1.16]$$

The backscattering coefficient in HV polarization is written as follows:

$$\begin{aligned}\sigma^0_{hv} &= \frac{k^2}{16\pi}e^{-2k^2Hrms^2\cos^2\theta}\sum_{n=1}^{+\infty}\sum_{m=1}^{+\infty}\frac{(k^2Hrms^2\cos^2\theta)^{n+m}}{n!m!}\\ &\iint\left[\left|F_{hv}(u,v)\right|^2+F_{hv}(u,v)F^*_{hv}(-u,-v)\right]W^{(n)}(u-k\sin\theta,v)W^{(m)}(u+k\sin\theta,v)\,du\,dv\end{aligned} \quad [1.17]$$

with:

$$f_{hh}=\frac{-2R_h}{\cos\theta}\ ;\ f_{vv}=\frac{2R_v}{\cos\theta} \quad [1.18]$$

$$F_{hh}=2\frac{\sin^2\theta}{\cos\theta}\left[4R_h-\left(1-\frac{1}{\varepsilon_r}\right)(1+R_h)^2\right] \quad [1.19]$$

$$F_{vv}=2\frac{\sin^2\theta}{\cos\theta}\left[\left(1-\frac{\varepsilon_r\cos^2\theta}{\mu_r\varepsilon_r-\sin^2\theta}\right)(1-R_v)^2+\left(1-\frac{1}{\varepsilon_r}\right)(1+R_v)^2\right] \quad [1.20]$$

$$F_{hv}(u,v)=\frac{uv}{k\cos\theta}\left[\frac{8R^2}{\sqrt{k^2-u^2-v^2}}+\frac{-2+6R^2+\dfrac{(1+R)^2}{\varepsilon_r}+\varepsilon_r(1-R)^2}{\sqrt{\varepsilon_rk^2-u^2-v^2}}\right] \quad [1.21]$$

$$R=\frac{R_v-R_h}{2} \quad [1.22]$$

$$R_h = \frac{\cos\theta - \sqrt{\varepsilon_r(1-\sin^2\theta)}}{\cos\theta + \sqrt{\varepsilon_r(1-\sin^2\theta)}}$$: Fresnel's coefficient for horizontal polarization

$$R_v = \frac{\cos\theta - \sqrt{\frac{1}{\varepsilon_r}(1-\sin^2\theta)}}{\cos\theta + \sqrt{\frac{1}{\varepsilon_r}(1-\sin^2\theta)}}$$: Fresnel's coefficient for vertical polarization

$$W^n(a,b) = \frac{1}{2\pi}\iint \rho^n(x,y)e^{-i(ax+by)}dxdy \qquad [1.23]$$

μ_r: permittivity

θ: incidence angle

k: radar wave number

pp: polarization of the incident wave and the reflected wave; h for horizontal and v for vertical

Re: real part of a complex number

f_{pp}^* : conjugate of the complex number f_{pp}

1.4.2. *Description of the Dubois model*

The Dubois model [DUB 95] is a semi-empirical model for modeling the backscattering coefficient in HH and VV polarizations (σ^0_{HH} and σ^0_{VV}) of bare surfaces as a function of the radar incidence angle (θ), the soil's dielectric constant (ε_r), the standard deviation of heights (*Hrms*), and the radar wavelength ($\lambda = 2\pi/k$, in cm):

$$\sigma^0_{HH} = 10^{-2.75}\left(\frac{\cos^{1.5}\theta}{\sin^5\theta}\right)10^{0.028\varepsilon_r \tan\theta}\left(k\,Hrms\sin\theta\right)^{1.4}\lambda^{0.7} \qquad [1.24]$$

$$\sigma^0_{VV} = 10^{-2.35}\left(\frac{\cos^3\theta}{\sin^3\theta}\right)10^{0.046\varepsilon_r \tan\theta}\left(k\,Hrms\sin\theta\right)^{1.1}\lambda^{0.7} \qquad [1.25]$$

The model's domain of validity is: $k\ Hrms \leq 2.5$, $mv \leq 35$ vol.% and $\theta \geq 30°$.

1.4.3. *Description of the Oh model*

Oh *et al.* developed a semi-empirical backscattering model with numerous versions between 1992 and 2004. The backscattering coefficients $p=\sigma°_{HH}/\sigma°_{VV}$ and $q=\sigma°_{HV}/\sigma°_{VV}$ were defined as a function of the incidence angle (θ), the radar wave number (k), *Hrms*, and the soil's volumetric water content (mv). The expressions of p and q are defined as follows [OH 02, OH 04]:

$$p = 1 - \left(\frac{\theta}{90^{\circ}}\right)^{0.35\,mv^{-0.65}} e^{-0.4(k\,Hrms)^{1.4}} \qquad [1.26]$$

$$q = 0.095\left(0.13 + \sin 1.5\theta\right)^{1.4}\left(1 - e^{-1.3(k\,Hrms)^{0.9}}\right) \qquad [1.27]$$

Furthermore, the backscattering coefficient in HV polarization can be written in the following way:

$$\sigma^{0}_{HV} = 0.11\,mv^{0.7}\cos^{2.2}\theta\left(1 - e^{-0.32(k\,Hrms)^{1.8}}\right) \qquad [1.28]$$

The Oh model's domain of validity is:

$0.13 < k\ Hrms \leq 6.98$, $4 < mv$ (vol.%) ≤ 29.1, and $10° \leq \theta \leq 70°$.

1.4.4. *Numerical modeling*

The equations that govern electromagnetic backscattering can only be solved analytically through the introduction of simplifications limiting the domain of validity of the obtained model. Exact numerical methods like the moment method [HAR 68, ZRI 00] are often used to simulate radar signals for complex soil surface conditions, such as a strong moisture profile or roughness heterogeneities, or to validate new analytical or semi-empirical models. Figure 1.7 illustrates the consistency of numerical simulations with radar data acquired for rough soil.

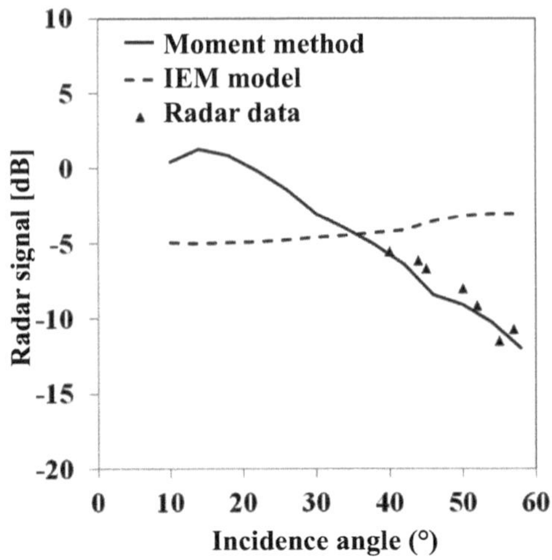

Figure 1.7. *Comparison of numerical simulations (moment method), the IEM, and radar data acquired by the SIR-C sensor (radar frequency = 5.3 GHz, HH polarization, Hrms = 2 cm, L= 9 cm, fractal dimension "D" =1.27, ε = 16.3-2.8i)*

Numerous studies have assessed the different radar backscattering models. Most of the time, strong differences have been observed between model simulations and real SAR data able to reach several decibels, making the results of radar signal inversion imprecise. This difficulty of accurately modeling the real radar signal is particularly linked to the complexity of describing and measuring the soil's roughness parameters (autocorrelation function, correlation length, and standard deviation of heights) precisely.

In recent years, different improvements have been made to radar backscattering models in order to correctly describe the measured signal's behavior as a function of surface parameters, in particular the soil moisture and roughness. Baghdadi *et al.* [BAG 06, BAG 11a, BAG 11b, BAG 15] proposed a semi-empirical calibration of the IEM backscattering model with the intention of better reproducing the radar backscattering coefficient of bare soils. This calibration replaced the measured correlation length, which is the least precise roughness parameter, with a forcing parameter. The results showed that the calibration parameter depends on *Hrms*, radar incidence angle (θ), and radar frequency.

The calibration parameter *Lopt* obtained with a Gaussian autocorrelation function is described as follows (Figure 1.8):

In the X-band:

$$\begin{cases} Lopt\,(Hrms,\theta,HH) = 18.102\,e^{-1.8910\theta}\ Hrms^{0.7644\,e^{0.2005\,\theta}} \\ Lopt\,(Hrms,\theta,VV) = 18.075\,e^{-2.17150\theta}\ Hrms^{1.2594\,e^{-0.8308\,\theta}} \end{cases} \quad [1.29]$$

In the C-band:

$$\begin{cases} Lopt\,(Hrms,\theta,HH) = 0.162 + 3.006\ (\sin 1.23\,\theta)^{-1.494}\ Hrms \\ Lopt\,(Hrms,\theta,HV) = 0.9157 + 1.2289\ (\sin 0.1543\,\theta)^{-0.3139}\ Hrms \\ Lopt\,(Hrms,\theta,VV) = 1.281 + 0.134\ (\sin 0.19\,\theta)^{-1.59}\ Hrms \end{cases} \quad [1.30]$$

In the L-band:

$$\begin{cases} Lopt\,(Hrms,\theta,HH) = 2.6590\,\theta^{-1.4493} + 3.0484\,Hrms\,\theta^{-0.8044} \\ Lopt\,(Hrms,\theta,VV) = 5.8735\,\theta^{-1.0814} + 1.3015\,Hrms\,\theta^{-1.4498} \end{cases} \quad [1.31]$$

θ is in radians; *Lopt* and *Hrms* are in centimeters. This calibration has been performed using numerous databases, acquired at numerous study sites, with different radar sensors (ERS, JERS, RADARSAT, ASAR, PALSAR, TerraSAR-X, COSMO-SkyMed, SIR-C/X), with incidence angles from 23° to 57°, and HH, HV and VV polarizations.

Using the IEM with the calibration parameter *Lopt* allows considerable improvement of the adequacy between the model and SAR data, regardless of the radar configuration and the study site (reduction of the root mean square errors). Furthermore, with this calibration, bare agricultural soil can be defined with only two surface parameters (standard deviation of heights "*Hrms*" and soil moisture "*mv*") instead of four (*Hrms*, *L*, autocorrelation function and *mv*).

1.5. Estimation of soil parameters at plot scale based on high and very high spatial resolution data

1.5.1. *Case of bare soil*

The estimation of soil parameters, particularly water content and roughness, is in most cases done by inverting either simple empirical relationships between the radar signal and soil parameters or a radar backscattering model (such as the IEM, Dubois or Oh).

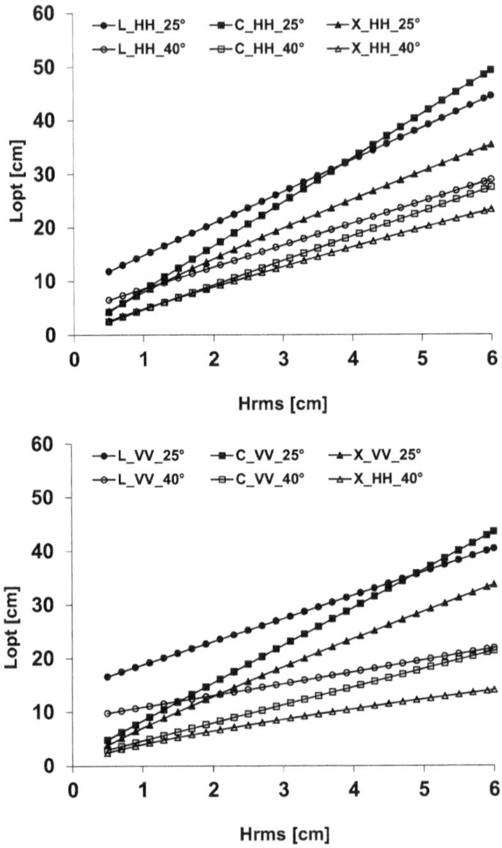

Figure 1.8. *The IEM's calibration parameter "Lopt" as a function of soil roughness "Hrms" for the L-, C- and X-bands and for the two polarizations HH and VV*

1.5.1.1. *Using a single SAR configuration*

In the case of a single radar layer (one date, one polarization, one incidence angle), simple empirical relationships, linear for the soil moisture and logarithmic (or exponential) for its roughness, are built using reference data collected on the study site's chosen plots. These reference data refer to *in situ* measurements of soil moisture and/or roughness, as well as the mean of the backscattering coefficient on each reference plot calculated using acquired radar images.

In this case, the radar backscattering coefficient (in decibels) is defined by a single soil parameter (soil moisture or roughness), while the other soil parameter is assumed to be constant or have little effect on the radar signal (this configuration only allows one soil parameter to be determined at a time). The SAR sensors' parameters are then chosen to increase the signal's sensitivity to the soil parameter to be estimated (e.g. moisture) and to minimize the effect of the other soil parameters (e.g. roughness). For the estimation of soil moisture, the following relation is used:

$$\sigma^0_{dB} = \alpha mv + b_1 \quad [1.32]$$

The following expression is used for mapping roughness:

$$\sigma^0_{dB} = \beta \log(k\,Hrms) + b_2 \quad [1.33]$$

For a given radar configuration (one polarization, one incidence angle, and one radar wavelength), b_1 depends on the soil's roughness, while b_2 depends on the soil moisture. The coefficients α (signal sensitivity to soil moisture) and β are dependent on the radar frequency, the polarization, and the incidence angle.

This monolayer approach was developed for the first spatial radars, which were mono-configuration (ERS-1/2, RADARSAT-1) [BAG 02], but which have also been used by second-generation multi-polarization sensors (ASAR, RADARSAT-2, TerraSAR-X, COSMO-SkyMed). Figure 1.9 shows examples of mapping the soil moisture and roughness using monolayer images. The relationships between the radar signal and soil parameters (aforementioned relations) strongly depend on the study site's characteristics, which makes calibration necessary for each site.

The precision obtained for the estimation of soil moisture differs from one site to another, but we could summarize the results in the following way: (1) by using a weak incidence (around 20°), the precision is roughly 5 vol.% in the C- and L-bands and roughly 3 vol.% in the X-band (stronger signal sensitivity to soil moisture). For a radar image with a strong incidence (>45°), the precision of the estimation of mv is roughly 10 vol.% in the C- and L-bands versus 4 vol.% in the X-band.

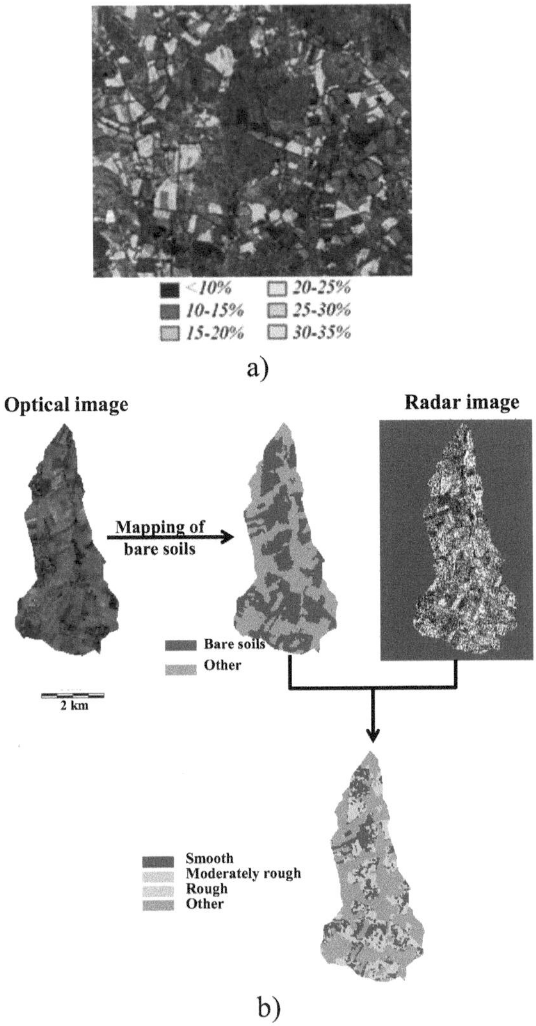

Figure 1.9. *a) Moisture map obtained for a study site in the Paris region (France) using TerraSAR-X data (HH-25°; $\sigma^0_{dB} = 0.411\,mv - 15.7$); b) soil roughness map according to three roughness classes on a site in Normandy (France) using a RADARSAT-1 image (HH-47°; $\sigma^0_{dB} = -9.81 - 9.46e^{-k\,Hrms}$): smooth surfaces with Hrms < 1 cm, rough surfaces with Hrms > 2 cm, and moderately rough surfaces with 1 cm < Hrms < 2 cm. For a color version of this figure, see www.iste.co.uk/baghdadi/4.zip*

1.5.1.2. *Using multi-incidence SAR data*

The launch of multi-incidence SAR satellites (ASAR, RADARSAT-2, TerraSAR-X and COSMO-SkyMed) has given the scientific community a greater ability to estimate the soil moisture and surface roughness simultaneously, even if multi-incidence acquisitions actually represent two acquisitions on two different dates. However, it is assumed that the soil parameters have not changed between the two acquisition dates. For this configuration, the radar images are acquired at two different incidence angles, generally one image with a weak incidence ($\theta_{weak} \sim 20°$) and one image with a strong incidence ($\theta_{strong} \sim 40°$).

In this case, the dependence between the radar signal and the two soil parameters is described by defining the radar backscattering coefficient σ^0_{dB} as the sum of the linear function that describes the dependence between the signal and the soil moisture, and the logarithmic function that illustrates the dependence between the signal and roughness:

$$\begin{cases} \sigma^0(\theta_{weak}) = \alpha_1\, mv + \beta_1\, \log(k\, Hrms) + c_1 \\ \sigma^0(\theta_{strong}) = \alpha_2\, mv + \beta_2\, \log(k\, Hrms) + c_2 \end{cases} \qquad [1.34]$$

By replacing the term $\log(k\, Hrms)$ in $\sigma^0(\theta_{weak})$ with one $\sigma^0_{dB}(\theta_{strong})$, the soil moisture can be estimated using the following relation:

$$mv = A\sigma^0(\theta_{weak}) + B\sigma^0(\theta_{strong}) + C \qquad [1.35]$$

Several studies used the relation $\sigma^0(\theta_{weak}) / \sigma^0(\theta_{strong})$ when formulating the inversion procedure because of this relation's maximum sensitivity to soil roughness [ZRI 03]:

$$mv = A\sigma^0\left(\theta_{weak}\ or\ \theta_{strong}\right) + B\left(\frac{\sigma^0\left(\theta_{weak}\right)}{\sigma^0\left(\theta_{strong}\right)}\right)_{dB} + C \qquad [1.36]$$

The roughness parameter *Hrms* can be estimated by using the relation:

$$Hrms = \frac{1}{k} 10^{D\sigma^0(\theta_{weak}) + E\sigma^0(\theta_{strong}) + F} \qquad [1.37]$$

The results of different studies show a clear improvement in the C-band of the precision of soil moisture estimates when two images at two incidence angles are used (one weak and one strong) in comparison with the case of a single image [ZRI 03]. The precision thus increases from around 6 vol.% with a single incidence angle to about 3.5 vol.% with two incidence angles. Unlike in the C-band, the precision in the X-band is not improved when two incidences are used instead of just one (an image with a weak or strong incidence angle) [BAG 12a]. This is especially due to the decreased sensitivity to soil roughness of measurements in the X-band (compared to the C-band).

1.5.1.3. *Using multi-polarization SAR data*

The arrival of multi-polarization satellite sensors (two polarizations simultaneously) since the 2002 launch of ASAR/ENVISAT encouraged scientists to analyze the potential of this acquisition mode – already tested occasionally from airborne measurements – for estimating soil parameters. In fact, the radar signal can exhibit different behaviors depending on the polarization. As an example, the ratio between the HH and VV polarizations is very sensitive to soil roughness. It increases with *Hrms* to reach a value near 1 for strong soil roughness values [OH 92].

Studies realized using SAR images have shown that using images with two polarizations does not improve the estimation of soil moisture, for example, in comparison with cases where a single polarization is used. In fact, the improvement is less than 1 vol.% [BAG 06]. [PAL 13] identified an improvement of the estimation of soil moisture when using several polarizations.

Oh's semi-empirical model is based on the ratio between the polarizations $\sigma°_{HH}/\sigma°_{VV}$ and $\sigma°_{HV}/\sigma°_{VV}$ for estimating the soil moisture and roughness. However, data in polarizations HH, HV and VV are needed for this purpose. Finally, the availability of images in HH and VV allows the soil parameters *Hrms* and *mv* to be estimated by inverting the Dubois model, which links the signals in HH and VV to the soil's parameters.

1.5.1.4. *Using polarimetric SAR data*

Radar polarimetry has also been assessed to study its potential for improving the estimation of soil moisture and roughness. Studies have generally been done using airborne data or the polarimetric SAR satellites RADARSAT-2 (C-band) and PALSAR-1 (L-band).

[HAJ 03] have shown that entropy and the alpha angle (derived from the Cloude decomposition) increase with soil moisture and that anisotropy is independent of the soil moisture. Furthermore, it has been shown that entropy increases with *Hrms*, that the alpha angle is independent of *Hrms*, and that when *k.Hrms* increases to 1, anisotropy decreases in an almost linear way.

Contrary to the results obtained in the L-band, the results obtained in the C-band show that the use of polarimetric parameters (entropy, alpha angle, anisotropy, etc.) does not significantly improve the estimation of the soil moisture and roughness in comparison with methods that only use the backscattering coefficients in polarizations HH, HV, or VV [BAG 12b]. In fact, the dynamic of the polarimetric parameters in C-band is too weak to allow their direct use in the process of estimating soil parameters. Finally, a recent study done using airborne and satellite SAR data in the X-, C-, L- and P-bands (RAMSES, SETHI, RADARSAT-2) revealed that the main polarimetric parameters (entropy, alpha angle, and anisotropy) are not particularly sensitive to variations of the soil's parameters [BAG 13].

1.5.1.5. *Using multi-date SAR data*

Exploiting the potential of time series of radar measurements on continental surfaces essentially started with the development of change detection techniques based on data from scatterometers for estimating the soil moisture [WAG 99]. For some years now, this approach has been more and more widely used to offer operational estimates of the soil moisture using SAR measurements [ZRI 07]. In order to estimate the soil moisture in semi-arid zones, where it can be assumed that the soil roughness is constant throughout the year, a methodology based on the difference between an image from the wet season and an image from the dry season has been developed. The difference between the two images ($\Delta\sigma°$) allows the effects of roughness to be eliminated and only to depend upon the soil moisture:

$$\Delta\sigma°_{dB} = \alpha\,(mv_{wet\ season} - mv_{dry\ season}) = \alpha\Delta mv \qquad [1.38]$$

Studies carried out in Sahelian areas using satellite radar images [BAG 11c, ZRI 07] have shown good results. The relationship between $\Delta\sigma°$

and soil moisture is firstly calibrated using auxiliary data on some reference plots, then inverted on the radar image (bare soil or with low-density vegetation cover) to map the soil moisture with a spatial resolution generally consisting of 100 m to 1 km (depending on the resolution of the images used). In fact, the average of the $\Delta\sigma°$ is calculated for each resolution unit (which is generally several pixels) and the map of Δmv is then produced using the relation $\Delta mv = \Delta\sigma°_{dB} / \alpha$. To calculate the soil moisture on a given date during the wet season, the mean of the soil moisture on the date of the image during the dry season (assumed to be known) is to be added to Δm_v.

Figure 1.10 shows that the contribution of roughness on the backscattered signal has been eliminated since there is not an overestimation of the soil moisture at the border between low lands and plateaus (strong soil roughness and large slope). Furthermore, the soil moisture estimated on the plateaus is close to 0, which is closer to field observations.

Figure 1.10. *Example of a soil moisture map on a study site in Niger (Banizoumbou) using TerraSAR-X images (1 m spatial resolution) on 100 m x 100 m grids. The areas with strong vegetation cover have been hidden, as well as residential areas. The image covers 10 km x 10 km. For a color version of this figure, see www.iste.co.uk/ baghdadi/4.zip*

1.5.2. *Case of soil with vegetation cover*

Spatial–temporal monitoring of the soil moisture on plots with vegetation cover is a key parameter for different applications, notably for managing irrigation and estimating surface fluxes. Different theoretical and semi-empirical models have been developed to eliminate the effect of vegetation cover and estimate soil moisture. The most-used model is the water cloud model (WCM), a very simple, but very effective semi-empirical model developed by [ATT 78]. It defines the backscattered radar signal (on a linear scale) as the sum of the contributions from the vegetation (σ°_{veg}), the soil attenuated by the vegetation ($T^2\ \sigma^{\circ}_{soil}$), and multiple soil-vegetation scatterings (often neglected):

$$\sigma^{o} = \sigma^{o}_{veg} + T^2 \sigma^{o}_{soil}$$

$$\sigma^{o}_{veg} = A\, V_1 \cos\theta \left(1 - T^2\right) \qquad [1.39]$$

– T^2 is the attenuation by the vegetation layer; V_1 and V_2 are the vegetation's descriptors: biomass (kg/m^2), vegetation water content (kg/m^2), vegetation height (m), or leaf area index "LAI" (m^2/m^2);

– θ is the radar incidence angle;

– A and B are parameters that depend on the chosen vegetation's descriptor(s) and the radar configuration.

The soil's contribution can be simulated using a physical model (IEM in particular), with a semi-empirical model (e.g. the Oh and Dubois models), or also even through a simplified expression of the soil's contribution taking into account the soil moisture and roughness (on a linear scale):

$$\sigma^{o}_{soil} = C\, e^{D\, mv} \qquad [1.40]$$

This expression assumes that the soil's signal is the sum of a contribution due to the soil moisture (in dB) and another due to its roughness (in dB):

– C is a parameter that depends on the soil's roughness and the radar configuration;

– D is the radar signal's sensitivity (linear unit) to the bare soil moisture (it depends on the radar configuration).

Backscattering coefficient $\sigma°$ data derived from SAR images, field measurements of the soil's parameters (moisture and roughness, or only moisture if the effect of roughness is neglected), and measurements of vegetation descriptors (*in situ* or from imaging, particularly optical) are necessary to adjust the WCM and estimate parameters A, B, C and D.

Studies on irrigated grasslands [ELH 16] show that soil moisture can be estimated using SAR data in the X-band with a precision in the order of 5 vol.% for LAI values under 3 m^2/m^2, and in the order of 7 vol.% for LAI values between 3 and 6 m^2/m^2. Equivalent results have also been obtained in the C-band with an error in the order of 6 vol.% in estimating soil moisture, even for plots with strong vegetation cover (LAI going up to 5 m^2/m^2, biomass 3 kg/m^2, vegetation water content 2.5 kg/m^2, and vegetation height of 1 m) [BAG 16]. These estimates of *mv* have been done using LAI – measured or estimated using optical data – as the vegetation descriptor.

This same approach is also very widely used for annual crops, particularly cereals. With ASAR/ENVISAT data in synergy with satellite optical measurements, [ZRI 11] obtained a precision in the order of 4 vol.% for LAI under 3m^2/m^2 when estimating soil moisture in semi-arid zones.

Inversion approaches are generally based on a direct inversion of the semi-empirical model developed by minimizing a cost function or applying other techniques, such as neural networks or look up tables [PAL 13].

1.6. Estimation of soil parameters with medium spatial resolution

The studies developed for estimating the soil moisture with SAR are essentially realized at the plot scale. For different regional hydrological applications, estimating the soil moisture is requested with a medium spatial resolution, consisting of between 100 m and several kilometers. The proposed methodologies are generally based on SAR data with a medium spatial resolution or the aggregation of high spatial resolution SAR data. Two approaches are generally used to retrieve the soil moisture:

– The first is based on the inversion of a semi-empirical model linking the radar signal with the soil moisture. [LEH 02] have developed an approach for estimating the mean soil moisture at the scale of a watershed with ERS data (SAR with 25 m spatial resolution) in several steps: (1) classification of the satellite images (optical and/or radar) to identify land use, (2) in the case of vegetation cover, correction of the effect of vegetation using a radiative transfer model, (3) aggregation of the radar signal estimated at the watershed scale. The effect of roughness is considered weak at that spatial scale. The obtained signal is linked to the mean soil moisture calculated at the watershed scale. A precision of 4 vol.% is obtained on the estimated mean soil moisture. The same approach has been applied on olive groves in semi-arid zones (Tunisia) with ASAR/ENVISAT data, producing the soil moisture at a spatial resolution of 1 km [ZRI 11]. Figure 1.11 illustrates the temporal evolution of the soil moisture for the 12 hours before and after the arrival of a rainfall event.

– A second is based on the use of a large multi-temporal database of SAR acquisitions. It considers the change detection technique frequently used for globally mapping soil moisture [WAG 99]. It is developed in several steps: (1) normalization of radar data at a single incidence, (2) elimination of the effect of vegetation cover on the signal by relying on the angular behavior of SAR data or using optical ancillary data, (3) estimation of a relative moisture between 0% (the driest conditions) and 100% (the wettest conditions) using the normalized radar signal corrected from the effect of vegetation cover (linear dependence between signal and soil moisture). [ZRI 14a] tested this approach on semi-arid zones using medium resolution ASAR/ENVISAT data (150 m) together with optical SPOT-VEGETATION data. The precision of the soil moisture estimation is in the order of 4.5 vol.%.

1.7. Prospects

Studies carried out in the last 20 years have allowed the development and validation of innovative methodologies for exploiting and enhancing radar data. Different approaches have been developed with the objective of monitoring soil surface characteristics from the intra-plot level to that of the regional basin.

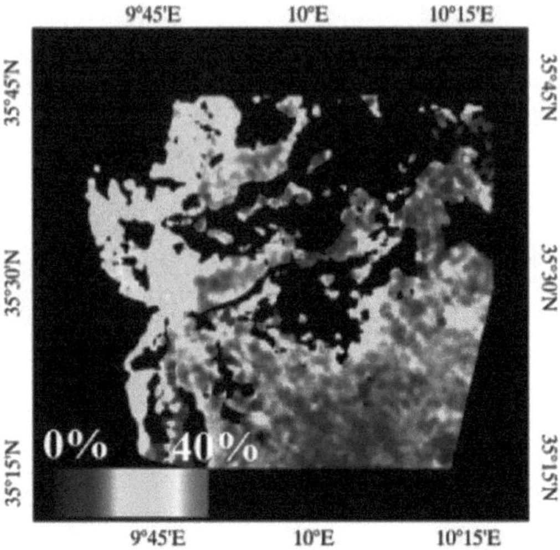

a)

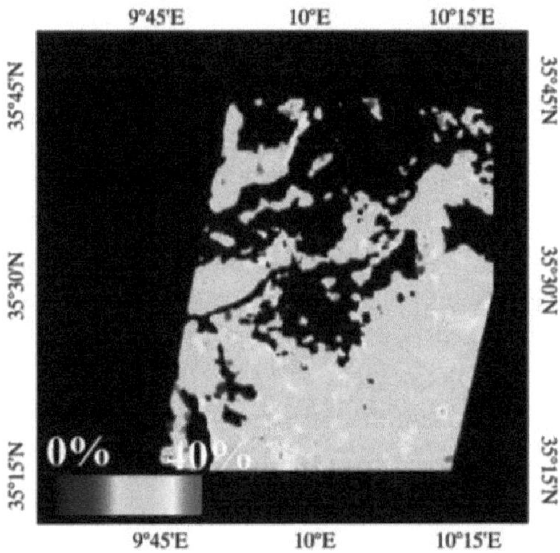

b)

Figure 1.11. *Mapping of the soil moisture on olive groves a) April 11, 2009, b) April 12, 2009 [ZRI 11]. For a color version of this figure, see www.iste.co.uk/baghdadi/4.zip*

SAR imaging is still not a tool commonly used by hydrologists and agronomists, even if certain applications have been developed for assimilating those surface parameters in agronomical or hydrological models [AUB 03, OUD 03]. This delay in use can be attributed to the lack of dedicated sensors, proposing configurations adapted to the needs, particularly in revisit time, as well as the lack of adequate algorithms allowing simple and precise estimates of soil parameters. Furthermore, SAR sensors allow the estimation of the soil moisture of superficial soil layers (max ~ 10 cm), while the root zone moisture profile (~ 1 m) is the essential parameter in agronomy and hydrology for understanding the functioning of soil-vegetation-atmosphere interactions. Numerous studies have recently been conducted in this field and this research axis remains a priority in numerous communities.

With the arrival of new SAR sensors such as the European Space Agency's Sentinel-1 constellation (free of charge products), the operational phase should be reached quickly, as for global products at a low spatial resolution, in order to offer satellite products with spatial and temporal resolutions adapted to the needs of assimilation in hydrological and surface models from the local to the regional scale.

It also seems essential to develop further the synergy with low-resolution microwave sensors (ASCAT, SMOS, etc.) for a disaggregation of low-resolution soil moisture products, but also the other spatial techniques, particularly optical, and the synergy between radar data with different wavelengths (X-, C- and L-bands). The arrival of new missions, such as BIOMASS (P-band), should expand the radar's potential towards the exploitation of the subsurface.

Significant efforts have been made in Europe (Copernicus) and in France in particular (Geosud, Theia, etc.) in recent years to create infrastructures and operational services for exploiting observation of Earth data. These initiatives focus on bringing about operational responses to user needs. They should also encourage the transfer of satellite techniques to the operational, particularly by offering new possibilities for accessing data, reducing costs and favoring image access for thematic research (hydrology, agronomy, biodiversity, etc.) and services for those in charge of public policy.

1.8. Key points

Soil surface characteristics play a key role in the understanding of different hydrological, meteorological, agronomical and similar processes. In the last 30 years, radar remote sensing has shown great potential in estimating soil surface characteristics. The strong sensitivity of the radar signal to the surface geometry (roughness) and the soil moisture has allowed the development of different inversion approaches for monitoring these soil parameters. These methods are based on multi-incidence, multi-polarization, or multi-date configurations. In the case of soil covered with vegetation, there are approaches based on radar/optical synergy to eliminate the effect of the vegetation and deduce the soil's characteristics. Aside from the two main parameters (soil moisture, surface roughness), there is also a scientific dynamic for estimating the soil's salinity and texture.

1.9. Bibliography

[ATT 78] ATTEMA E.W.P., ULABY F.T., "Vegetation modelled as a water cloud", *Radio Science*, vol. 13, no. 2, pp. 357–364, 1978.

[AUB 03] AUBERT D., LOUMAGNE C., OUDIN L., "Sequential assimilation of soil moisture and streamflow data in a conceptual rainfall-runoff model", *Journal of Hydrology*, vol. 280, pp. 145–161, 2003.

[AUB 11] AUBERT M., BAGHDADI N., ZRIBI M. *et al.*, "Analysis of TerraSAR-X data sensitivity to bare soil moisture, roughness, composition and soil crust", *Remote Sensing of Environment*, vol. 115, pp. 1801–1810, 2011.

[BAG 02] BAGHDADI N., KING C., BOURGUIGNON A. *et al.*, "Potential of ERS and RADARSAT data for surface roughness monitoring over bare agricultural fields: application to catchments in Northern France", *International Journal of Remote Sensing*, vol. 23, no. 17, pp. 3427–3442, 2002.

[BAG 06] BAGHDADI N., HOLAH N., ZRIBI M., "Calibration of the Integral Equation Model for SAR data in C-band and HH and VV polarizations", *International Journal of Remote Sensing*, vol. 27, no. 4, pp. 805–816, 2006.

[BAG 08] BAGHDADI N., CERDAN O., ZRIBI M. *et al.*, "Operational performance of current synthetic aperture radar sensors in mapping soil surface characteristics: application to hydrological and erosion modelling", *Hydrological Processes*, vol. 22, no. 1, pp. 9–20, 2008.

[BAG 11a] BAGHDADI N., ABOU CHAAYA J., ZRIBI M., "Semi-empirical calibration of the Integral equation Model for SAR data in C-band and cross polarization using radar images and field measurements", *IEEE Geoscience and Remote Sensing Letters*, vol. 8, no. 1, pp.14-18, 2011.

[BAG 11b] BAGHDADI N., SABA E., AUBERT M. *et al.*, "Comparison between backscattered TerraSAR signals and simulations from the radar backscattering models IEM, Oh, and Dubois", *IEEE Geoscience and Remote Sensing Letters*, vol. 8, no. 6, pp.1160–1164, 2011.

[BAG 11c] BAGHDADI N., CAMUS P., BEAUGENDRE N. *et al.*, "Estimating surface soil moisture from TerraSAR-X data over two small catchments in the Sahelian part of western Niger", *Remote Sensing*, vol. 3, pp. 1266–1283, 2011.

[BAG 12a] BAGHDADI N., AUBERT M., ZRIBI M., "Use of TerraSAR-X data to retrieve soil moisture over bare soil agricultural fields", *IEEE Geoscience and Remote Sensing Letters*, vol. 9, no. 3, pp. 512–516, 2012.

[BAG 12b] BAGHDADI N., CRESSON R., POTTIER E. *et al.*, "A potential use for the C-band polarimetric SAR parameters to characterise the soil surface over bare agriculture fields", *IEEE Transactions on Geoscience and Remote Sensing*, vol. 50, no. 10, pp. 3844–3858, 2012.

[BAG 13] BAGHDADI N., DUBOIS-FERNANDEZ P., DUPUIS X. *et al.*, "Sensitivity of multi-frequency (X, C, L, P and UHF-band) polarimetric SAR data to soil moisture and surface roughness over bare agricultural soils", *IEEE Geoscience and Remote Sensing Letters*, vol. 10, no. 4, pp. 731–735, 2013.

[BAG 15] BAGHDADI N., ZRIBI M., PALOSCIA S. *et al.*, "Semi-empirical calibration of the Integral Equation Model for co-polarized L-band backscattering", *Remote Sensing*, vol. 7, pp. 13626–13640, 2015.

[BAG 16] BAGHDADI N., EL HAJJ M., ZRIBI M. *et al.*, "Coupling SAR C-band and optical data for soil moisture and leaf area index retrieval over irrigated grasslands", *IEEE Journal of Selected Topics in Applied Earth Observations and Remote Sensing*, vol. 9, no. 3, pp. 1229–1243, 2015.

[BEA 90] BEAUDOIN A., LE TOAN T., GWYN Q.H.J., "SAR observations and modeling of the C-band backscatter variability due to multiscale geometry and soil moisture", *IEEE Transactions on Geoscience and Remote Sensing*, vol. 28, pp. 886–895, 1990.

[BEC 63] BECKMAN P., SPIZZICHINO A., *The Scattering of Electromagnetic Waves from Rough Surfaces*, Pergammon Press, New York, 1963.

[CHE 03] CHEN K.S., WU T.D., TSANG L. *et al.*, "Emission of rough surfaces calculated by the Integral Equation Method with comparison to three-dimensional moment method simulations", *IEEE Transactions on Geoscience and Remote Sensing*, vol. 41, pp. 90–101, 2003.

[DAV 00] DAVIDSON M.W., LE TOAN T., MATTIA F. *et al.*, "On the characterization of agricultural soil roughness for radar remote sensing studies", *IEEE Transactions on Geoscience and Remote Sensing*, vol. 38, pp. 630–640, 2000.

[DOB 85] DOBSON M.C., ULABY F.T., HALLIKAINEN M.T. *et al.*, "Microwave dielectric behavior of wet soil- Part II: Dielectric mixing models", *IEEE Transactions on Geoscience and Remote Sensing*, GE-23, pp. 35–46, 1985.

[DUB 95] DUBOIS P.C., VAN ZYL J., ENGMAN T., "Measuring soil moisture with imaging radars", *IEEE Transactions on Geoscience and Remote Sensing*, vol. 33, pp. 915–926, 1995.

[ELH 16] EL HAJJ M., BAGHDADI N., ZRIBI M. *et al.*, "Soil moisture retrieval over irrigated grassland using X-band SAR data", *Remote Sensing of Environment*, vol. 176, pp. 202–218, 2016.

[FUN 92] FUNG A.K., LI Z. CHEN K.S., "Backscattering from a randomly rough dielectric surface", *IEEE Transactions on Geoscience and Remote Sensing*, vol. 30, pp. 356–369, 1992.

[HAJ 03] HAJNSEK I., POTTIER E., CLOUDE S., "Inversion of surface parameters from polarimetric SAR", *IEEE Transactions on Geoscience and Remote Sensing*, vol. 41, no. 4, pp. 727–744, 2003.

[HAL 85] HALLIKAÏNEN M.T., ULABY F.T., DOBSON F.T. *et al.*, "Microwave dielectric behavior of wet soil. Part I: Empirical models and experimental observations", *IEEE Transactions on Geoscience and Remote Sensing*, vol. 23, pp. 25–34, 1985.

[HAR 68] HARRINGTON R.F., *Field Computation by Moment Method*, IEEE Press, 1968.

[LAS 08] LASNE Y., PAILLOU P., FREEMAN A. *et al.*, "Effect of salinity on the dielectric properties of geological materials: Implication for soil moisture detection by means of radar remote sensing", *IEEE Transactions on Geoscience and Remote Sensing*, vol. 46, no. 6, pp. 1674–1688, 2008.

[LEH 02] LE HÉGARAT-MASCLE S., ZRIBI, M., ALEM, F. *et al.*, "Soil moisture estimation from ERS/SAR data: Toward an operational methodology", *IEEE Transactions on Geoscience and Remote Sensing*, vol. 40, no. 12, pp. 2647–2658, 2002.

[LET 82] LE TOAN T., "Active microwave signatures of soil and crops- Significant results of three years of experiments", *Proceeding of the International Geoscience and Remote Sensing Symposium, IGARSS 82*, Munich, Germany, New York, pp. 25–32, 1–4 June 1982.

[LIE 09] LIEVENS H., VERNIEUWE H., ALVAREZ-MOZOS J. *et al.*, "Error in radar-derived soil moisture due to roughness parameterization: an analysis based on synthetical surface profiles", *Sensors Journal*, vol. 9, pp. 1067–1093, 2009.

[LOU 91] LOUMAGNE C., MICHEL C., NORMAND M., "Soil water conditions and forecasting rain water runoff", *Journal of Hydrology*, vol. 123, pp. 1–17, 1991.

[MAT 99] MATTIA F., LE TOAN T., "Backscattering properties of multi-scale rough surfaces", *J. Electro. Waves Appl.*, vol. 13, pp. 491–526, 1999.

[MIR 04] MIRONOV V.L., DOBSON M.C., KAUPP, V.H. *et al.*, "Generalized refractive mixing dielectric model for moist soils", *IEEE Transactions on Geoscience and Remote Sensing*", vol. 42, no. 4, pp. 773–785, 2004.

[NAR 15] NARVEKAR P.S., ENTEKHABI D., KIM S. B. *et al.*, "Soil moisture retrieval using L-band radar observations", *IEEE Transactions on Geoscience and Remote Sensing*, vol. 53, no. 6, pp. 3492–3506, 2015.

[OH 92] OH Y., SARABANDI K., ULABY F.T., "An empirical model and an inversion technique for radar scattering from bare soil surfaces", *IEEE Transactions on Geoscience and Remote Sensing*, vol. 30, no. 2, pp. 370–381, 1992.

[OH 98] OH Y., KAY Y., "Condition for precise measurement of soil surface roughness", *IEEE Transactions on Geoscience and Remote Sensing*, vol. 36, no. 2, pp. 691–695, 1998.

[OH 02] OH Y., SARABANDI K., ULABY F.T., "Semi-empirical model of the ensemble-averaged differential Mueller matrix for microwave backscattering from bare soil surfaces", *IEEE Transactions on Geoscience and Remote Sensing*, vol. 40, pp. 1348–1355, 2002.

[OH 04] OH Y., "Quantitative retrieval of soil moisture content and surface roughness from multipolarized radar observations of bare soil surfaces", *IEEE Transactions on Geoscience and Remote Sensing*, vol. 42, pp. 596–601, 2004.

[OUD 03] OUDIN L, WEISS A., LOUMAGNE C. *et al.*, "Assimilation of soil moisture into hydrological models for floods forecasting a variational approach", *Canadian Journal of Remote Sensing*, vol. 29, no. 6, pp. 679–686, 2003.

[PAL 08] PALOSCIA S., PAMPALONI P., PETTINATO S. *et al.*, "A comparison of algorithms for retrieving soil moisture from ENVISAT/ASAR images", *IEEE Transactions on Geoscience and Remote Sensing*, vol. 46, pp. 3274–3284, 2008.

[PAL 13] PALOSCIA S., PETTINATO S., SANTI E. *et al.*, "Soil moisture mapping using Sentinel-1 images: Algorithm and preliminary validation", *Remote Sensing of Environment*, vol. 134, pp. 234–248, 2013.

[RIC 51] RICE S.O., "Reflection of electromagnetic waves from slightly rough surfaces", *Communications in Pure and Applied Mathematics*, vol. 4, pp. 361–378, 1951.

[SHA 03] SHAO Y., HU Q., GUO H. *et al.*, "Effect of dielectric properties of moist salinized soils on backscattering coefficients extracted from RADARSAT image", *IEEE Transactions on Geoscience and Remote Sensing*, vol. 41, no. 8, pp. 1879–1888, 2003.

[TOP 80] TOPP G.C., DAVIS J.L., ANNAN A.P., "Electromagnetic determination of soil water content: Measurement in coaxial transmission lines", *Water Resources Research*, vol. 16, pp. 547–582, 1980.

[ULA 78] ULABY F.T., BATLIVALA P.B., DOBSON M.C., "Microwave backscatter dependence on surface roughness, soil moisture and soil texture: Part I – Bare soil", *IEEE Transactions on Geoscience and Remote Sensing*, vol. 16, pp. 286–295, 1978.

[ULA 86] ULABY F.T., MOORE R.K., FUNG A.K., *Microwave Remote Sensing: Active and Passive*, Norwood, Artech House Inc., 1986.

[VER 08] VERHOEST N.E.C., LIEVENS H., WAGNER W. *et al.*, "On the soil roughness parameterization problem in soil moisture retrieval of bare surfaces from Synthetic Aperture Radar", *Sensors*, vol. 8, no. 7, pp. 4213–4248, 2008.

[WAG 99] WAGNER W., LEMOINE G., ROTT H., "A method for estimating soil moisture from ERS scatterometer and soil data", *Remote Sensing of Environment*, vol. 70, pp. 191–207, 1999.

[ZRI 98] ZRIBI M., Développement de nouvelles méthodes de modélisation de la rugosité pour la rétrodiffusion hyperfréquence de la surface du sol, PhD Thesis, University of Toulouse, France, 1998.

[ZRI 00] ZRIBI M., CIARLETTI V., TACONET O., "Validation of a rough surface model based on fractional Brownian geometry with SIRC and ERASME radar data over Orgeval site", *Remote Sensing of Environment*, vol. 73, pp. 65–72, 2000.

[ZRI 02] ZRIBI M., CIARLETTI, V., TACONET, O. *et al.*, "Effect of rows structure on radar microwave measurements over soil surface", *International Journal of Remote Sensing*, vol. 23, no. 24, pp. 5211–5224, 2002.

[ZRI 03] ZRIBI M., DECHAMBRE M., "A new empirical model to retrieve soil moisture and roughness from radar data", *Remote Sensing of Environment*, vol. 84, pp. 42–52, 2003.

[ZRI 05] ZRIBI M., BAGHDADI N., HOLAH N. "New methodology for soil surface moisture estimation and its application to ENVISAT-ASAR multi-incidence data inversion", *Remote Sensing of Environment*, vol. 96, pp. 485–496, 2005.

[ZRI 07] ZRIBI M., SAUX-PICART S., ANDRÉ C. *et al.*, "Soil moisture mapping based on ARSAR/ENVISAT radar data over a sahelian site", *International Journal of Remote Sensing*, vol. 28, no. 16, pp. 3547–3565, 2007.

[ZRI 11] ZRIBI M., CHAHBI A., LILI Z. *et al.*, "Soil surface moisture estimation over a semi-arid region using ENVISAT ASAR radar data for soil evaporation evaluation", *Hydrology and Earth System Sciences*, vol. 15, pp. 345–358, 2011.

[ZRI 12] ZRIBI M., KOTTI F., LILI-CHABAANE, Z. *et al.*, "Soil texture mapping over a semi-arid area using TERRASAR-X radar data over a semi-arid area", *IEEE Transactions on Geoscience and Remote Sensing Letters*, vol. 9, no. 3, pp. 353–357, 2012.

[ZRI 13] ZRIBI M., BAGHDADI N., LE HÉGARAT S. *et al.*, "Suivi de l'état hydrique du sol par télédétection radar", in LOUMAGNE C., TALLEC G. (eds), *L'observation long terme en environnement, Exemple du bassin de l'Orgeval*, Editions QUAE, Versailles, 2013.

[ZRI 14a] ZRIBI M., GORRAB A., BAGHDADI N., "A new soil roughness parameter for the modelling of radar backscattering over bare soil", *Remote Sensing of Environment*, vol. 152, pp. 62–73, 2014.

[ZRI 14b] ZRIBI M., GORRAB A., BAGHDADI N. *et al.*, "Influence of radar frequency on the relationship between bare surface soil moisture vertical profile and radar backscatter", *IEEE Geoscience and Remote Sensing Letters*, vol. 11, no. 4, pp. 848–852, 2014.

2

Estimation of Soil Water Conditions Using Passive Microwave Remote Sensing

2.1. General introduction

The water content of soil, or soil moisture, is a key element in several fields, such as hydrology [ANC 04], meteorology [SEN 10, KOS 11], agriculture [CRO 10], and forestry [LAV 05]. This statement is explained by the determining role of soil moisture in the processes (infiltration, runoff, evaporation, etc.) governing the water cycle and the global energy balance [JUN 10]. This role has resulted in the recognition of soil moisture as an essential climate variable [GCO 06] by the Global Climate Observing System (GCOS). However, as a result of its dependence on several factors, the spatial and temporal variability of soil moisture is very complex [FAM 08]. In addition to precipitation and evapotranspiration, it is linked to surface characteristics such as the type of soil and vegetation, the topography, the surface roughness and so on [GOW 02]. In order to analyze the variability of soil moisture, dense network stations for soil moisture observation have been temporarily installed during short field campaigns, among them the Cold Land Processes Field Experiment (CLPX) in northern Colorado [ELD 09], the Soil Moisture Experiment in 2002 (SMEX02) in Iowa [FAM 08, LAK 13], the Canadian Experiment for Soil Moisture in 2010 (CanEx-SM10) and the SMAP Validation Experiment in 2012

Chapter written by Ramata MAGAGI, Yann KERR and Jean-Pierre WIGNERON.

(SMAPVEX12) in the Canadian Prairies [MAG 13, MCN 15]. For continuous soil moisture data, the installation of dense networks capable of providing historical series of soil moisture is ideal, but expensive [CHA 10, COS 13]. As a result, the International Soil Moisture Network has been developed through international cooperation in order to put together and archive in a database soil moisture measurements collected during field campaigns, and to make this database available to researchers [DOR 11]. At the watershed or regional scale, hydrological models are used as an alternative for determining soil moisture [BRO 09, MES 06]. At the global scale, information about soil moisture can be obtained from passive or active microwave satellite measurements [GRU 10, KER 10, ALY 14a, ALY 14b]. In an effort to help soil moisture users, a database of soil moisture estimated at the global scale from passive and active microwave satellites has been developed as part of the Climate Change Initiative (CCI) program launched in 2010 [HOL 13].

However, despite the importance of soil moisture in the water cycle and efforts devoted to its ground measurements and its estimation from microwave data [KER 07, PAL 06], determining the spatial–temporal variability of soil moisture in an operational context still remains a challenge for several reasons:

– the difficulty of establishing and maintaining permanent measurement networks on a large scale [COS 13];

– the influence of some biophysical and geophysical variables of the soil (e.g. surface roughness, surface temperature) and vegetation (e.g. plant's water content, height, density) on microwave signals' sensitivity to soil moisture [WIG 95, BIN 06, SAL 09, CAL 11].

In this chapter, we will discuss the surface soil moisture, commonly called soil moisture, obtained only from passive microwave measurements in a top soil layer of ~ 0–3 cm from the soil surface [ESC 10]. Although passive microwave satellite measurements benefit from better temporal resolution than active measurements (see Chapter 1 of this book) and are less affected by soil and vegetation parameters [LAK 13, NAR 15], their weak spatial resolution is a handicap for applications at the local or regional scale, including, for example, monitoring crops (growth, water stress) and flood risks. To diversify the use of soil moisture estimated from passive microwave satellite data for local scale applications, in connection with the societal demand, some authors have developed disaggregation methods that make it possible to improve the spatial resolution of soil moisture [MAS 10, PIL 11, MER 12]. Moreover, in certain conditions, passive microwave measurements can be strongly disrupted by radio frequency interferences (RFI) [WIG 16], which leads to erroneous values of soil moisture estimates. It is important to note that other moisture products derived from passive microwaves exist, such as moisture index and the root zone moisture.

This chapter will present the principle of soil moisture estimation using passive microwaves (section 2.2), the methods for estimating soil moisture using passive microwaves and inversion problems (section 2.3), the products of soil moisture derived from passive microwave spatial remote sensing as well as their evaluation (section 2.4), the disaggregation methods of these products (section 2.5), moisture indicators as well as root zone moisture (section 2.6), and finally, some applications of soil moisture (section 2.7).

2.2. Principle of passive microwave soil moisture estimation

Dry soil is a mixture of air and soil particles in a ratio that depends on the type of soil. In the case of wet soil, water is added to these air and soil particles, which will increase the dielectric constant of the medium, which strongly impacts passive or active microwave measurement [ULA 86]. However, in the microwave domain, the dielectric constants of the water, air, and soil are ~80, ~1 and ~4, respectively. This strong contrast between the dielectric properties of water and those of the other soil components is the basis of the principle for estimating soil moisture using microwave measurements. By increasing soil moisture, the soil dielectric constant increases while the brightness temperature, which represents passive microwave measurement, decreases. All investigations into passive microwave soil moisture estimation are based on this relationship between soil moisture, the dielectric constant, and the brightness temperature measurements. However, it has been shown that surface roughness, vegetation cover, and the atmosphere more or less perturb this relationship depending on the sensor's characteristics, such as the frequency, incidence angle, and polarization [ULA 86].

2.2.1. *Effect of surface roughness on the brightness temperature of bare soil*

The brightness temperature of bare soil, T_{bs}^{p}, is written as:

$$T_{bs}^{p} = e_s^p\, T_s = \left(1 - r_s^p\right) T_s \qquad [2.1]$$

where,

p is the horizontal (h) or vertical (v) polarization of the measurement,

T_s is the soil's physical temperature,

e_s^p is the soil's emissivity,

r_s^p is the Fresnel reflectivity if the soil is smooth, or the reflectivity of a rough soil expressed by an empirical relationship taking into account the roughness height of the observed surface [WAN 81, MON 15].

The brightness temperature of rough bare soil being higher than that of smooth soil, its response to moisture is weaker than that of smooth soil [MON 15].

2.2.2. *Effect of vegetation on the brightness temperature of vegetation-covered soil*

When the sensor observes a target covered with vegetation, at a frequency ≤ 10 GHz, the brightness temperature measured by the sensor is given by (tau-omega model, τ-ω):

$$T_{bs}^{p} = e_s^p T_s \exp(-\tau_v^p) + T_v (1 - w^p)\left\{1 - \exp(-\tau_v^p)\right\}\left\{1 + r_s^p \exp(-\tau_v^p)\right\} \qquad [2.2]$$

where,

T_v is the temperature of the vegetation,

τ_v^p and ω^p are the radiative parameters of the vegetation for a polarization p. They represent the vegetation's optical depth and its simple scattering albedo, respectively. Both depend on the water content and the type of vegetation [NJO 99, WIG 04a, WIG 07, NJO 06].

In [2.2], the first term represents the brightness temperature coming from the soil, but attenuated by the vegetation, and the second term, the vegetation's contribution. The attenuation factor of vegetation ($\exp(-\tau_v^p)$) depends on the sensor's parameters, and for a given vegetation cover, it increases with the vegetation's density and water content. If the sensor operates at high frequency and/or high incidence angle, for dense vegetation, the second term of [2.2] will be the dominant contributor and the soil will be masked, which leads to a decrease in the signal's sensitivity to soil moisture, as shown in Figure 2.1.

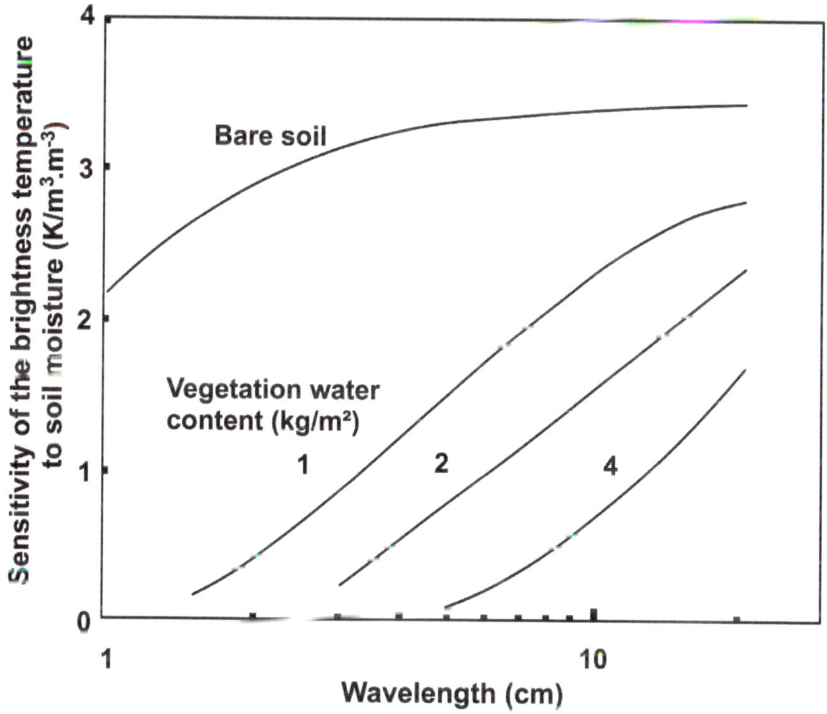

Figure 2.1. *Sensitivity of the brightness temperature (polarization h, incidence angle 10°) to soil moisture as a function of the wavelength for bare soil and vegetation-covered soil with various vegetation water content values (1, 2, and 4 kg/m²). Adapted from [JAC 93]*

For a heterogeneous target composed of a mixture of soil and vegetation, the brightness temperature measurement, T_B^p, is given as:

$$T_B^p = (1 - C_v) T_{bs}^p + C_v T_{bv}^p \qquad [2.3]$$

where C_v is the vegetation fractional cover and T_{bs}^p and T_{bv}^p are the brightness temperatures of bare soil [2.1] and of a surface covered with vegetation [2.2], respectively.

2.2.3. *Effect of soil moisture on the signal's sensitivity to surface roughness and vegetation*

From sections 2.2.1 and 2.2.2, it is clear that vegetation and soil roughness affect the response of microwave signals to soil moisture. The intensity of this effect is modulated by soil moisture, which, according to its magnitude, acts more or less on the interaction between the signal and the soil and vegetation's parameters. In fact, Figure 2.2 shows a strong sensitivity of the passive microwave signal to surface roughness and vegetation for very wet soils. Therefore, it is necessary to take into account surface roughness and vegetation in order to determine soil moisture in such wet conditions. In contrast, for relatively dry soil, uncertainties about the surface roughness and vegetation will have a less noticeable impact on the quality of soil moisture estimation than for wet conditions (Figure 2.2).

2.2.4. *Penetration depth of the signal*

The penetration depth of a given wave is defined as the depth from which emanates 63% of the energy emitted by a medium [ULA 86]. It results from the propagation processes in the medium and is expressed according to the medium's dielectric properties and the propagation wavelength:

$$\delta_p \sim \frac{\lambda\sqrt{\varepsilon'}}{2\pi\varepsilon''} \qquad [2.4]$$

Where λ is the electromagnetic wavelength, ε' and ε'' are the real and imaginary parts of the dielectric constant.

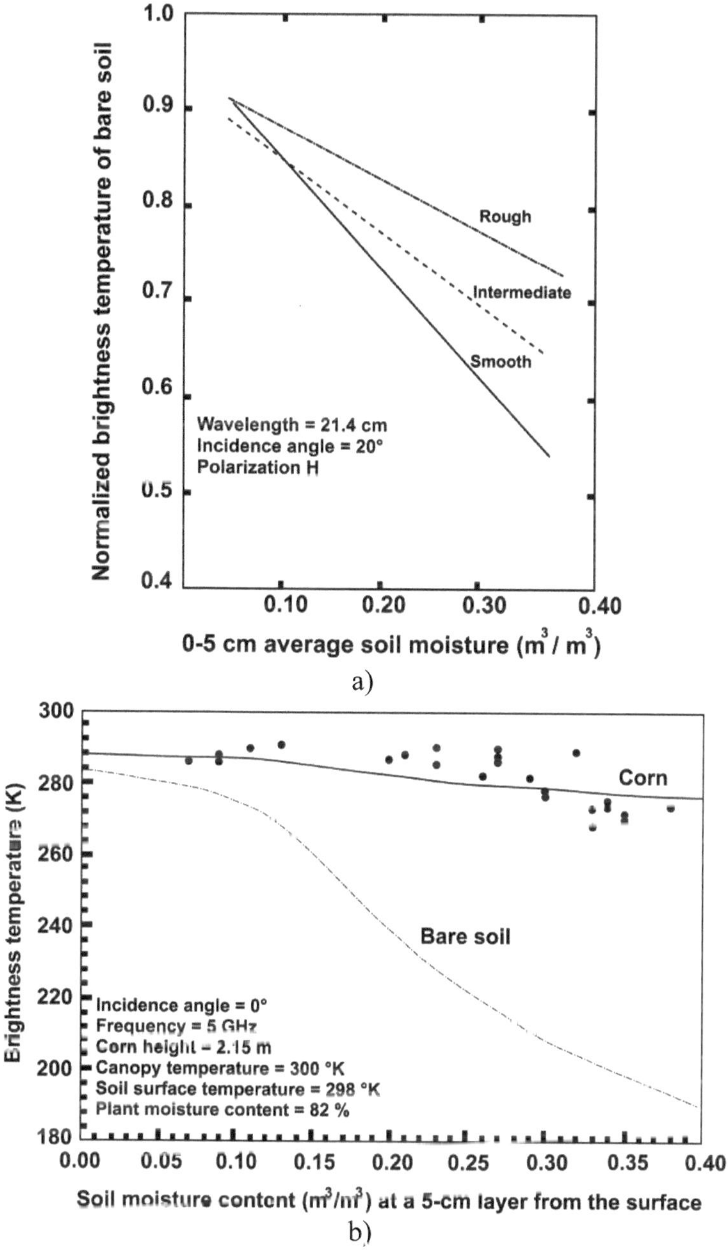

Figure 2.2. *Relationships between passive microwave measurements and soil moisture for a) bare soil (adapted from [NEW 80]); b) soil covered with vegetation (adapted from [ULA 86])*

Figure 2.3 shows the relationships between the penetration depth and the wavelength for different soil moisture conditions. Unlike dry soils, the signal only slightly penetrates wet soils, while a high wavelength provides a better penetration than a short wavelength [CHA 97, ESC 10]. Due to this higher penetration depth, microwave measurements based on long wavelengths allow soil moisture estimation over a greater depth. Furthermore, these long wavelengths lead to small atmospheric attenuation [ULA 86] and important vegetation transmissivity (Figure 2.4). As a result, long wavelengths, particularly L-band (wavelength of ~21.5 cm), are most suitable for soil moisture estimation.

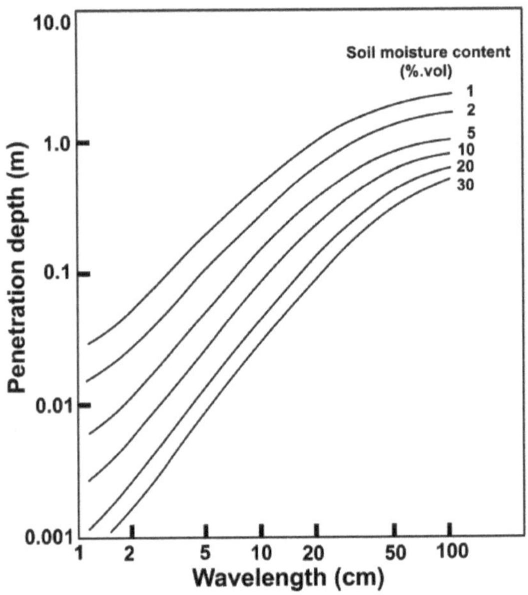

Figure 2.3. *The penetration depth of the signal as a function of the wavelength for several soil moisture values. Adapted from [NJO 96]*

2.3. Methods for surface soil moisture estimation

Here, we will adopt the classification used by Wigneron *et al.* [WIG 03] to distinguish the different methods for estimating soil moisture. Three categories are distinguished:

– methods based on the inversion of a direct model;

– statistical methods;

– explicit inverse methods.

Afterwards, we will briefly present assimilation methods.

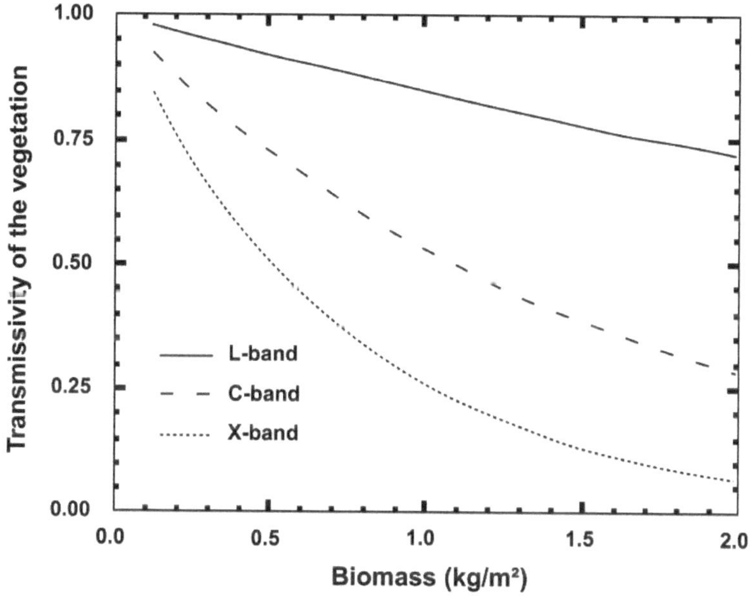

Figure 2.4. *Transmissivity of the vegetation as a function of the biomass in L (1.4 GHz), C (6 GHz), and X (10 GHz) bands. Adapted from [JAC 93]*

2.3.1. *Inversion of direct models: 1-P, 2-P, and N-P inversion of the L-MEB model*

This method consists of inverting a direct model that calculates observation values (T_B^p) as a function of the surface variables soil moisture (SM), τ, and T_s. Inversion allows the surface variables of interest to be calculated as a function of T_B^p. This is the method used to estimate SM on a global scale using Advanced Microwave Scanning Radiometer (AMSR-E, AMSR2), Soil Moisture and Ocean Salinity (SMOS) and Soil Moisture Active and Passive (SMAP) satellites. It is generally based upon an iterative minimization approach of a cost function, estimating the difference between the measurement and the modeling of T_B, for a set of measurement configurations in terms of incidence angle (θ), polarization (p) and frequency (f). It is also possible to use lookup tables (tables containing a very large number of simulations obtained from the use of a large variety of input parameters in the direct model) or neural networks.

The inversions 1-P, 2-P,…. N-P are different from one another depending on whether one variable (SM) is estimated for the approach 1-P, or two variables (generally SM and τ) for the approach 2-P, or N variables (SM, τ, and other variables such as the temperature or surface roughness) for the approach N-P.

In 1-P inversion, only soil moisture is estimated. In this case, observations for one single measurement configuration are necessary. This is typically the case for the Single Channel Approach, or SCA method [ONE 14], proposed for SMAP, in which the effect of vegetation is parameterized with the Normalized Difference Vegetation Index (NDVI). The method relies on the measurement of T_B^v (SCA-V) or T_B^h (SCA-H).

In 2-P inversion, SM and τ are estimated based on bipolarization and/or multi-angular observations. In the algorithm proposed for SMOS, SM and τ are estimated based on bipolarization and multi-angular observations for a range of angles that varies as a function of the pixel's position in the swath. The accuracy of the SM inversion thus depends upon the SMOS measurement configuration and surface conditions (dry or wet soil, weak or strong vegetation biomass, etc.) [WIG 00]. In the Land Parameter Retrieval Model (LPRM) algorithm or the Dual Channel Approach (DCA) algorithm, SM and τ are estimated from bipolarization observations, relying on an iterative approach or the minimization of a cost function, respectively [ONE 14]. Miernecki *et al.* [MIE 14] assessed these different approaches based on long-term data acquired on the VAS site (Valencia region in Spain) in Mediterranean conditions.

In N-P inversion, SM, τ, and other surface variables are estimated based on bipolarization, multi-angular, or multi-frequency observations. For example, Njoku and Li [NJO 99] developed 3-P inversions (SM, τ, and *Ts*) for AMSR-E using bipolarization and multi-frequency measurements (6.9, 10.7, and 18.7 GHz). Pardé *et al.* [PAR 04] assessed N-P inversions using bipolarization and multi-angular observations in L-band.

2.3.2. *Statistical approaches*

The general principle here is the development of an empirical relationship, or one based on physical principles, between soil moisture, which is the parameter of interest, and sensor measurements conducted in several measurement configurations for θ, p, or f. Two categories of approaches can be distinguished:

– those based on classifications;

– most common, those based on numerical relationships.

For the latter, the simplest statistical relationship is the linear relation, which is generally written between the soil moisture (SM) and the surface emissivity:

$$SM = a\ e_s^p + b \qquad [2.5]$$

The coefficients a and b are obtained by regression and are mainly a function of the sensor's measurement configuration (θ, p, f) and of the soil's characteristics, in terms of surface texture, structure, and roughness [WAN 83].

The following step, much more complex, is to move away from bare soil and to account for the vegetation's effect. In most cases, this effect is taken into account through vegetation index (NDVI, Normalized Difference Water Index (NDWI), Enhanced Vegetation Index (EVI), Leaf Area Index (LAI), etc.) derived from observations in the optical domain, which allow to dissociate the effects of soil moisture and surface roughness from those of the vegetation [CHO 88, AHM 95, PEL 03].

Another approach consists of separating the effects of soil moisture (SM) and vegetation water content (effects parameterized by the vegetation's optical thickness, τ_v^p), relying on a mathematical calculation which is based on the equations of the tau-omega model, τ-ω [2.2]. Wigneron *et al.* [WIG 04b] and Saleh *et al.* [SAL 06] thus show that by making certain realistic hypotheses, SM and τ can be written as a linear combination of the measured biangular (r_{s1}^p, r_{s2}^p) or bipolarizations (r_{s1}^p, r_{s1}^q) microwave reflectivities. The principle is to linearize the equations of the τ-ω model by considering the logarithm of the reflectivity, and to reach a simple system of

2 equations and 2 unknowns (SM and τ). For biangular microwave reflectivity, the system's resolution leads to [WIG 04b]:

$$\text{Log (SM)} = a_1 \log(r_{s1}^{p}) + a_2 \log(r_{s2}^{p}) + a_3 \quad [2.6]$$

$$\tau = b_1 \log(r_{s1}^{p}) + b_2 (\log r_{s2}^{p}) + b_3 \quad [2.7]$$

For bipolarization microwave reflectivities, it will be necessary to replace r_{s2}^{p} with r_{s1}^{q} in [2.6] and [2.7]. Coefficients a_1, a_2, a_3, b_1, b_2, and b_3 are best fit coefficients calculated by a regression approach. Adjustment by regression requires a reference value of SM. This value can be a ground measurement, a surface model simulation, or an estimation using a microwave sensor. In this way, Al-Yaari *et al.* [ALY 16] adjust this relationship using AMSR-E satellite observations with the SM data from SMOS satellite. This technique allows the development of a homogeneous SM dataset between SMOS and AMSR-E. In the same way, relationship using SMOS data (SM and T_B^p and $\theta = 40°$) can also be directly applied to the SMAP's bipolarization measurements.

2.3.3. *Explicit inversions*

2.3.3.1. *Neural networks*

The use of neural networks (NN) generally corresponds to an explicit inversion approach: the NN inputs are the brightness temperature measurements for different sensor configurations and the outputs are the variables of interest. For example, Liu *et al.* [LIU 02] find τ and SM using measurements at 1.4 and 10.65 GHz over agricultural crops. Rodríguez-Fernández *et al.* [ROD 15] apply NN to find SM using multi-angular SMOS measurements at L-band. They also test the potential contribution of complementary input data: maps of soil or vegetation types, or ancillary remote sensing data (observations from the Advanced Scatterometer satellite ASCAT, vegetation index, etc.). To train the algorithm, SM reference data are required. They can either be ground data or model simulation data, or SM data derived from an inversion algorithm; the NN then makes a sort of copy of the inversion algorithm. Fortunately, this copy is not perfect (as it does not reproduce the algorithm exactly, with its imperfections). Instead, the NN optimizes it by only selecting the pairs of relationships "observation – variables" that are consistent for a large number of cases. For a large sample of data, the NN can thus perform better than the algorithm that it copies.

If NN methods exhibit the disadvantage of being a sort of "black box" (there is not necessarily physics in NN modeling), they are extremely powerful and quick and thus well adapted to operational approaches, such as *Near Real Time*, for example.

2.3.3.2. *Assimilation of data*

Land Surface Models (LSM) are another approach, very complimentary to satellite observations, of estimating soil moisture on large scales. These models simulate the exchanges of matter and energy at the surface-atmosphere interface. However, there also remains a large degree of uncertainty in both the SM estimations derived from LSM simulations (uncertainties about climate forcings, parameterizations, soil texture and land use, conceptual errors of modeling, etc. [LIE 15]) and those derived from satellite observations (instrumental errors, uncertainties in the algorithms, perturbations of the measurements by RFI, etc.). The combination of LSM simulations (hydrological models, Numerical Weather Prediction models (NPW), etc.) and satellite observations by means of assimilation approaches allows the improvement of large scale SM estimates and the predictions of LSM in terms of runoff, river flow [DRA 11] or weather predictions [DER 13]. It is possible to assimilate SM data derived from inversion models or brightness temperatures (a modeling platform of T_B^p as a function of SM is thus necessary in the assimilation system). For example, the assimilation of brightness temperature data in the global system NWP is conducted at the European Center for Medium range Weather Forecasting (ECMWF) on the basis of the extended Kalman filter integrated into the assimilation system. This system has also been tested on SM data from the Meteorological Operational satellite programme/Advanced Scatterometer (METOP/ ASCAT). For details on data assimilation techniques, the reader can refer to Chapters 11 and 12 of this book, and to [OTT 16].

2.3.4. *Inversion problems*

The difficulties related to soil moisture estimation from microwave data are the result of several sources of error. These include the characteristics of the soil (surface roughness, texture, topography, temperature), vegetation cover (type of cover, density, water content), as well as the sensor radiometric parameters (spatial resolution, incidence angle, frequency, polarization), instrumentation (calibration problems), radio frequency interferences (RFI) at low frequencies, and atmospheric conditions, specifically at frequencies above 10 GHz. With regard to the sources of error related to the target (soil, vegetation), several studies have contributed to the understanding and consideration of the effects of surface roughness [NJO 06, WIG 11, MAR 15, MON 15] and vegetation [WIG 95, WIG 07, JAC 91, NJO 06] when estimating soil moisture. These two factors have similar effects on the brightness temperatures observed. As a result, they are difficult to dissociate [NJO 06, PAT 13, FER 15], hence the complexity of estimating soil moisture.

In recent years, with the launch of satellites operating at low frequencies (AMSR, SMOS, Aquarius, SMAP), more attention was devoted to RFI [LI 04, OLI 12, PIE 14]. Figure 2.5 shows an example of RFI probability occurrence on SMOS brightness temperature data. The zones whose brightness temperatures are largely corrupted by RFI appear in red. The corresponding data are then filtered and not used in the SMOS algorithm for estimating soil moisture.

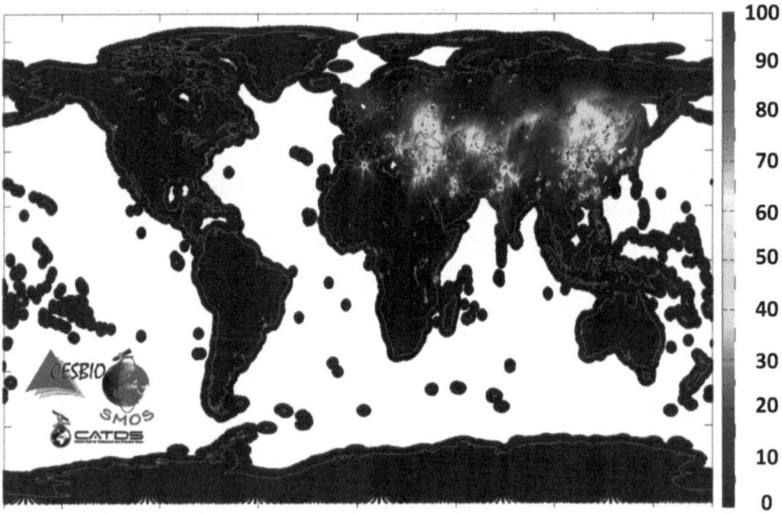

Figure 2.5. *Average RFI probability occurrence in SMOS observations (L-band) acquired on October 25, 2015 in descending mode over ~ one week. For a color version of this figure, see www.iste.co.uk/baghdadi/4.zip*

2.4. Soil moisture products derived from passive microwave space-borne observations

2.4.1. *Spatial passive microwave remote sensing evolution for soil moisture estimation*

The potential of passive microwave measurements for determining soil moisture on a global scale has been shown with data from the Scanning Multichannel Microwave Radiometer (SSMR) satellite, launched in 1978 [NJO 98], and those from its successor, Special Sensor Microwave Imager (SSM/I), launched in 1987 [HOL 90]. Afterwards, several satellite missions were launched (Table 2.1), proposing global scale soil moisture estimation with a targeted accuracy of 0.04 m^3/m^3 or better. The various algorithms developed for soil moisture estimation (explained in section 2.3) take into account the specificities of the different sensors described in Table 2.1.

For example, for the AMSR-E sensor, the algorithm uses brightness temperatures at X-band (10.65 GHz [NJO 04]) or a combination of brightness temperatures at C-band (6.91 GHz) and X-band [OWE 01, OWE 08] in the inversion process of a radiative transfer model [2.2]. However, due to interactions with vegetation and radio frequency interferences (RFI), which perturb the C-band data, these frequencies are not optimal for obtaining soil moisture [CAL 11].

With the launch of SMOS (November 2009), the first passive microwave satellite operating at L-band (1.41 GHz) and at incidence angles of 0–55°, new soil moisture products are now available on a global scale and over the top 0–5 cm soil surface [KER 10]. Due to its weak sensitivity to surface roughness, vegetation cover (biomass up to 5 kg/m^2), and atmosphere, L-band is more suited for assessing soil moisture conditions (Figures 2.3 and 2.4). It is, however, perturbed by RFI (Figure 2.5, section 2.3.4) in Asia, the Middle East and Europe to a lesser extent. The algorithm allowing the estimation of SMOS soil moisture is based on the inversion of a microwave radiative transfer model [2.2] that uses multi-angular brightness temperatures acquired at L-band, at H and V polarizations. The principle is based on the simultaneous inversion (2-P inversion) of the vegetation's optical depth and soil moisture using multi-angular and bipolarization measurements [WIG 95, WIG 00, KER 11]. Figure 2.6 presents a global map of the SMOS monthly average soil moisture obtained for August 2014. Desert and semi-arid zones appear with low soil moisture values (< 0.1 m^3/m^3); tropical zones show higher values of SM.

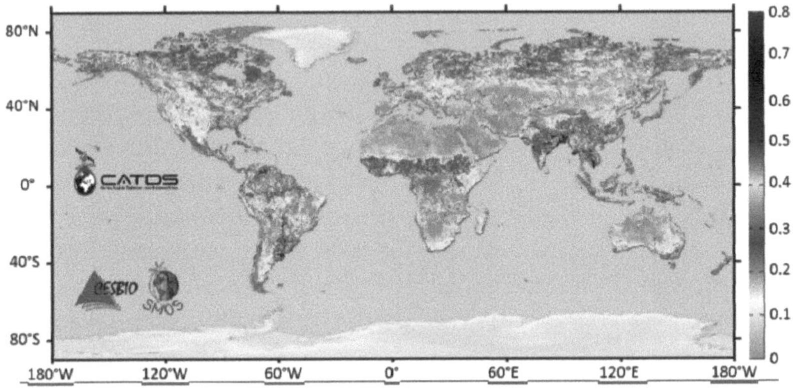

Figure 2.6. *SMOS monthly average soil moisture (m^3/m^3) obtained from brightness temperature measurements acquired in ascending mode from August 1 to August 31, 2014. For a color version of this figure, see www.iste.co.uk/baghdadi/4.zip*

In late January 2015 the Soil Moisture Active and Passive (SMAP) satellite was launched, which also operates at L-band (1.41 GHz), but at only one incidence angle (40°), unlike SMOS [ENT 10]. The particularity of SMAP is that it carries both L-band radiometer and radar sensors. The idea is to take advantage of the high temporal frequency of the passive measurements (2–3 days) and the good spatial resolution of the SMAP radar measurements (~ 3 km) to provide soil moisture at a better resolution (~ 9 km) than the one obtained with only passive measurements (Table 2.1). Unfortunately, the SMAP radar sensor stopped functioning in July 2015. Since then, only passive measurements at a spatial resolution of ~ 40 km are available. The four SMAP algorithms used to estimate soil moisture consist of inverting a radiative transfer model whose input variables differ from one algorithm to another [ONE 14]. Like SMOS and Aquarius brightness temperature data and the low frequency data from AMSR, RFI perturb the SMAP measurements, thereby reducing the success rate of the soil moisture inversion algorithm.

Sensor/Space mission	Operation period	Frequency (GHz)	Resolution (km)	Incidence angle (°)
SSMR/Nimbus-7	1978 – 1988	6.6, 10.7, 18, 21, 37	150 at 6.6 GHz	50
SSM/I/DMSP	1987 –	19.3, 22.3, 37, 85.5	25 at 19.3 GHz	53
TMI/TRMM	1997 – 2015	10.65, 19.35, 21.3, 37, 85.5	50 at 10.65 GHz	53
AMSR-E/EOS PM-1	2002 – 2011	6.9, 10.7, 18.7, 23.8, 36.5, 89	56 at 6.9 GHz	55
AMSR2/GCOM-W1	2012 –	6.92, 7.3, 10.65, 18.7, 23.8, 36.5, 89	10 at 6.92 GHz	55
MIRAS/SMOS	2009 –	1.42	25 – 40	0 – 55
Aquarius/SAC-D	2011 – 2015	1.41	76 × 94, 84 × 120, 96 × 156	29, 38, 46
SMAP/SMAP	2015 –	1.43	40	40

Table 2.1. *Characteristics of the SSMR, SSM/I, TMI, AMSR-E, AMSR2, SMOS, Aquarius and SMAP sensors*

2.4.2. *Validation and representativeness of the large scale AMSR-E, SMOS, and SMAP SM retrievals*

Several field campaigns have been developed to validate the soil moisture products estimated from microwave satellites [ONE 14], and also to understand the physical processes that govern the interaction between the signal and the target (soil, vegetation). The field campaign sites include different landscapes and diverse soil moisture and soil type conditions in order to identify the sources of error of inversion algorithms. These errors are difficult to estimate, especially with the heterogeneity of natural surfaces at a spatial resolution of ~ 40 km. The field campaigns aim to produce databases of ground measurements of the target's characteristics (moisture, surface roughness, vegetation, etc.) and/or airborne measurements of brightness temperatures, which will be acquired simultaneously or almost simultaneously to the satellite overpasses.

The validation of the retrieval algorithm generally relies on a comparison between the estimated and measured soil moisture values. Due to the low spatial resolution of passive microwaves spatial sensors, a direct comparison between the estimated and the measured moisture is not reliable. Therefore, the authors generally use a weighted average of the measured soil moisture [GHE 12, DJA 15a]. This weighted average takes into account the distance between the measurement point and the center of the satellite's pixel according to the equation:

$$SM = \sum_{i=1}^{n} p_i\, SM_i \quad with \quad \sum_{i=1}^{n} p_i = 1 \qquad [2.8]$$

where SM is the weighted average of the soil moisture measurements (SMi) calculated at the satellite's spatial resolution:0

– p_i are the weights attributed to each soil moisture measurement (SM_i). They depend on the distance between the location of SM_i and the center of the considered pixel [KER 11]. The weight p_i is higher as the measurement point is closer to the center of the considered pixel.

– n is the number of measurement points located at a distance of one spatial demi-resolution or less from the center of the considered pixel.

To ensure that this weighted average (SM) is representative of the satellite's spatial resolution, it is necessary to collect a certain number of ground soil moisture measurements [FAM 08]. At a spatial resolution of 40 km, which corresponds to that of satellites designed for estimating soil moisture at a global scale (SMOS and SMAP), about twenty spatially distributed ground measurements would be necessary [FAM 08] and would also make it possible to satisfy the targeted accuracy of 0.04 m^3/m^3 set by the space missions for estimating soil moisture.

2.4.3. *Example of SMOS soil moisture evaluation over the agricultural sites of CanEx-SM10 and SMAPVEX12*

The field campaigns CanEx-SM10 and SMAPVEX12 were led in the Canadian Prairies from May 31 to June 16, 2010, and from June 6 to July 17, 2012, respectively, in order to validate the SMOS data and prepare for the launch of SMAP. Detailed information about the collection of soil moisture validation data by SMOS are presented in Magagi *et al.* [MAG 13] and McNairn *et al.* [MCN 15].

SMOS moisture evaluation was made on the basis of statistical parameters (correlation, bias, mean square error, slope, etc.), evaluating the agreement/disagreement with the ground measurements. For the sites sampled during CanEx-SM10 and SMAPVEX12, Table 2.2 shows an underestimation of soil moisture by SMOS (L2 product, version 551), which is also observed on several SMOS validation sites [LAC 12]. As a result, it is important to correct this negative bias before any use of SMOS soil moisture data. While a strong correlation is observed between SMOS soil moisture and the mean values of ground measurements of SM [2.8], the mean square error is higher than the targeted accuracy of 0.04 m^3/m^3 [KER 10].

	CanEx-SM10				SMAPVEX12	
SMOS Pixels	147226		147228		Several	
Orbits	Asc.	Desc.	Asc.	Desc.	Asc.	Desc.
Correlation	0.58	0.58	0.59	0.64	0.78	0.59
Bias (m^3/m^3)	-0.1	-0.03	-0.11	-0.02	-0.1	-0.08
MSE (m^3/m^3)	0.11	0.07	0.12	0.06	0.11	0.1
Slope	0.46	0.59	0.34	0.51	0.84	0.9
N	30	40	34	39	178	125

Table 2.2. *Regression statistics (correlation, bias, mean square error (MSE), slope) between the SM values estimated by SMOS (product L2 version 551) and the SM ground measurements over the agricultural sites of CanEx-SM10 and SMAPVEX12. The results are given according to the ascending (Asc., 6 am) and descending (Desc., 6 pm) satellite overpasses. N is the number of SM measurements used in the regression [DJA 15a, DJA 15b]*

2.4.4. *Qualitative analysis of the soil moisture values estimated by SMOS and AMSR-E over the agricultural and forest sites of CanEx-SM10*

In order to analyze the influence of precipitation and vegetation on the soil moisture (SM) estimated by SMOS and AMSR-E, agricultural (Figures 2.7(a), (b) and (c)) and forested (Figures 2.7(d), (e) and (f)) sites of CanEx-SM10 are considered. Figure 2.7 shows the temporal profiles of the SM estimated from SMOS (product L2, version 551), AMSR-E/NSIDC, and AMSR-E/VUA, as well as those from the ground measurements of SM, daily cumulative rainfall corresponding to the satellites' ascending and descending overpasses over the two sites. Due to the high vegetation cover fraction in forested sites, soil moisture retrievals are more difficult in forested regions, but much less with SMOS (Figures 2.7(d)) than with AMSR-E (Figure 2.7(e) and (f)). This is a consequence of signal attenuation problems that increase with the signal frequency (section 2.2.2). The same consequence is observed over the agricultural site, where AMSR-E/NSIDC soil moisture values, primarily estimated from X-band data [NJO 04], are nearly stable when the vegetation is well developed (Figure 2.7(b)). In contrast, over the agricultural site, the SMOS estimations seem to respond better to precipitation and they better match the ground measurements (Figure 2.7(a)).

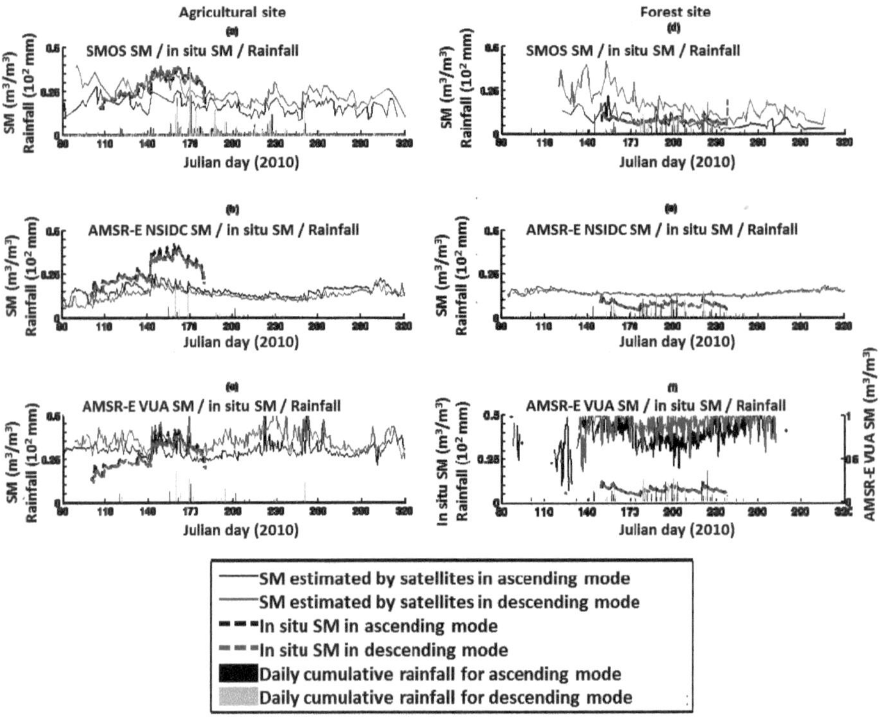

Figure 2.7. *Comparison of the soil moisture values (SM) estimated from SMOS (product L2, version 551), AMSR-E/NSDIC, and AMSR-E/VUA over agricultural (a-b-c, respectively) and forested (d-e-f, respectively) sites of CanEx-SM10. For the agricultural and forest sites, the considered SMOS pixels are DGG ID 147228 and DGG ID 138528, respectively. Adapted from [DJA 15a]*

At the regional or global scale, satellite soil moisture products are evaluated in comparison with moisture data derived from assimilation processes (section 2.3.3.2) or collected by ground measurements of network stations (Table 2.3). The statistical parameters (correlation, bias, mean square error, slope, etc.) obtained are interpreted using:

– the potential of passive microwave signal to interact with the observed surface according to the characteristics of different regions (desert, arid, forest, etc. zones);

– the perturbations caused by RFI;

– the characteristics of moisture measuring network stations, including the measurement sampling depth and the number of stations [ALY 14b, ALB 12].

Soil moisture measuring network stations	Measurement depth (cm)	Number of stations used
SMOSMANIA (France)	5, 10, 20, 30	12
SMOSMANIA-E (France)	5, 10, 20, 30	9
OZNET (Australia)	0–5 or 0–8, and 0–30	38
NCRS-SCAN (US)	~5, ~ 20	154
AMMA (West Africa)	5, 10, [illegible], 120	10
REMEDHUS (Spain)	5	21
UMSUOL (Italy)	10	1
UDC-SMOS (Germany)	5	7

Table 2.3. *Specifications of some measuring network stations used to validate remote sensing soil moisture products. Adapted from [ALB 12]*

2.5. Methods for disaggregating satellite soil moisture products derived from passive microwave observations

Global products of soil moisture estimated from passive microwave data (AMSR, SMOS, and SMAP) are available at coarse spatial resolutions (~25–40 km) that limit their application at a local scale. The disaggregation methods allowing the spatial resolution of these global soil moisture products to be improved (for example, increasing the spatial resolution from 25–40 km to 1 km), generally integrate high spatial resolution data as ancillary data [LAK 13]. The latter can be synthetic aperture radar data or optical and thermal infrared (TIR) remote sensing data. Figure 2.8 outlines the general disaggregation process for soil moisture.

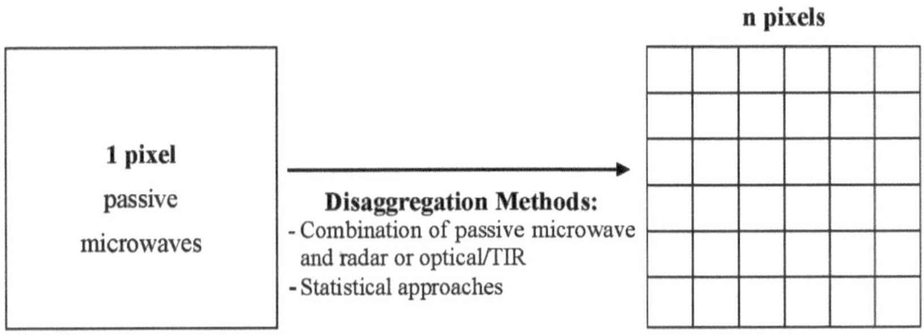

Figure 2.8. *Schematic representation of the disaggregation process*

In the literature, most disaggregation methods combining passive and active microwave measurements apply a change detection approach governed by soil moisture variation in the radar signal [LAK 13, NAR 06]. Others directly use the radar data to disaggregate the microwave brightness temperature, the result of which is then integrated into an inversion algorithm (section 2.3.1) to compute soil moisture at a finer spatial resolution [ENT 14b]. As for disaggregation methods based on passive microwave and optical/TIR/data, they depend on the relationships between soil moisture, surface temperature, and vegetation through the "universal triangle" concept [PIL 11, MER 12]. These empirical or physical methods differ from one another according to the input data and the expression used to link soil moisture, surface temperature, and vegetation. For example, Merlin *et al.* [MER 12] use the soil evaporation efficiency estimated using surface temperature and NDVI from the *MODerate resolution Imaging Spectoradiometer* (MODIS) satellite. As for Piles *et al.* [PIL 11], in addition to NDVI and MODIS surface temperature, they integrate the SMOS brightness temperature into their disaggregation algorithm at spatial resolutions of 10 km and 1 km. Figure 2.9 shows SMOS soil moisture retrieved on January 19, 2010 over the Murrumbidgee basin in Australia, before and after disaggregation [PIL 11]. The spatial variability of soil moisture is much better represented at a resolution of 1 km (Figure 2.9(c)) than at 10 km (Figure 2.9(b)) and 40 km (Figure 2.9(a)). In addition, the Murrumbidgee River clearly appears in the disaggregation results at a resolution of 1 km (Figure 2.9(c)). The major disadvantage of these disaggregation methods based on optical/TIR data is that they are not applicable in cloudy conditions.

Other more marginal disaggregation methods are based on statistical methods. They take into account the spatial variability of soil moisture, and most use the fractal or multi-fractal behavior of this variability [KIM 02, MAS 10].

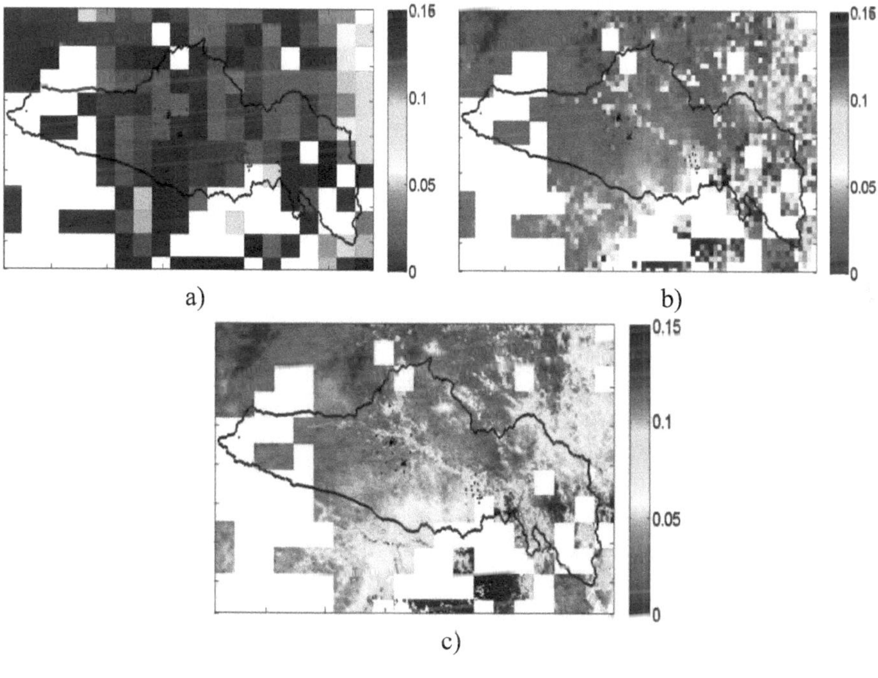

Figure 2.9. *a) SMOS soil moisture at a resolution of 40 km, acquired on January 10, 2010 for the Murrumbidgee basin (Australia) and the disaggregation results obtained at b) a resolution of 10 km and c) a resolution of 1 km with Piles et al.'s algorithm [PIL 11]. For a color version of this figure, see www.iste.co.uk/baghdadi/4.zip*

2.6. Other moisture products derived from passive microwave observations

2.6.1. *Moisture index*

Before the launch of passive microwave satellites operating at low frequencies, surface moisture indexes were developed with data from the SMMR and SSM/I sensors whose frequencies are [6.6, 10.7, 18, 21, and 37 GHz] and [19.35, 22.2, 37, and 85.5 GHz], respectively. We will present the Fraction of Water Surface (FWS), the Bassist Wetness Index (BWI), and the Antecedent Precipitation Index (API).

FWS [FIL 03] is mainly based on the large decrease in emissivity due to the presence of water extents in the SSMR and SSM/I pixels, as well as to moist soils. It is given by:

$$FWS = \frac{e^p - e_{dry}}{e_w - e_{dry}} \qquad [2.9]$$

where e^p, e_{dry}, and e_w are the emissivity of the pixel observed at a polarization p, the emissivity of a dry surface, and the emissivity of the water, respectively.

BWI [BAS 98, BAS 01] integrates two concepts: the large decrease in emissivity due to liquid water (including that at the soil surface, the canopy, etc.) and the dynamic of the increasing relationships between emissivity and measurement frequency [BAS 98]. This dynamic is strongly influenced by the percentage of water extents in the observed pixel; it is very weak for small percentage of water extents and gradually increases with the increase in this percentage. BWI is given by:

$$BWI = \Delta e\, Ts = \beta_0 \left[T_B^v(f_2) - T_B^v(f_1) \right] + \beta_1 \left[T_B^v(f_3) - T_B^v(f_2) \right] \qquad [2.10]$$

Where Δe is the decrease in emissivity due to water, Ts the surface temperature, T_B^v the surface brightness temperature at vertical polarization [2.3]; f_1, f_2, and f_3 are the 19, 37, and 85 GHz frequencies of SSM/I, respectively; β_0 and β_1 are parameters taking into account the frequency behavior of the emissivity in the presence of water extents.

API [KOH 51], which is the weighted sum of daily precipitation, is considered as an interesting soil moisture index [BLA 81]. For a day j, it is given by:

$$API(j) = k\ API(j-1) + P(j) \qquad [2.11]$$

where $P(j)$ is the precipitation of day j and k is a depletion coefficient.

Several authors have found a relationship between the API index and the microwave brightness temperatures (acquired at different frequencies) that depends on vegetation [BLA 81, WAN 95]. For example, the relationship developed by Choudhury and Golus [CHO 88] including the vegetation index NDVI is:

$$API = \frac{a + b\ NDVI - T_B^h(6.6GHz)}{c + d\ NDVI} \qquad [2.12]$$

2.6.2. *Root zone moisture*

From the knowledge of the surface moisture with an adequate temporal frequency (better than three days in most cases), it is possible to estimate the soil moisture in deeper layers [ALB 13]. The input data is surface moisture and ancillary data such as vegetation indexes. For example, in the case of SMOS, root zone moisture is obtained by using the *double bucket* method. It is based on a modified exponential filter for the first layer and a physical infiltration model for the second [ALB 13]. This model is a linear approximation of Richards' 1-D equations [RIC 31], which govern flow in an unsaturated porous medium. This model also takes into account vegetation transpiration, infiltration in the deep layer using gravitational percolation. Hourly surface moisture values of the first layer are used by the model of the second layer to estimate deep moisture. Figure 2.10 shows an example of soil moisture simulation results in the root zone obtained using surface SM values measured on-site [TRA 10]. The statistical evaluation of the comparison between the observations and the simulations (R^2, mean square error, bias) show the feasibility of the approach despite overestimation episodes observed in Niger and Mali.

Other root zone moisture estimation approaches use a 3-D assimilation method based on the ensemble Kalman filter (EnKF). This is the method applied to the SMAP SM data to generate the root zone soil moisture [ENT 14a].

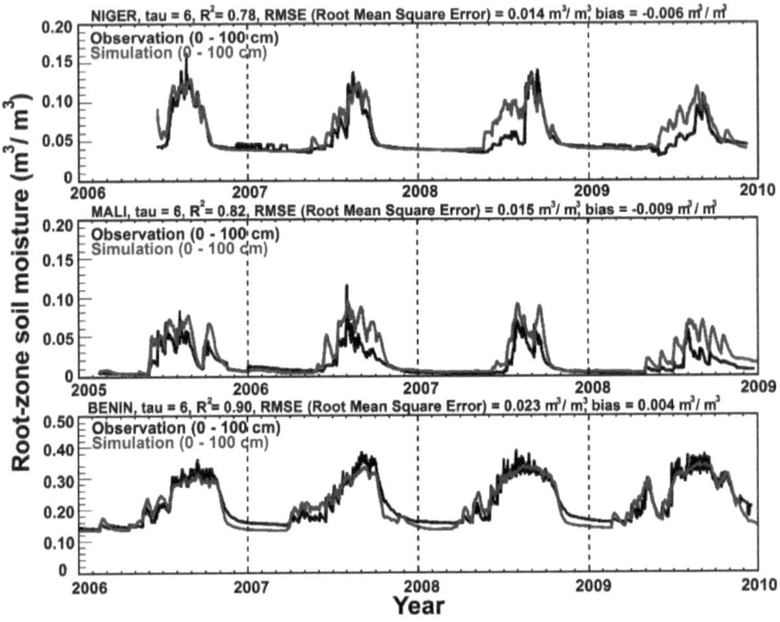

Figure 2.10. *Root zone soil moisture observed and simulated using surface soil moisture data measured over three sites in West Africa located in Niger, Mali and Benin from 2005 to 2010. Adapted from [TRA 10]*

2.7. Principal applications

There is wide range of soil moisture applications in environmental science, research, etc. The following subsections present an overview of some applications. For more details on the applications of soil moisture, the reader can refer to the other chapters of this volume.

2.7.1. *Applications in meteorology (NWP), climate modeling, drought*

The first use of global moisture data is its application in meteorological forecasting (see Chapter 12 of this book). With the relatively recent availability of these data, models are not yet ready to assimilate them, but numerous studies are in progress, and to date the most advanced are in Europe (ECMWF), United States (NOAA), and Canada (Environment Canada). These studies show that the assimilation of the surface soil moisture data allow forecasts to be improved, with the only limitation being that the models themselves should be refined beforehand in order to fully benefit from the contribution of surface SM data [CAR 15, MUN 15]. By studying climatologies obtained from L-band observations, it is possible to establish drought indexes. These studies also show that these data allow – at least in some cases – better anticipation of extreme rainfall.

2.7.2. *Applications in hydrology, flood risks, agriculture, forestry, etc.*

In hydrology, the assimilation of moisture data allows river flows to be improved [LIE 15]. Numerous studies focus on the estimation of open water extents and their fluctuations (section 2.6.1), the monitoring of floods, and the prediction of flood risks (Leroux and Pellarin, Chapter 11 of this volume). Similarly, the simultaneous monitoring of surface moisture and of the vegetation water content allows L-band radiometry to provide useful information about fire risks, particularly in forests (30% of land surfaces). In forest zones, an extreme increase in soil moisture can lead to paludification [LAV 05] and reduces the quality of the woods. It has also been shown that these data could be used to characterize vegetation phenological cycles, and probably to predict agricultural yields [PAT 13].

2.7.3. *Applications in Nordic environments*

In high latitude areas, radiometry at L-band makes it possible to monitor freezing and thawing conditions [RAU 12, RAU 14, ROY 15], and additional active microwave measurements make it possible to improve the monitoring. It is also possible to follow thin sea ice [BEI 15]. More recently, various studies aim to better characterize rich organic soils [BIR 14] in order to improve their monitoring and the understanding of their climatic impacts.

2.8. Conclusion

Using data from the first passive microwave satellites (SMMR, SSM/I), indicators have been developed in order to follow the temporal variations in surface soil moisture. The estimation of soil moisture from passive microwave spatial sensors has seen great advances, with the launch of SMOS (November 2009), which operates at the L-band. Global soil moisture products are available today at a spatial resolution of 40 km and a temporal frequency of 2–3 days. Some soil moisture products of the new SMAP satellite, launched in 2015, have the same characteristics and could also ensure the continuity of the SMOS data. The frequency band (1.4 GHz) of SMOS and SMAP is more sensitive to soil moisture than the higher frequency of AMSR-E (6.9 GHz). However, the data at L-band are strongly affected by RFI effects, which reduce or nullify the inversion rate of the algorithms, especially in Asia, the Middle East and Europe to a lesser extent. Also, over the SMOS validation sites presented in this chapter, it seems that the SMOS and AMSR-E estimation algorithms face difficulties in the presence of forests, but much less with SMOS due to its lower frequency (1.4 GHz). Over agricultural areas where vegetation is less abundant, SMOS responds to precipitation and better matches ground SM measurements.

In order to develop the use of these global soil moisture products in more local applications, several disaggregation algorithms have been developed. While methods integrating radar data are perturbed by surface roughness, those integrating optical/TIR data are affected by clouds so that disaggregated products are not yet continuously available, or at least not at the same temporal frequency as SMOS or SMAP (2–3 days). As a result, these products still remain inadequate for operational applications at the local scale. On a global or regional scale, numerous studies in progress integrate global soil moisture or brightness temperature data for a better evaluation of different processes involved in hydrological or climate models. Additionally, applications are being developed over specific geographic zones, such as the monitoring of freezing/thawing in high northern latitudes, and to forest ecosystems, such as the monitoring of forest fires.

Finally, algorithms have also been developed to estimate root zone moisture from the SMOS and SMAP data, which is very useful for crops monitoring.

2.9. Key points

Due to its important role in several environmental fields (hydrology, climatology, agriculture, etc.), several methods have been developed to estimate surface soil moisture (an extremely variable parameter in time and space). Indeed, several factors perturb the process of soil moisture estimation and limit the accuracy of the estimated soil moisture products. Throughout this chapter, we have mentioned soil characteristics and those of vegetation cover as well as sensor parameters, RFI at low frequencies, and atmospheric conditions, particularly at frequencies above 10 GHz. With the new satellites operating at L-band (SMOS and SMAP), improvements have been made, but areas with high vegetation density still remain problematic in SM retrieval studies. Considering the coarse spatial resolution (~25–40 km) of the soil moisture products estimated using passive microwave satellite sensors, disaggregation methods (for example, increasing the spatial resolution from 25–40 km to 1 km) are being developed in order to compute refined products better suited for applications at more local scale such as flood risks, monitoring crops, droughts, etc.

2.10. Acknowledgments

The authors would like to thank Najib Djamaï and Amirouche Benchall, both doctoral students in the field of remote sensing at the Department of Applied Geomatics (University of Sherbrooke), for their contribution to the writing of this chapter.

2.11. Bibliography

[AHM 95] AHMED N.U., "Estimating soil moisture from 6.6 GHz dual polarization, and/or satellite derived vegetation index", *International Journal of Remote Sensing*, vol. 16, no. 4, pp. 687–708, 1995.

[ALB 12] ALBERGEL C., DE ROSNAY P., GRUHIER C. *et al.*, "Evaluation of remotely sensed and modelled soil moisture products using global ground-based *in situ* observations", *Remote Sensing of Environment*, vol. 118, pp. 215–226, 2012.

[ALB 13] AL BITAR A., KERR Y. H., MERLIN O. *et al.*, "Global drought index from SMOS soil moisture", *International Geoscience and Remote Sensing Symposium (IGARSS 2013)*, 22–26 July, Melbourne, Australia, 2013.

[ALY 14a] AL-YAARI A., WIGNERON J.P., DUCHARNE A. *et al.*, "Global-scale evaluation of two satellite-based passive microwave soil moisture datasets (SMOS and AMSR-E) with respect to Land Data Assimilation System estimates", *Remote Sensing of Environment*, vol. 149, pp. 181–195, 2014.

[ALY 14b] AL-YAARI A., WIGNERON J.P., DUCHARNE A. *et al.*, "Global-scale evaluation of two satellite-based passive (SMOS) and active (ASCAT) microwave soil moisture retrievals using MERRA-land soil moisture estimates", *Remote Sensing of Environment*, vol. 152, pp. 614–626, 2014.

[ALY 16] AL-YAARI A., WIGNERON J.P. , KERR Y. *et al.*, "Testing regression equations to derive long-term global soil moisture datasets from passive microwave observations", *Remote Sensing of Environment*, vol. 180, pp. 453–464, 2016.

[ANC 04] ANCTIL F., MICHEL C., PERRIN C. *et al.*, "A soil moisture index as an auxiliary ANN input for stream flow forecasting", *Journal of Hydrology*, vol. 286, pp. 155–167, 2004.

[BAS 98] BASIST A., GRODY N.C., PETERSON T.C. *et al.*, "Using the Special Sensor Microwave/Imager to monitor land surface temperatures, wetness, and snow cover", *Journal of Applied Meteorology*, vol. 37, pp. 888–911, 1998.

[BAS 01] BASIST A., WILLIAMS C., GRODY N. *et al.*, "Using the Special Sensor Microwave Imager to monitor surface wetness", *Journal of Hydrometeorology*, vol. 2, pp. 297–308, 2001.

[BEI 15] BEITSCH A., KERN S., KALESCHKE L., "Comparison of SSM/I and AMSR-E sea ice concentrations with ASPeCt ship observations around Antarctica", *IEEE Transactions on Geoscience and Remote Sensing*, vol. 53, no. 4, pp. 1985–1996, 2015.

[BIN 06] BINDLISH R., JACKSON T.J., GASIEWSKI A.J. *et al.*, "Soil moisture mapping and AMSR-E validation using the PSR in SMEX02", *Remote Sensing of Environment*, vol. 103, no. 2, pp. 127–139, 2006.

[BIR 14] BIRCHER S., DEMONTOUX F., HØGH JENSEN K. *et al.*, "SMOSHiLat - Microwave L-band emission from organic-rich soils in the northern cold climate zone in the framework of the SMOS mission", *International Geoscience and Remote Sensing Symposium (IGARSS)*, July 2014, Québec, Canada, 2014.

[BLA 81] BLANCHARD B.J., MCFARLAND M.J., SCHMUGGE T.J. *et al.*, "Estimation of soil moisture with API algorithms and microwave emission", *Water Resources Bulletin*, vol. 17, no. 5, pp. 767–774, 1981.

[BRO 09] BROCCA L., MELONE F., MORAMARCO, T. *et al.*, "Assimilation of observed soil moisture data in storm rainfall-run off modelling", *Journal of Hydrologic Engineering*, vol. 14, pp. 153–165, 2009.

[CAL 11] CALVET J.C., WIGNERON J.P., WALKER J. *et al.*, "Sensitivity of passive microwave observations to soil moisture and vegetation water content: L-Band to W-Band", *IEEE Transactions on Geoscience and Remote Sensing*, vol. 49, pp. 1190–1199, 2011.

[CAR 15] CARRERA M.L., BÉLAIR S., BILODEAU B., "The Canadian Land Data Assimilation System (CaLDAS): Description and synthetic evaluation study", *Journal of Hydrometeorology*, vol. 16, pp. 1293–1314, 2015.

[CHA 10] CHAMPAGNE C., BERG A., BELANGER J. *et al.*, "Evaluation of soil moisture derived from passive microwave remote sensing over agricultural sites in Canada using ground-based soil moisture monitoring networks", *International Journal of Remote Sensing*, vol. 31, no. 14, pp. 3669–3690, 2010.

[CHA 97] CHANZY A., RAJU S., WIGNERON J.P., "Estimation of soil microwave effective temperature at L and C bands", *IEEE Transactions on Geoscience and Remote Sensing*, vol. 35, no. 3, pp. 570–580, 1997.

[CHO 88] CHOUDHURY B.J., GOLUS R.E., "Estimating soil wetness using satellite data", *International Journal of Remote Sensing*, vol. 9, pp. 1251–1257, 1988.

[COS 13] COSH M.H., JACKSON T.J., SMITH C. *et al.*, "Validating the BERMS in situ soil water content data record with a large scale temporary network", *Vadose Zone Journal*, vol. 12, no. 2, 2013.

[CRO 10] CROW W.T., XIWU Z., JACKSON T.J. *et al.*, "Evaluating the utility of remotely sensed soil moisture retrievals for operational agricultural drought monitoring", *IEEE Transactions on Geoscience and Remote Sensing*, vol. 3, pp. 57–66, 2010.

[DER 13] DE ROSNAY P., DRUSCH M., VASILJEVIC D. *et al.*, "A simplified Extended Kalman Filter for the global operational soil moisture analysis at ECMWF", *Quarterly Journal of the Royal Meteorological Society*, vol. 139, no. 674, pp. 1199–1213, 2013.

[DJA 15a] DJAMAI N., MAGAGI R., GOÏTA K. *et al.*, "Evaluation of SMOS soil moisture products over the CanEx-SM10 area", *Journal of Hydrology*, vol. 520, pp. 254–267, 2015.

[DJA 15b] DJAMAI N., MAGAGI R., GOÏTA K. *et al.*, "A new approach to downscale SMOS soil moisture estimates during cloudy days", *International Geoscience and Remote Sensing Symposium (IGARSS)*, 26–31 July, Milan, Italy, 2015.

[DOR 11] DORIGO W.A., WAGNER W., HOHENSINN R. *et al.*, "The International Soil Moisture Network: a data hosting facility for global in situ soil moisture measurements", *Hydrology and Earth System Sciences*, vol. 15, pp. 1675–1698, 2011.

[DRA 11] DRAPER C., MAHFOUF J.F., CALVET J.C. *et al.*, "Assimilation of ASCAT near-surface soil moisture into the SIM hydrological model over France", *Hydrology and Earth System Sciences*, vol. 15, pp. 3829–3841, 2011.

[ELD 09] ELDER K., CLINE D., LISTON G.E. *et al.*, "NASA cold land processes experiment (CLPX 2002/03): Field measurements of snowpack properties and soil moisture", *Journal of Hydrometeorology*, vol. 10, pp. 320–329, 2009

[ENT 10] ENTEKHABI D., NJOKU E, O'NEILL P. *et al.*, "The Soil Moisture Active Passive (SMAP) Mission", *Proceedings of the IEEE*, vol. 98, no. 5, pp. 704–716, 2010.

[ENT 14a] ENTEKHABI D. *et al.*, "SMAP Handbook", California Institute of Technology, JPL CL#14-2285, JPL 400-1567 07/14, 2014.

[ENT 14b] ENTEKHABI D., DAS N., NJOKU E. *et al.*, "Algorithm theoretical basis document L2 & L3 radar/radiometer soil moisture (Active/Passive) data products", Revision A, Jet Propulsion Laboratory, California Institute of Technology, 2014.

[ESC 10] ESCORIHUELA M.J., CHANZY A., WIGNERON J.P. *et al.*, "Effective soil moisture sampling depth of L-band radiometry: A case study", *Remote Sensing of Environment,* vol. 114, no. 5, pp. 995–1001, 2010.

[FAM 08] FAMIGLIETTI J., RYU D., BERG A.A. *et al.*, "Field observations of soil moisture variability across scales", *Water Resources Research.,* vol. 44, *W01423,* 2008.

[FER 15] FERNANDEZ-MORAN R., WIGNERON J.P., LOPEZ-BAEZA E. *et al.*, "Roughness and vegetation parameterizations at L-band for soil moisture retrievals over a vineyard field", *Remote Sensing of Environment,* vol. 170, pp. 269–279, 2015.

[FIL 03] FILY M., ROYER A., GOÏTA K. *et al.*, "A simple retrieval method for land surface temperature and fraction of water surface determination from satellite microwave brightness temperatures in sub-arctic areas", *Remote Sensing of Environment*, vol. 85, no. 3, pp. 328–338, 2003.

[GCO 06] GCOS-107, "Systematic observation requirements for satellite-based products for climate – Implementation plan for the global observing system for climate in support of the UNFCCC", *WMO/TD* No. 1338, September 2006.

[GHE 12] GHERBOUDJ I., MAGAGI R., GOÏTA K. *et al.*, "Validation of SMOS data over agricultural and boreal forest areas in Canada", *IEEE Transactions on Geoscience and Remote Sensing*, vol. 50, no. 5, pp. 1623–1635, 2012.

[GOW 02] GOWARD S.N., XUE Y., CZAJKOWSKI K.P., "Evaluating land surface moisture conditions from the remotely sensed temperature/vegetation index measurements: an exploration with the simplified simple biosphere model", *Remote Sensing of Environment*, vol. 79, pp. 225–242, 2002.

[GRU 10] GRUHIER C., DE ROSNAY P., HASENAUER S. *et al.*, "Soil moisture active and passive microwave products: intercomparison and evaluation over a Sahelian site", *Hydrology and Earth System Sciences,* vol. 14, pp. 141–156, 2010.

[HOL 13] HOLLMANN R., MERCHANT C.J., SAUNDERS R. *et al.*, "The ESA climate change initiative: Satellite data records for essential climate variables", *American Meteorological Society*, vol. 94, pp. 1541–1552, 2013.

[HOL 90] HOLLINGER J.P., PIERCE J.L., POE G.A., "SSM/I instrument evaluation", *IEEE Transactions on Geoscience and Remote Sensing,* vol. 28, no. 5, pp. 781–790, 1990.

[JAC 91] JACKSON T.J., SCHMUGGE T.J., "Correction for the effects of vegetation on the microwave emission of soils", *IEEE Geoscience and Remote Sensing Proceedings*, Helsinki, Finland, pp. 753–756, 3–6 June 1991.

[JAC 93] JACKSON T.J., "Measuring surface soil moisture using passive microwave remote sensing", *Hydrological Processes*, vol. 7, no. 2, pp. 139–152, 1993.

[JUN 10] JUNG M., REICHSTEIN M., CIAIS P. *et al.*, "Recent decline in the global land evapotranspiration trend due to limited moisture supply", *Nature Letter*, vol. 467, no. 7318, pp. 951–954, 2010.

[KER 07] KERR Y.H., "Soil moisture from space: Where are we?", *Hydrogeology Journal,* vol. 15, pp. 117–120, 2007,

[KER 10] KERR Y.H., WALDTEUFEL P., WIGNERON J.P. *et al.*, "The SMOS Mission: New tool for monitoring key elements of the global water cycle", *IEEE Transactions on Geoscience and Remote Sensing,* vol. 98, pp. 666–687, 2010.

[KER 11] KERR Y.H., WALDTEUFEL P., RICHAUME P. *et al.*, "Algorithm theoretical basis document (ATBD) for the SMOS Level 2 soil moisture processor development continuation project", *SMOS level 2 Processor for Soil Moisture, SO-TN-ARR-L2PP-0037,* nos. 3–6, 2011.

[KIM 02] KIM G., BARROS A.P., "Downscaling of remotely sensed soil moisture with a modified fractal interpolation method using contraction mapping and ancillary data", *Remote Sensing of Environment,* vol. 83, pp. 400–413, 2002.

[KOH 51] KOHLER M.A., LINSLEY R.K., "Predicting the runoff from storm rainfall", U.S. Weather Bureau Research Paper, 1951.

[KOS 11] KOSTER R.D., MAHANAMA S.P.P., YAMADA T.J. *et al.*, "The second phase of the global Land-Atmosphere coupling experiment: Soil moisture contributions to subseasonal forecast skill", *Journal of Hydrometeorology,* vol. 12, no. 5, pp. 805–822, 2011.

[LAC 12] LACAVA T., MATGEN P., BROCCA L. *et al.*, "A first assessment of the SMOS soil moisture product with in-situ and modeled data in Italy and Luxembourg", *IEEE Transactions on Geoscience and Remote Sensing,* vol. 50, pp. 1612–1622, 2012.

[LAK 13] LAKSHMI, V., "Remote sensing of soil moisture", *ISRN Soil Science,* Article ID 424178, 2013.

[LAV 05] LAVOIE M., PARÉ D., FENTON N. *et al.*, "Paludification and management of forested peatlands in Canada: a literature review", *Environmental Reviews*, vol. 13, no. 2, pp. 21–50, 2005.

[LI 04] LI L., NJOKU E.G., IM E. *et al.*, "A preliminary survey of radio-frequency interference over the U.S. in Aqua AMSR-E Data", *IEEE Transactions on Geoscience and Remote Sensing,* vol. 42, no. 2, pp. 380–390, 2004.

[LIE 15] LIEVENS H., TOMER S.K., AL BITAR A. *et al.*, "SMOS soil moisture assimilation for improved hydrologic simulation in the Murray Darling Basin, Australia", *Remote Sensing of Environment*, vol. 168, pp. 146–162, 2015.

[LIU 02] LIU S. F., LIOU Y.A., WANG W.J. *et al.*, "Retrieval of crop biomass and soil moisture from measured 1.4 and 10.65 brightness temperatures", *IEEE Transactions on Geoscience and Remote Sensing*, vol. 40, no. 6, pp. 1260–1268, 2002.

[MAG 13] MAGAGI R., BERG A., GOÏTA K. *et al.*, "CanEX-SM 10 (Canadian Experiment for Soil Moisture in 2010): Overview and preliminary results", *IEEE Transactions on Geoscience and Remote Sensing,* vol. 51, pp. 347–363, 2013.

[MAR 15] MARTENS B., LIEVENS H., COLLIANDER A. *et al.*, "Estimating effective roughness parameters of the L-MEB model for soil moisture retrieval using passive microwave observations from SMAPVEX12", *IEEE Transactions on Geoscience and Remote Sensing*, vol. 53, no. 7, pp. 4091–4103, 2015.

[MAS 10] MASCARO G., VIVONI E.R., DEIDDA R., "Downscaling 2015. soil moisture in the southern Great Plains through a calibrated multifractal model for land surface modeling applications", *Water Resources Research,* vol. 46, 2010.

[MCN 15] MCNAIRN H., JACKSON T.J., WISEMAN G. *et al.*, "The Soil Moisture Active Passive Validation Experiment 2012 (SMAPVEX12): Pre-Launch calibration and validation of the SMAP soil moisture algorithms", *IEEE Transactions on Geoscience and Remote Sensing,* vol. 53, no. 5, pp. 2784–2801, 2015.

[MER 12] MERLIN O., RÜDIGER C., AL BITAR A. *et al.*, "Disaggregation of SMOS soil moisture in Southeastern Australia", *IEEE Transactions on Geoscience and Remote Sensing*, vol. 50, pp. 1557–1571, 2012.

[MES 06] MESINGER F., DIMEGO G., KALNAY E. *et al.*, "North American Regional Reanalysis (NARR)", *Bulletin of the American Meteorological Society*, vol. 87, pp. 343–360, 2006.

[MIE 14] MIERNECKI M., WIGNERON JP., LOPEZ-BAEZA E. *et al.*, "Comparison of SMOS and SMAP soil moisture retrieval approaches using tower-based radiometer data over a vineyard field", *Remote Sensing of Environment*, vol. 154, pp. 89–101, 2014.

[MON 15] MONTPETIT B., ROYER A., WIGNERON J.P. *et al.*, "Evaluation of multi-frequency bare soil microwave reflectivity models", *Remote Sensing of Environment*, vol. 162, pp. 186–195, 2015.

[MUN 15] MUÑOZ-SABATER J., "Incorporation of passive microwave brightness temperatures in the ECMWF soil moisture analysis", *Remote Sensing,* vol. 7, pp. 5758–5784, 2015.

[NAR 06] NARAYAN U., LAKSHMI V., JACKSON T.J., "High resolution estimation of soil moisture using L-band radiometer and radar observations made during the SMEX02 experiments", *IEEE Transactions on Geoscience and Remote Sensing,* vol. 44, pp. 1545–1554, 2006.

[NAR 15] NARVEKAR P.S., ENTEKHABI D., KIM S.B. *et al.*, "Soil moisture retrieval using L-Band radar observations", *IEEE Transactions on Geoscience and Remote Sensing,* vol. 53, no. 6, pp. 3492–3506, 2015.

[NEW 80] NEWTON R.W., ROUSE J.W., "Microwave radiometer measurements of soil moisture", *IEEE Transaction on Antennas propagation*, vol. AP-28, no. 5, pp. 680–686, 1980.

[NJO 96] NJOKU E.G., ENTEKHABI D., "Passive microwave remote sensing of soil moisture", *Journal of Hydrology*, vol. 184, no. 1-2, pp. 101–129, 1996.

[NJO 98] NJOKU E., RAGUE B., FLEMING K., "The Nimbus-7 SMMR pathfinder brightness data set", Jet Propulsion Laboratory, California Institute of Technology, 1998.

[NJO 99] NJOKU E.G., LI L., "Retrieval of land surface parameters using passive microwave measurements at 6 to 18 GHz", *IEEE Transactions on Geoscience and Remote Sensing*, vol. 37, no. 1, pp. 79–93, 1999.

[NJO 04] NJOKU E.G, "AMSR-E/Aqua daily L3 surface soil moisture, interpretive parameters, & QC EASE-Grids", *Version 2*, Boulder, Colorado USA: NASA DAAC at the National Snow and Ice Data Center, 2004.

[NJO 06] NJOKU E.G., CHAN S.K., "Vegetation and surface roughness effects on AMSR-E land observations", *Remote Sensing of Environment*, vol. 100, no. 2, pp. 190–199, 2006.

[OLI 12] OLIVA R., DAGANZO E., KERR Y.H. *et al.*, "SMOS radio frequency interference scenario: Status and actions taken to improve the RFI environment in the 1400–1427-MHz passive band", *IEEE Transactions on Geoscience and Remote Sensing*, vol. 50, pp. 1427–1439, 2012.

[ONE 14] O'NEILL P., CHAN S., NJOKU E. *et al.*, "Soil Moisture Active Passive (SMAP) algorithm theoretical basis document Level 2 & 3 soil moisture (passive) data products", Revision A, Jet Propulsion Laboratory, California Institute of Technology, 2014.

[OTT 16] OTTLÉ C., MAHFOUF J-F., "Data Assimilation of Observations from Space", in BAGHDADI N.N., ZRIBI M. (eds), *Microwave Remote Sensing of Land Surfaces*, ISTE Ltd, London and Elsevier Ltd, Oxford, 2016.

[OWE 01] OWE M., DE JEU R., WALKER J.P., "A methodology for surface soil moisture and vegetation optical depth retrieval using the microwave polarization difference index", *IEEE Transactions on Geoscience and Remote Sensing*, vol. 39, pp. 1643–1654, 2001.

[OWE 08] OWE M., DE JEU R., HOLMES T.R.H., "Multi-sensor historical climatology of satellite derived global land surface moisture", *Journal of Geophysical Research*, vol. 113, p. 17, 2008.

[PAL 06] PALOSCIA S., MACELLONI G., SANTI E., "Soil moisture estimates from AMSR-E brightness temperatures by using a dual-frequency algorithm", *IEEE Transactions on Geoscience and Remote Sensing*, vol. 44, no. 11, pp. 3135–3144, 2006.

[PAR 04] PARDÉ M., WIGNERON J.P., WALDTEUFEL P. *et al.*, "N-Parameter retrievals from L-band microwave measurements over a variety of agricultural crops", *IEEE Transactions on Geoscience and Remote Sensing*, vol. 42, no. 6, pp. 1168–1178, 2004.

[PAT 13] PATTON J., HORNBUCKLE B., "Initial validation of SMOS vegetation optical thickness over Iowa", *IEEE Geoscience and Remote Sensing Letters,* vol. 10, pp. 647–651, 2013.

[PEL 03] PELLARIN T., CALVET J.C., WIGNERON J.P., "Surface soil moisture retrieval from L-band radiometry: a global regression study", *IEEE Transactions on Geoscience and Remote Sensing,* vol. 41, no. 9, pp. 2037–2051, 2003.

[PIE 14] PIEPMEIER J.R., JOHNSON J.T., MOHAMMED P.N. *et al.*, "Radio-frequency interference mitigation for the Soil Moisture Active Passive microwave radiometer", *IEEE Transactions on Geoscience and Remote Sensing,* vol. 52, no. 1, pp. 761–775, 2014.

[PIL 11] PILES M., CAMPS A., VALL-LLOSSERA M. *et al.*, "Downscaling SMOS-derived soil moisture using MODIS visible/infrared data", *IEEE Transactions on Geoscience and Remote Sensing,* vol. 49, pp. 3156–3166, 2011.

[RAU 12] RAUTIAINEN K., LEMMETYINEN J., PULLIAINEN J. *et al.*, "L-Band radiometer observations of soil processes in Boreal and Subarctic environments", *IEEE Transactions on Geoscience and Remote Sensing*, vol. 50, pp. 1483–1497, 2012.

[RAU 14] RAUTIAINEN K., LEMMETYINEN J., SCHWANK M. *et al.*, "Detection of soil freezing from L-band passive microwave observations", *Remote Sensing of Environment,* vol. 147, pp. 206–218, 2014.

[RIC 31] RICHARDS L.A., "Capillary conduction of liquids through porous mediums", *Physics,* vol. 1, no. 5, pp. 318–333, 1931.

[ROD 15] RODRÍGUEZ-FRENÁNDEZ N.J., AIRES F., RICHAUME P. *et al.*, "Soil moisture retrieval using neural networks: Application to SMOS", *IEEE Transactions on Geoscience and Remote Sensing*, vol. 53, no. 11, pp. 5991–6007, 2015.

[ROY 15] ROY A., ROYER A., DERKSEN C. *et al.*, "Evaluation of spaceborne L-Band radiometer measurements for terrestrial freeze/thaw retrievals in Canada", *IEEE Journal of Selected Topics in Applied Earth Observations and Remote Sensing,* vol. 8, no. 9, pp. 4442–4459, 2015.

[SAL 06] SALEH K., WIGNERON J.P., DE ROSNAY P. *et al.*, "Semi-empirical regressions at L-band applied to surface soil moisture retrievals over grass", *Remote Sensing of Environment,* vol. 101, pp. 415–426, 2006.

[SAL 09] SALEH K., KERR Y.H., RICHAUME P. *et al.*, "Soil moisture retrievals at L-band using a two-step inversion approach (COSMOS/NAFE'05 Experiment)", *Remote Sensing of Environment*, vol. 113, pp. 1304–1312, 2009.

[SEN 10] SENEVIRATNE S. I., CORTI T., DAVIN E.L. *et al.*, "Investigating soil moisture–climate interactions in a changing climate: A review", *Earth Science Reviews*, vol. 99, pp. 125–161, 2010.

[TRA 10] TRAN X.T., "Estimation de l'état hydrique des sols en Afrique de l'Ouest par télédétection spatiale", PhD Thesis, University of Grenoble, 2010.

[ULA 86] ULABY F.T., MOORE R.K., FUNG A.K., *Microwave Remote Sensing: Active and Passive, Vol. III, From Theory to Application*, Artech House, Dedham, MA, 1986.

[WAN 81] WANG J.R., CHOUDHURY B.J., "Remote sensing of soil moisture content over bare fields at 1.4 GHz frequency", *Journal of Geophysical Research*, vol. 86, pp. 5277–5282, 1981.

[WAN 83] WANG J.R., O'NEILL P.E., JACKSON T.J. *et al.*, "Multifrequency measurements of the effects of soil moisture, soil texture, and surface roughness", *IEEE Transactions on Geoscience and Remote Sensing*, vol. 21, no. 1, pp. 44–51, 1983.

[WAN 95] WANG J.R., CHOUDHURY B.J., "Passive microwave radiation from soil", in CHOUDHURY B.J., KERR Y.H., NJOKU E.G. *et al.* (eds), in *Passive Microwave Remote Sensing of Land-Atmosphere Interactions*, VSP Utrecht, The Netherlands, 1995.

[WIG 95] WIGNERON J.P., CHANZY A., CALVET J.C. *et al.*, "A simple algorithm to retrieve soil moisture and vegetation biomass using passive microwave measurements over crop fields", *Remote Sensing of Environment*, vol. 51, pp. 331–341, 1995.

[WIG 00] WIGNERON J.P., WALDTEUFEL P., CHANZY A. "Two-D microwave interferometer retrieval capabilities of over land surfaces (SMOS Mission)", *Remote Sensing of Environment*, vol. 73, pp. 270–282, 2000.

[WIG 03] WIGNERON J.P., CALVET J.C., PELLARIN T. *et al.*, "Retrieving near surface soil moisture from microwave radiometric observations: Current status and future plans", *Remote Sensing of Environment*, vol. 85, pp. 489–506, 2003.

[WIG 04a] WIGNERON J.P., PARDÉ M., WALDTEUFEL P. *et al.*, "Characterizing the dependence of vegetation parameters on crop type, view angle and polarization at L-Band", *IEEE Transactions on Geoscience and Remote Sensing,* vol. 42, no. 2, pp. 416–425, 2004.

[WIG 04b] WIGNERON J.P., CALVET J.C., DE ROSNAY P. *et al.,* "Soil moisture retrievals from bi-angular L-band passive microwave observations", *IEEE Geoscience and Remote Sensing Letters,* vol.1, no. 4, pp. 277–281, 2004.

[WIG 07] WIGNERON J.P., KERR Y., WALDTEUFEL P. *et al.*, "L-band Microwave Emission of the Biosphere (L-MEB) model: Description and calibration against experimental data sets over crop fields", *Remote Sensing of Environment*, vol. 107, pp. 639–655, 2007.

[WIG 11] WIGNERON J.P., CHANZY A., KERR H. *et al.*, "Evaluating an improved parameterization of the soil emission in L-MEB", *IEEE Transactions on Geoscience and Remote Sensing,* vol. 49, no. 4, pp. 1177–1189, 2011.

[WIG 16] WIGNERON J-P., KERR Y., "Passive Low Frequency Microwaves: Principles, Radiative Transfer, Physics of Measurements", in BAGHDADI N., ZRIBI M. (eds), *Microwave Remote Sensing of Land Surfaces*, ISTE Ltd, London and Elsevier Ltd, Oxford, 2016.

3

Using Satellite Scatterometers to Monitor Continental Surfaces

3.1. Introduction

Scatterometers are sidelooking radar sensors designed to precisely evaluate the radar backscatter coefficient σ^0 of the surfaces being observed. However, as we will aim to demonstrate, this high radiometric resolution is obtained at the cost of spatial resolution. Scatterometers are therefore radar sensors complementary to synthetic aperture radars (SAR). These promote spatial resolution at the cost of radiometric resolution. Thus, while current SAR are characterized by a spatial resolution of about 1 m, with a strong radiometric variation linked to the speckle effect, satellite scatterometers have a spatial resolution accurate to about 10 km, with an precision on the estimation of σ^0 within a tenth of a decibel. Originally designed in order to evaluate the speed and direction of wind across ocean surfaces, the potential of these sensors for the monitoring of continental surfaces on a regional or global scale has been demonstrated [KEN 89a, KEN 89b, MOU 95a, FRI 96a, FRI 96b, MAG 97, WAG 99a, WAG 99b, WIS 00, MES 97].

The first two spaceborne scatterometers were put into orbit by NASA: the first as part of the SKYLAB (SkyLaboratory) mission in 1973–74 and the second, given the name SASS (SEASAT Advanced Scatterometer System), during the three-month-long mission of the SEASAT satellite. Nevertheless, it is only since 1991 that a continuous observation of Earth has been in place,

Chapter written by Pierre-Louis FRISON, Lionel JARLAN and Eric MOUGIN.

using various satellite scatterometers. The first of these, called ESCAT (ERS Scatterometer), was put into orbit onboard the European satellite ERS-1. These satellites are summarized in Table 3.1. It can be seen here that the two frequency bands used are C-band ($f = 5.4$ GHz, $\lambda = 5.6$ cm) for the European sensors and Ku band ($f = 13.4$ GHz, $\lambda = 2.2$ cm) for the other sensors. These most recent sensors, developed in order to maintain observations in Ku-band (since 1999), were largely put into orbit by NASA, with the exception of OSCAT (OceanSat-2 Scatterometer), which was put into orbit by the ISRO (Indian Space Research Organization).

Satellite	Scatterometer	Carrier frequency	Polarization	Spatial resolution	Swath	Duration of the mission
SKYLAB	RADSCAT	Ku	VV or HV or HH	11 – 30 km		1973–1974
SEASAT	SASS	Ku	HH–VV	~50 km	2 × 600 km	Jul – Oct 1978
ERS-1	ESCAT	C	VV	~50 km	500 km	Jul 91 – Mar 00
ERS-2						Apr 1995 –
ADEOS-1	NSCAT	Ku	HH–VV	25 and 50 km	2 × 600 km	Aug 96 – June 97
QuikSCAT	SeaWinds	Ku	HH–VV	~ 25 km	1800 km	June 99 – Nov 09
ADEOS-2						Dec 02 – Oct 03
METOP-A	ASCAT	C	VV	25 – 50 km	2 × 550 km	Oct 06 –
METOP-B						Sept 2012 –
OceanSat2	OSCAT	Ku	HH–VV	12 – 50 km	1800 km	Sept 09 – Apr 14
ISS	RapidScat	Ku	HH–HV	~ 15 km	900 km	Sept 2014 –

Table 3.1. *The characteristics of various satellite scatterometers operating in frequency bands Ku (λ = 2.1 cm) and C (λ = 5.6 cm)*

3.2. Principle of acquisition for scatterometers

Until now, two types of configuration for satellite scaterometers have been in use: the SASS, ESCAT, ASCAT and NSCAT scatterometers use multiple and fixed beams. The antennae are characterized by long, narrow footprints, which cover one or two swaths parralel to the satellite sub-track (Figures 3.1(a) and 3.1(b)). The NSCAT scatterometer was a successor to the SASS scatterometer: the same principle and the same Ku carrier frequency. Several scatterometers followed on from NSCAT: SeaWinds, on board the ADEOS-2, QuikSCAT, RapidScat and OSCAT, all operating in the Ku-band. These scatterometers employ a rotating antenna which emits two beams, both of which sweep a cone around the vertical which passes alongside the satellite (Figure 3.1(c)). Using similar configurations, these sensors have made possible an almost continuous global observation in the Ku band since 1999.

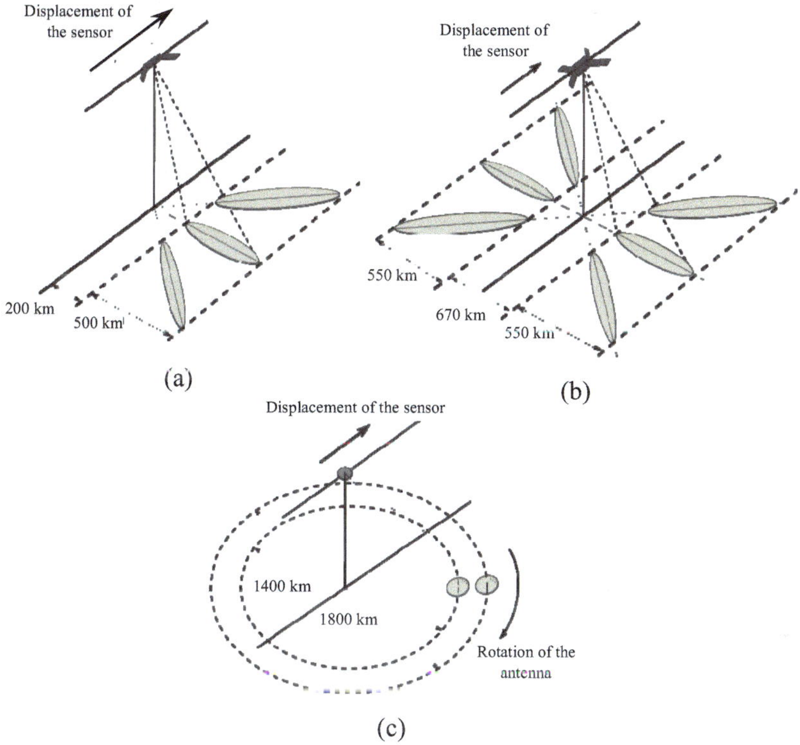

Figure 3.1. *Configurations of scatterometers using multiple fixed antennae: a) ERS and b) ASCAT; c) SeaWinds using a conically scanning pencil-beam*

The various configurations of scatterometers make it possible to view a point at the centre of the swath from different azimuthal angles as the spacecraft moves on its orbit. For oceanic surfaces, this azimuthal signature makes it possible to estimate wind direction, with the signal intensity making it possible to estimate their speed.

Owing to the speed $\vec{v}$ of the sensor (positioned at *O*, Figure 3.2(a)), a point M is characterized by a shift in the frequency domain linked to the Doppler effect. This shift is given by the expression [MAS 13]:

$$f_D = -\frac{2}{\lambda}\vec{v}\,.\overrightarrow{u_R} \qquad [3.1]$$

with $\overrightarrow{u_R} = \frac{\vec{R}}{R}$, $\vec{R} = \overrightarrow{OM}$, designating the vector between the sensor and the point M being taken into consideration at the surface of Earth, and $R = \|\vec{R}\|$.

For a given Doppler frequency, the set of points of the space having the same frequency is located on along straights supported by $\overrightarrow{u_R}$, such that $\vec{v}\,.\overrightarrow{u_R} = cste$. From these are generated two cones of an angle $\alpha = acos\left(-\frac{\lambda f_D}{2\,v}\right)$, of an apex which is the sensor, of an axis which is the satellite orbit, one open towards the front and the other open towards the back. The point where these cones intersect with the terrestrial surface (presumed to be flat) positioned at an altitude z = h beneath the sensor, form hyperboles. An antenna's footprint on the terrestrial surface can therefore be subdivided into several cells in accordance with a division which observes the equidistant and/or isodoppler lines.

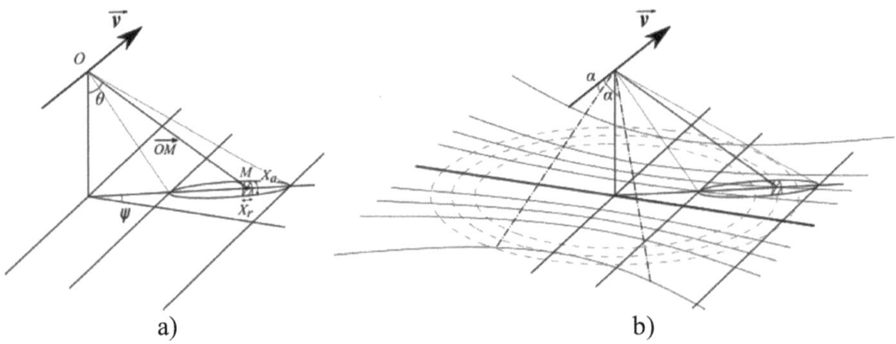

Figure 3.2. *a) Acquisition geometry of a radar sensor; X_a and X_r correspond to the dimensions of the resolution cell including the point M; b) equidistant circles (magenta, dashed lines) and isodoppler lines (blue) on the terrestrial surface, taken to be flat. For each aperture angle (α), there are 2 cones and therefore 2 hyperboles. On the figure are also shown the two generators of the cone associated to the hyperbole the most backward in Figure 3.2(b)*

Echoes received by radar sensors are ranged according to their distance from the sensor. Echoes from targets which belong to the same sphere will reach the sensor at the same time and cannot be separated. The points where these spheres intersect with the terrestrial surface (thought to be flat) form concentric circles, centred at the point on the nadir of the satellite (Figure 3.2(b)).

The radar backscatter coefficient σ^0 [FER 16] is estimated by measuring the power P_rd of the wave backscattered by the surface observed, and the system configurations, using the radar equation:

$$P_r = \frac{\lambda^2}{(4\pi)^3} P_e \iint_{Surf.obs.} \frac{G_e(\vec{R})G_r(\vec{R})}{L_e L_r} \frac{1}{R^4} \sigma^0(\vec{R}) dS \qquad [3.2]$$

where G_e, L_e, G_r, L_r are the antenna's gains and the loss factors of the emission and reception systems, λ is the wavelength of the carrier frequency, and P_e is the power of the emitted pulse. The integration focuses on the surface under observation, bounded by the principal lobe of the antenna. If we are to suppose that L_e, L_r, $\vec{R}$, and $\sigma^0(\vec{R})$ remain constant, and we note that $\overrightarrow{R_0}$ the vector between the sensor and the centre of the cell under observation and S_{eff} the effective surface, verifying $\iint_{Surf.obs.} G_e(\vec{R})G_r(\vec{R})dS = G_e(\overrightarrow{R_0})G_r(\overrightarrow{R_0})S_{eff}$, we obtain:

$$\sigma^0\ (m^2/m^2) = \frac{(4\pi)^3 R^4}{\lambda_0^2} \frac{L_e L_r}{G_e(\vec{R})G_r(\vec{R})} \frac{P_r}{P_e} \frac{1}{S_{eff}} \qquad [3.3]$$

Given the dynamic range of σ^0, the latter is often converted into decibels : $\sigma^0_{dB} = 10 \log_{10} \sigma^0$.

Owing to the fading affect (also called speckle) typical of coherent radar measurements carried out on natural surfaces, the power P_r, recorded by multiple measurements corresponding to different resolution cells acquired on a homogeneous natural surface, varies greatly. Consequently, the power P_r from different resolution cell over an homogeneous area follows exponential law probability with an average of $\langle P_r \rangle = m$ (proportional to σ^0) and at a variance of $var(P_r) = m^2$. In order to improve radiometric resolution, the aim is therefore to reduce the variance without altering the average. This can be done efficiently by taking an average of independent samples. The average intensity $\langle P_r \rangle = \frac{1}{N}\sum_{q=0}^{N} P_{r_q}$ of the variable which is obtained by taking an average of N independent samples, described as "look", remains unchanged $(\langle P_r \rangle) = E(P_r)$, while its variance $var(\langle P_r \rangle) = \frac{var(P_r)}{N}$ is reduced by a factor of N. The radiometric resolution is commonly fixed at the standard deviation of the measurement of the backscatter coefficient [ULA 86]. This can be evaluated using the K_p coeeficient followng:

$$K_p = \frac{\sqrt{var(\langle P_r \rangle)}}{\langle P_r \rangle} = \frac{\sqrt{var(\langle \sigma^0 \rangle)}}{\langle \sigma^0 \rangle} = \frac{1}{\sqrt{N}} \quad [3.4]$$

which is a result of the relation:

$$\sigma^0_{dB} = \langle \sigma^0 \rangle_{dB} + 10 \log_{10} \left(1 \pm \frac{\sqrt{var(\langle \sigma^0 \rangle)}}{\langle \sigma^0 \rangle} \right) \quad [3.5]$$

The aim of a scatterometer is to take the average from a large number of samples in order to obtain a very small signal variance. The average of these samples is taken at the cost of spatial resolution, given that the looks used are resolution cells obtained for adjacent impulses[1] during an integration time during the sensor's displacement.

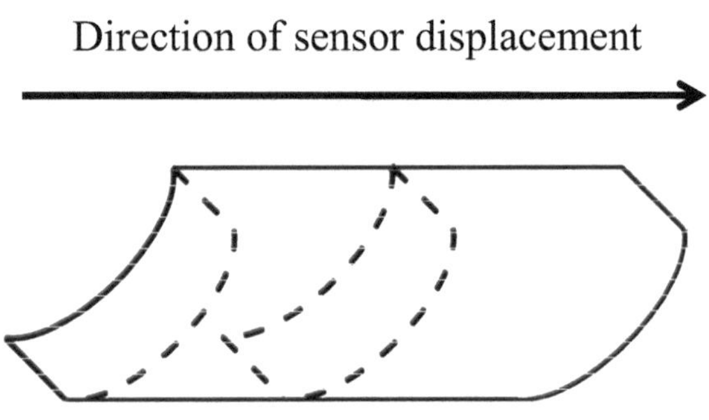

Figure 3.3. *Resolution cell for an impulse illustrated by a dotted line, corresponding to the hatched surface in Figures 3.2(a), and (b) resolution cell after averaging during the integration time (continuous line) for a fixed antenna scatterometer*

For a given signal with a passband B, and measured during a period of integration T, the valuation of the number of resultant looks is $N \approx BT$

1 σ^0 is estimated by replacing P_r in [3.2] with $\langle P_r \rangle$.

[ULA 86]. In the appropriate azimuthal direction, the "Doppler" bandwidth B_D corresponds to the Doppler frequency variation during the sensor's displacement. Here it is shown that, for an antenna with an oblique view (Figure 3.2(a))[2]:

$$B_D = \frac{2v}{\lambda}\frac{\cos\theta}{h}(X_r \cos^2\theta \, \sin\psi + X_a \cos\psi) \qquad [3.6]$$

X_r and X_a are the resolutions in terms of distance and azimuth[3] and θ and ψ are incidence and azimuth angles. For a given period of integration, B_D has to be compared with the pulse repetition frequency (PRF). For a PRF $\leq B_D$, pulses will be statistically independent and the number of looks will be equal to the number of pulses received during the integration period.

3.3. The main scatterometers

This section will look in more detail at the ESCAT and ASCAT scatterometers which have obtained data in the C-band on a continuous basis since 1991. Other scatterometers with fixed antennae, on board the SEASAT and ADEOS-1 satellites and operating on the Ku-band, only had a short acquisition period (4 and 10 months respectively). We will then turn to look at the conically rotating pencil-beam scatterometer SeaWinds which enabled acquisitions for 10 consecutive years.

3.3.1. *Fixed antenna scatterometers*

3.3.1.1. *ESCAT scatterometer*

The satellites ERS-1 and 2 were put into orbit by the ESA (the European Space Agency) in 1991 and 1997 respectively. They carried similar

---

2 From [3.1] we are able to conclude that $f_D = \frac{2v}{\lambda}\sin\theta\sin\psi$ leading to the Doppler dispersion relation: $B_D = \Delta f_D = \frac{2v}{\lambda}(\cos\theta\sin\psi\,\Delta\theta + \sin\theta\cos\psi\,\Delta\psi)$. In addition $\Delta\theta = \frac{X_r\cos^2\theta}{h}$ and $\Delta\psi = \frac{X_a}{h\tan\theta}$.

3 X_r depends on the passband, B_c, on the emitted pulse, and X_a, on the 3dB aperture, $\beta_a = \frac{\lambda}{L}$, of the antenna in the azimuthal direction (λ being the wavelength of the pulse, and L the length of the antenna). They verify: $X_r = \frac{c}{2\,B_c\sin\theta}$ and $X_a = \frac{h\,\beta_a}{\cos\theta}$, where c is the speed of light.

scatterometers onboard which were acquiring data at VV polarization over a period of 20 years, from July 1991 to 2011. The radar signal is continuously delivered and analyzed by three sideways antennas: one pointing normal to the satellite flight path (midbeam antenna) and the two others pointing 45° forward (forebeam antenna) and 45° backward (aftbeam antenna). Along the swath, incidence angles range from 18° to 47° for the the Midbeam antenna and from 25° to 59° for the two others antennae. The footprint of a beam is divided according to the equidistant circles in the radial direction while the sampling in the azimuthal direction follows the movement of the sensor. A total of 384 pulses are averaged, which roughly corresponds to an integration period of 12 seconds and a distance along the ground of 85 km[4]. Given that radiometric precision depends on the number N of looks, it becomes necessary to estimate the one associated to the 384 pulses. When we consider the midbeam antenna ($\psi = 0$) for example, of a length L = 3.5 m, [3.6] gives us $B_D = \frac{2v}{L} \approx$ 6kHz (for a satellite speed of $\approx 7.7\ \mathrm{km/s}$). This frequency, much higher than PRF ≈ 115 Hz, means that the received echoes are statistically independent. It results a total look number of N = 384, leading to $K_p = \frac{1}{\sqrt{384}} \approx 5\%$ [3.4].

The swath is divided up according to a square grid 25 km on one side. The apexes of each grid, termed nodes, represent the centre of a surface integral of roughly 50 × 50 km^2 while the contribution of each pulse within this surface is weighted by a Hamming window. The result of this is a measurement of the backscatter coefficient σ^0 which represents an almost circular resolution cell around 50 km in diameter.

3.3.1.2. *The ASCAT scatterometer*

In October 2006, the ESA launched the first of a series of three satellites, known as METOP, which each had on board the same type ofscatterometer, ASCAT (Advanced Scatterometer). The METOP satellites are controlled by the organisation EUMETSAT (European Meteorological Satellite) as part of the EPS (European Polar System) program, which was developed for operational monitoring in meteorology and climate domains. ASCAT scatterometers operate in the C-band and with VV polarization, which allows

---

4 The integration is carried out by taking the average from 12 FMA cycles (Fore-Mid-Aft). An FMA cycle is made up of 32 pulses emitted by each of the three antennae [LEC 98].

for continuous observations with respect to ESCAT onboard ERS satellites. In terms of design, they are similar to those previously mentioned, featuring three antennae which illuminate the same swath: the midbeam antenna, with incidence angles between 25° and 53°, and fore and aftbeam antennae with incidence angles between 34° and 64°. Compared to their predecessors, these scatterometers feature significant improvements, such as:

– double swath: each is comprised of two sets of three antennae (Fore, Mid and Aft) which make it possible to illuminate two swaths of 550 km (divided into 21 nodes) on either side of the satellite's track;

– improved spatial resolution: in addition to 50 km resolution ASCSZ0 products, EUMETSAT also provide ASCSZR products which feature a spatial resolution of roughly 30 km for nodes 12.5 km apart (41 nodes across each swath). These products possess a radiometric precision in the same order of magnitude as 50 km ERS scatterometers.

3.3.1.3. *Angular effects*

Owing to the size of the resolution cell, measurements of σ^0 carried out on the majority of terrestrial surfaces are independent from the azimuthal angle [FRI 96a, ZIN 04][5]. On the other hand, the configuration of fixed antenna scatterometers means that measurements will be acquired across a wide range of incidence angles. These variations of incidence angles lead to significant change in σ^0 measurements, and depend on the surface being observed, as is illustrated in Figure 3.4. We observe a linear decrease of σ^0, particularly for angles above 30° [KEN 89a, KEN 89b, FRI 96a]. The slope of the linear regression is reduced depending on the vegetation density. Thus, tropical forests provide a relatively weak slope (–0.06 dB/°) while for deserts the slope can be as high as –0.24 dB/°. Therefore, it is necessary to carry out angular normalization in order to make use of different measurements obtained at different angles of incidence. Essentially, in the case of natural surfaces, denser vegetation results in more isotropic backscatter. It also means that the dependency of σ^0 at the incidence angles is less significant. In the case of tropical forests, backscatter is typical of volumic diffusion, while for deserts it is typical of surface diffusion. In this last case, the gradient is primarily a function of the roughness of the Earth

5 Significant variations were observed for the Sahara desert, characterized by vast expanses of dunes angled in a set direction [STE 07].

(see Chapter 1 of this book). Measurements at higher angles of incidence are preferred for monitoring terrestrial vegetation, given that, in this case, the underneath surface contribution is minimized and the intensity of the backscattered signal is primarily dependent on the vegetation. Conversely, for monitoring soil moisture, measurements at lower incidence angles are preferred in order to minimize the disruptive effect of the vegetal covering. By way of illustration, [FRI 98], whose work aims to monitor the herbaceous biomass of the Sahelian steppes, opt for data where the angle of incidence is greater than 45°, while [ZRI 08] only retain data obtained at an incidence angles less than 35° for estimating soil moisture in Western Africa.

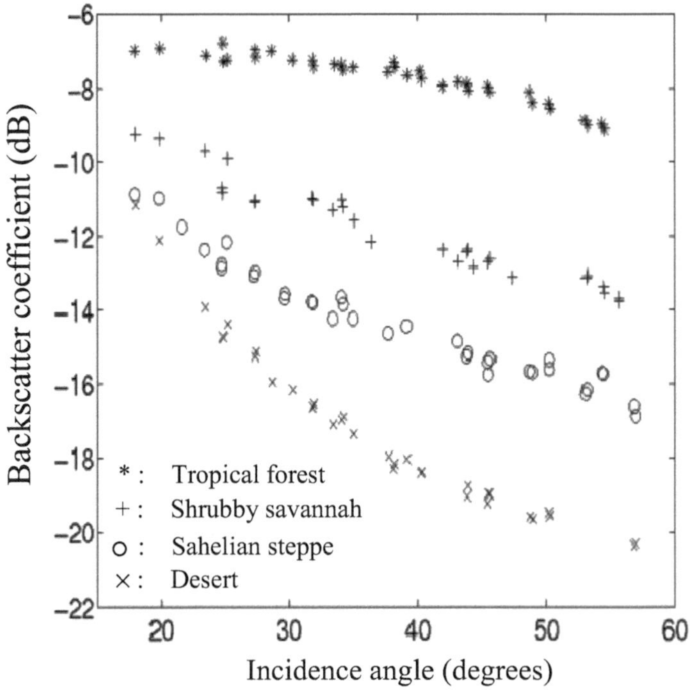

Figure 3.4. *Showing the development of the radar backscatter coefficient σ^0 as a function of the angle of incidence across different types of surface (ASCAT data)*

3.3.2. *Rotating antennae scatterometers: SeaWinds*

The scatterometer SeaWinds on board the satellite QuikSCAT (Figure 3.1(c)) was in use between 1999 and 2009. It was rapidly developed in order to compensate for the failure of the NSCAT scatterometer. A similar scatterometer, on board the satellite ADEOS-2, acquired data at the Ku-band between December 2002 and October 2003. It is important to note here that Ku-band measurements are sensitive to water present in the atmosphere (including water vapour and rain). These effects are taken into account according to the relation: $\sigma^0_{cor} = \tau\,\sigma^0_{mes} + \sigma^0_{rain}$, where σ^0_{mes}, σ^0_{cor}, σ^0_{rain} are the respective backscatter coefficients measured, corrected for atmosphere and for rain, and with τ as the absorption coefficient. SeaWinds is a conical pencil-beam scatterometer, which operates at constant incidence for each of its two beams. The interior beam operates at VV polarization with an incidence of 46° while the exterior beam operates in HH polarization with an incidence of 54°. This configuration ensures a very large swath, of 1400 km and 1800 km for the interior and exterior beams respectively, and makes it possible to cover 90% of the Earth's surface in a period of 24 hours. With the exception of the points on the satellite's track, four measurements are carried out for the points within the interior swath (1400 km): 2 in VV polarization when it will be viewed from the front and then the rear of the satellite and 2 in HH polarization. The rotation of the antenna (18 rotations/min) combined with the PRF (92.5 Hz) enables the superposition of the footprints on the ground [SPE 97]. The footprint measures roughly 25 km × 35 km in azimuth and in the direction of the view angle respectively. It is subdivided into five cells of roughly 7 km × 25 km (the concentric lines in Figure 3.2(a)) according to the viewing direction [SPE 01]. The configuration of the conical pencil-beam entails a smaller number of resolution cells (and therefore a weaker precision of σ^0) involved when calculating the average, compared to a configuration using fixed antennae. Nevertheless, the values of K_p are smaller than 17%. On the other hand, measurements carried out by QuikSCAT are also combined with those carried out by OSCAT during the period where they are both operational in order to significantly improve the precision of the measurements [SEE 14].

3.4. Thematic applications

The purpose of this section is to review the main results of the application of data obtained from scatterometers for the monitoring of surface parameters. Primarily designed for wind retrieval above oceans, it quickly became apparent that this data could be used for monitoring continental surfaces [KEN 89a, KEN 89b, MOU 95a]. For illustration purposes, Figure 3.5 shows the backscatter coefficient σ^0 obtained by ESCAT in December 1992 for the entire Earth. Here, it is possible to identify a close link between this image and the main mountainous areas as well as the main types of vegetation. The highest responses can be observed in the glacial regions of the Arctic and Antarctica while desert surfaces (Sahara, Gobi, Takli Makan) are characterized by the lower backscattering (≈ –26 dB). The strong backscatter observed in the Sahara desert correspond to the mountain ranges (Aïr, Tibesti and Hoggar). The highest response for surfaces covered by vegetation can be observed in tropical forests in the basins of the Amazon and the Congo (–7 dB). Intermediate values correspond to different types of vegetation depending on their density, as shown by the responses of herbaceous steppes, shrubland and open or closed savannahs along a north–south gradient in West Africa.

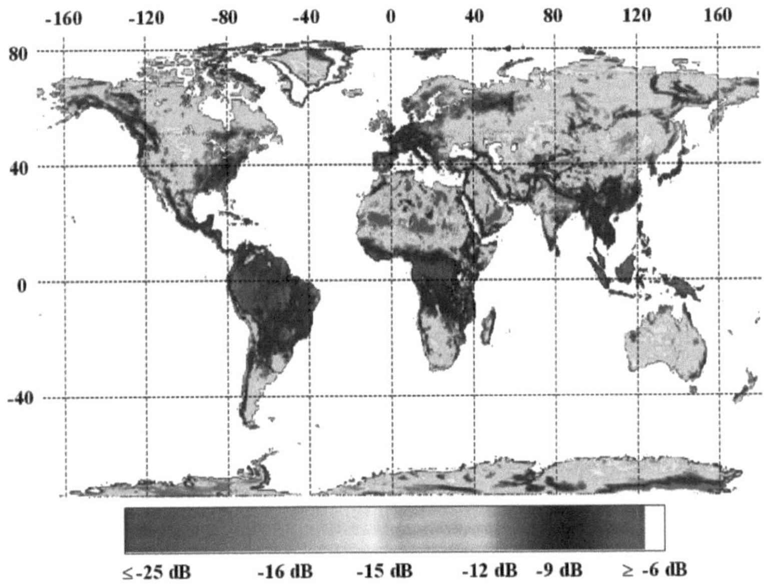

Figure 3.5. *Backscatter coefficient σ^0 (incidence angle: 40°) obtained by the ESCAT scatterometer in December 1992 (adapted from [FRI 96b]). For a color version of this figure, see www.iste.co.uk/baghdadi/4.zip*

Such potential of scatterometers for land surface monitoring results mainly from their high temporal frequency of acquisition (on an almost daily basis for current scatterometers), which is especially well-adapted for relating signal variations to seasonal change of the surface biophysical parameters. The first studies undertaken following the launch of ERS-1 focused on geographical locations displaying temporal signals of the backscatter coefficient with a pronounced seasonality. This is particularly true for the semi-arid areas of West Africa [FRI 96b] and the high-latitude regions of the Northern Hemisphere [FRI 96b, WIS 00]. These two regions are characterized by a significant change of the surface's dielectric properties. For Sahelian regions, the first rainfall events after a long dry season, during which the soil are dry and bare, dampens the soil and herbaceous plants begin to grow. This results in a significant increase of the backscattering coefficient which is used to retrieve the superficial soil moisture and the vegetation aerial biomass [MAG 97, FRI 98, WOO 00, JAR 03]. At the end of the rainy season, the backscattering coefficient decreases as the soil and the vegetation dry out before reaching a low and constant level during the following dry season (Figure 3.10, section 3.4.2.1). For northern regions, the water change of phase from solid to liquid when thaw occurs causes an increase of backscattering coefficient. This sudden changes occuring during freezing or thawing makes it possible to identify the freeze and thaw dates. Scatterometer data have then been used for numerous applications. Figure 3.6 is a spectacular illustration of the potential of scatterometers for the monitoring of the seasonal variations of surface parameters for a desertic area. It displays the time series of the backscattering coefficient obtained by ASCAT in the Chott El-Jerid region in Tunisia between 2007 and 2010 [FRI 13]. This saline lowland located in an arid environment is subject to regular floods causing a specular reflection of the radar signal (and therefore a low backscatter around –22 dB). High rates of evaporation lead to the formation of evaporites and a salt crust, increasing the surface's roughness as well as creating a strong variation in the dielectric constant. This results in a drastic increase of the radar response (> 10 dB, among the highest observed on terrestrial surfaces) with values as high as –12 dB. The high time frequency of ASCAT makes it possible to detect floods at Chott [CHO 07, CHO 09], emphasized by sharp falls in the signal observed. Because of its weaker observation frequency (roughly one measurement every 20 days in this range of angle of incidence) these events are more difficult to detect using ASAR radar imaging.

The main applications of these observations are described below and divided in two sections; the first one deals with the monitoring of the parameters of the surface hydrological cycle and the second is dedicated to the monitoring of vegetation.

Interestingly enough, other studies have been published, exploring the potential to retrieve variables indirectly related to the backscattering coefficient. Among others, [JIM 09] develop an approach to estimate turbulent flux at the surface–atmosphere interface on a monthly and global level from scattermeter observations used in combination with other satellite data. [PRI 05], based on *in situ* measurements from arid and semi-arid zones, and the backscattering coefficient measured by ESCAT retrieved the aerodynamic roughness length, a key variable for land surface models.

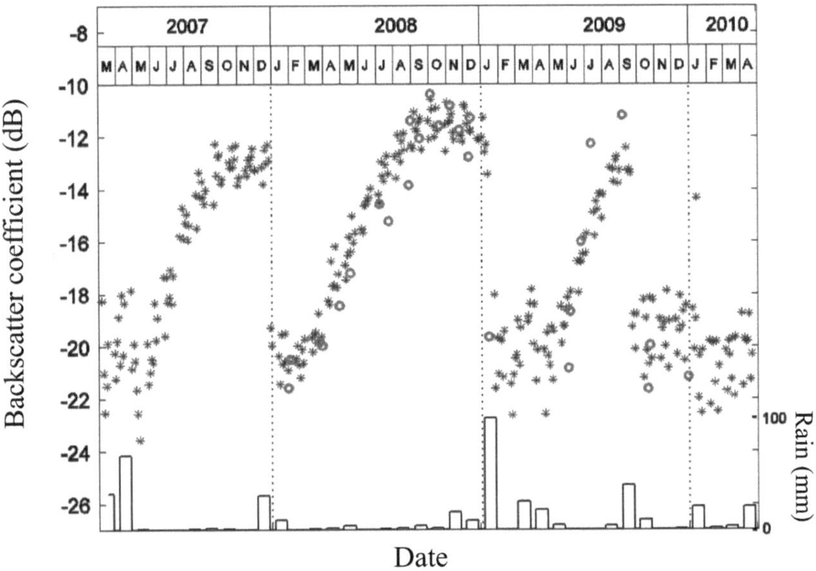

Figure 3.6. *Temporal profiles of radar backscatter coefficient σ^0, observed at Chott El-Jerid in Tunisia, by ASCAT (*) and by ASAR (o) at an angle of incidence of 40°. The average levels of precipitation observed at the neighboring stations of Tozeur and Kebili are also shown here [FRI 13]*

3.4.1. *Surface water hydrology*

3.4.1.1. *Surface soil moisture*

The potential of scatterometers for the monitoring the surface soil moisture has been the major contributing factor towards the promotion of scatterometers for observing continental surfaces. This is the first time that such estimations on a global scale were distributed among the scientific community, responding to a pressing need for meteorological applications. The soil moisture is a key variable in the terrestrial hydrological cycle as it partly determines the partition of available energy between sensible and latent heat at the soil–vegetation–atmosphere interface and therefore impacts weather and climate [SHU 82]. Since then, serious efforts have been made to provide products derived from passive radiometers sensors which have been specifically designed for this purpose (AMSR, SMOS, SMAP)[6].

The approaches developed until now aim to separate the effects of the surface roughness and of the vegetation in order to extract information on the soil water content. The first application of scatterometers data for monitoring soil moisture was described initially in Wagner *et al.* [WAG 99a, WAG 99b]. It led to the production of operational soil moisture products derived from ERS and ASCAT data which quickly became intensively used, in hydrology and meteorology in particular. Several studies have used these products in order to force the trajectory of land surface models within weather forecast models through data assimilation (see Chapter 12 of this book), in particular at Météo France [BAR 14], at the European Forecasting Centre [SCI 08, DER 13], and at the Met Office [DHA 11]. In hydrology, stream flow and flood risk are partly determined by the initial conditions of the hydric state of the soil, particularly in semi-arid zones [TRA 11]. Even if the maturity is not the same as for weather forecasting, the integration of these products in hydrological models has been successfully tested by several teams [BRO 09, DRA 11].

In practice, the approach taken by Wagner *et al.* [WAG 99a, WAG 99b] is based on the hypothesis that the seasonal variations observed by C-band scatterometerss (Figure 3.10, section 3.4.2.1) are mainly caused by variations in surface soil moisture, while the effect of the seasonal variation of vegetation is only taken into account implicitly when the data is standardized. The inter annual variability is simply ignored. This leads to an estimation of the relative value of the soil's superficial moisture, m_s, using a linear interpolation between the extreme values of σ^0 observed over a period

6 AMSR: Advanced Microwave Scanning Radiometer. SMOS: Soil Moisture and Ocean Salinity. SMAP: Soil Moisture Active and Passive.

of several years: the minimal value is associated with the driest conditions $\sigma^0_{dry}(t)$ and the maximal value is associated with wettest conditions $\sigma^0_{wet}(t)$. The value is thus taken to be between 0 for water content at its residual value and 1 for saturated soil. In the case of soil that is not frozen and where there has been no snowfall, the surface soil moisture is estimated according to the relation:

$$m_s(t) = \frac{\sigma^0\,(t) - \sigma^0_{dry}(t)}{\sigma^0_{wet}(t) - \sigma^0_{dry}(t)} \qquad [3.7]$$

Provided that sufficiently long ime series are made available, it is possible to obtain the extreme values $\sigma^0_{wet}(t)$ and $\sigma^0_{dry}(t)$. Estimations of the surface soil moisture products were successfully validated in various regions around the world: in France by Météo-France's studies with ERS [PEL 06] and ASCAT [ALB 09, ALB 12] products, Amri *et al.* [AMR 12] in Tunisia using ASCAT, Gruhier *et al.* [GRU 09] in the Sahel in West Africa and most recently Cho *et al.* [CHO 15] in Korea. These studies have been able to demonstrate the coherence of these products on a global scale. Nevertheless, certain divergences with measurements of moisture (*in situ*, products derived from radiometers, reanalyses, etc.) have been observed in the Sahelian region [GRU 09] and in four drainage basins in North America [LER 14]. In semi-arid regions, there has been evidence of abnormally high levels of retrieved surface soil moisture when the soil is particularly dry. These errors have been attributed to a variation of the backscatter coefficient resulting in a deeper level of wave penetration when the soil dries [WAG 13]. However, the absence of an explicit consideration of vegetation effects and of its interannual variability are the main limitations of these products.

[ZRI 08] put forward an alternative empirical approach which explicitly takes into account the effects of vegetation. Humidity m_s is estimated according to the relationship:

$$m_s(t) = \alpha\,\Delta\sigma \qquad [3.8]$$

where $\Delta\sigma$ represents the difference between the value of the backscattering coefficient corresponding to dry soil σ^0_{dry} and the observed actual value,

$\sigma^0(t,)$; α is a coefficient which takes into account the properties of the soil (texture and roughness) and the properties of the vegetation as follows:

$$\alpha = \alpha_{soil} \cdot f(veg) \qquad [3.9]$$

α_{soil} can be simply estimated using a linear regression established on bare soil between *in situ* soil moisture measurements and the observed $\Delta\sigma$ ($f(veg) = 1$ for this particular case). The main added value of the approach is the decoupling of the effects of the soil roughness from those of the vegetation based on the use of the vegetation index NDVI from the NOAA-AVHRR (National Oceanic and Atmospheric Organization – Advanced Very High Resolution Radiometer) sensors. Figure 3.7 provides an example a surface soil moisture map for a period of five days towards the end of August 1999. The presence of the monsoon front to the north of the zone on the Sahelian strip caused heavy spells of rain which led to the soil becoming damp (red marks).

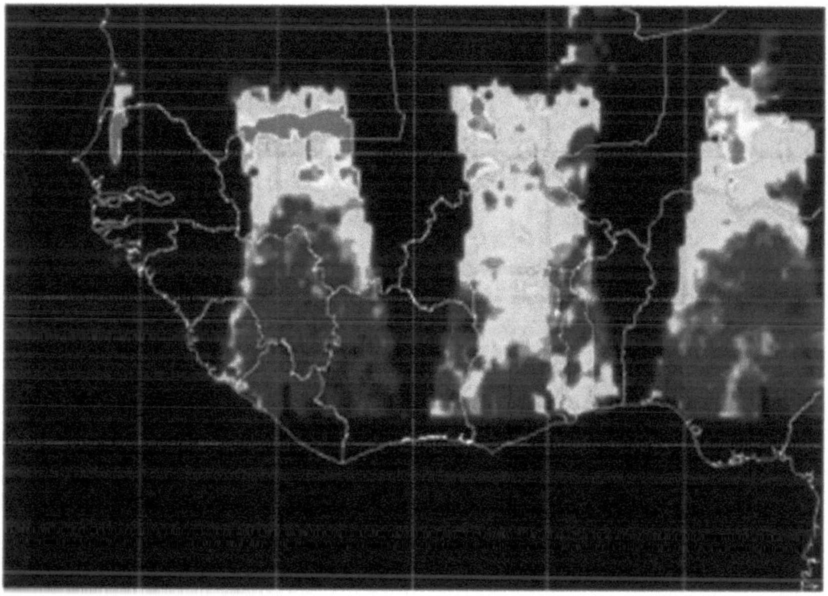

Figure 3.7. *Map of the superficial soil moisture derived from Zribi et al.'s approach [ZRI 08] over a period of 5 days between the 20th and the 25th August 1999 in West Africa. The level ranges from 0 m^3/m^3 (dark purple) to 0.25 m^3/m^3 (red). For a color version of this figure, see www.iste.co.uk/baghdadi/4.zip*

Moreover, from the surface soil moisture value, it is possible to retrieve a normalized water content of the entire root zone by applying an exponential filter which is far more relevant for agronomists than superficial soil moisture. This normalized value, called the Soil Wetness Index (SWI) [DOU 99] ranges from 0 (for water content at wilting point) and 1 (for water content at saturation). This exponential filter is a result of the resolution of a simple soil water budget equation with the assumption that the transfer of water between the superficial layer and the root reservoir is proportional to the difference in water content between the two layers [WAG 99b, PAR 08, MAN 14]. Once this equation has been solved, the SWI can be expressed as follows:

$$SWI(t) = \frac{\sum_i m_s(t_i) e^{-(t-t_i)/T}}{\sum_i e^{-(t-t_i)/T}} \qquad [3.10]$$

with $m_s(t_i)$ the value of superficial soil moisture estimated using scatterometer data at a time of t_i. This evaluation can be written recursively, making it easier to implement [ALB 08]. The only parameter requiring calibration is T, a time characteristic of the temporal dynamic of the root zone moisture which includes the effect of a number of factors, including the hydrodynamic characteristics of the soil, the redistribution of water between evaporation, runoff and drainage. Amri *et al.* [AMR 12] have developed a normalized anomaly index of the SWI which proves how useful this new index is for the detection of drought.

3.4.1.2. *Dates of freezing and thawing in northern regions*

The dates of freezing and thawing are important informations when studying the global carbon and methane cycle given that northern regions discharge high quantities of these greenhouse gases during the summer. These dates also provide a good indicator of climate change because regions at such high latitudes are scarcely anthropized and are particularly sensitive to reheating that will have a direct impact on the length and phasing of the summer. The principle for detecting the dates of freezing and thawing was described by Wismann [WIS 00] with ERS, by Kimball *et al.* [KIM 01] with NSCAT, and Bartsch *et al.* [BAR 07] with QuikSCAT. This principle is illustrated in Figure 3.8, which shows two time series of the backscatter coefficient for two study sites in Siberia.

Thaw, for the point furthest south, is accompanied by a sudden increase of σ^0_{40} due to the phase change of water from solid to liquid. For the point furthest north, which is probably covered by snow, the liquid water contained in the melting snow causes a drop in the signal before a sudden increase for the same reasons listed above. At the end of summer, we notice a sharp decrease in the level of backscatter when the water is freezing. An algorithm for the detection of change makes it possible to automatically estimate the dates of thawing and freezing. More complex situations can arise, such as when several successive periods of thawing and freezing have taken place at the beginning of the season when the temperature hovers around 0°C. In order to remove any ambiguities which could come about as a result of this, the detection approach of freeze-thaw dates was refined by Zwieback *et al.* [ZWI 15] taking into account data on the air temperature obtained by reanalysis.

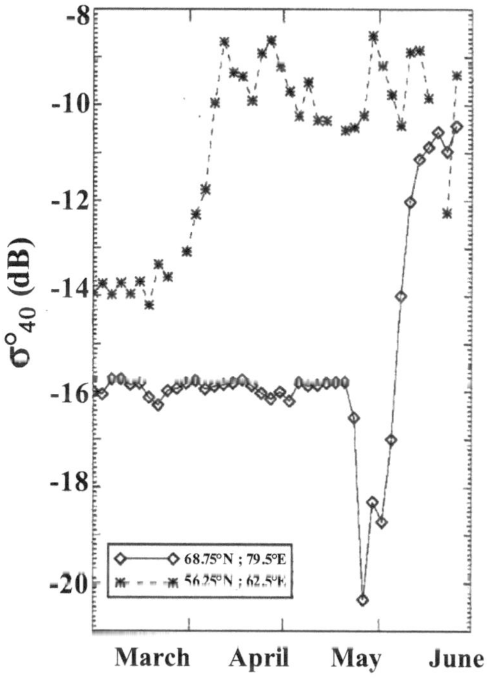

Figure 3.8. *Temporal series of the backscatter coefficient standardized at 40° of incidence (σ^0_{40}) for two pixels in Siberia [WIS 00]*

This information relating to the dates and the length of the thawing season for high northern latitudes has been explored in a number of different contexts. Kimball *et al.* verified the consistency between the dates and the length of the season estimated based on the SeaWinds data and the distinct primary productivity, measured simultaneously *in situ* at several sites and from maps produced by a carbon model. A few studies have also used the information available since the beginning of the 1990s in order to constrain hydrological and carbon cycles in surface models through data assimilation technics. The work of Farhadi *et al.* [FAR 15] shows how the assimilation of the dates of freezing and thawing can be used in order to reduce errors on the surface temperatures and soil temperatures simulated by a land surface model.

3.4.1.3. *Other uses*

Data from scatterometers have also been used in order to extract other information essential to the understanding of the continental hydrological cycle. Prigent *et al.* [PRI 01a] as well as Papa *et al.* [PAP 10] were able to map flooded zones on a global scale from 1993 to 2004 using a combination of several sets of satellite data, including data from the ERS scatterometer. Mapping flooded zones is directly linked to the capacity of scatterometers to distinguish open water, based on the strong specular reflection (and therefore weak backscatter) caused by these flat surfaces. Through analysis of these monthly maps, it was been possible to demonstrate a decrease of 5.7% of flooded surface between 1993 and 2004.

3.4.2. *Monitoring vegetation*

3.4.2.1. *Estimating aerial biomass*

Unlike the studies discussed previously, the aim here is to target the effect the vegetation canopy by minimising those effects caused by variations in the soil's properties (water content and roughness). As radar acquisitions in the C-band saturates for relatively small quantities of biomass (> 50 tonnes of dry matter per hectare), these studies were focused on arid and semi-arid regions (Mediterranean and Sahelian). For the most part, the methods used are based on the inversion of a radiative transfer model in order to achieve a more precise deconvolution of the different contributions. The advantage of this type of approach, as opposed to empirical approaches, is to be able to quantify the contributions and thus better understand the sensitivity of the radar signal to the different surface parameters.

Because of the pixel size (tens of kilometers), studies which are based on the modeling of the interaction between the radar wave and a random terrestrial surface generally tend to split the study area into one component representing bare soil and another composed of the vegetation canopy [MAG 97, FRI 98, WOO 00]. The backscatter coefficient of the area can be expressed as the weighted sum of the soil contribution $\sigma^0{}_{soil}$ and the canopy contribution $\sigma^0{}_{canopy}$, according to their respective fraction cover:

$$\sigma^0_{area}(t) = C_v\ \sigma^0_{canopy}(t) + (1 - C_v)\sigma^0_{soil} \qquad [3.11]$$

with C_v representing the fraction cover of vegetation.

The vegetation contribution σ^0_{canopy} is computed using a radiative transfer model [ULA 88, KAR 92, ATT 78]. The soil's contribution σ^0_{soil} is calculated using a surface backscatter model [OH 92, FUN 94]. Vegetation is characterized by one or several horizontal layers of randomly distributed scatterers, assuming an azimuthal isotropic distribution.

The canopy contribution σ^0_{canopy} corresponds to the incoherent sum of three contributions [ULA 82]: the contribution of the vegetation representing the volume diffusion, σ^0_{veg}, the contribution of the soil below the vegetation layer (surface diffusion), σ^0_{soil}, which is attenuated by a transmission factor of the canopy, L, and the interaction between the canopy and the soil σ^0_{int}:

$$\sigma^0{}_{conopy} = L^2\,\sigma^0{}_{soil} + \sigma^0{}_{veg} + \sigma^0{}_{int} \qquad [3.12]$$

These three contributions are represented in Figure 3.9. The term $\sigma^0{}_{int}$ will generally never exceed 1 dB for areas of low diffusion in VV polarization [ULA 82]. This is particularly true when the vegetation cover fraction is relatively low and in these cases this term can be discarded.

The only differences between the approaches cited here lie in the way in which the terms σ^0_{veg} are calculated. Woodhouse and Hoekman [WOO 00] assume that in C-band, σ^0_{veg} is at a level of saturation for all ecosystems observed (σ^0_{veg} = constant = –8 dB at 45° incidence) and the vegetation contribution only varies through the vegetation fraction cover C_v.

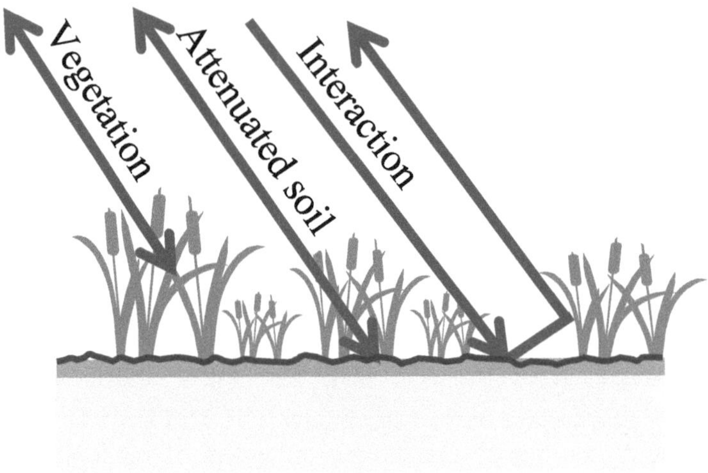

Figure 3.9. *First order solution for radiative transfer equations*

Magagi and Kerr [MAG 97] and Frison *et al.* [FRI 98] employ a radiative transfer model which provides an expression similar to the water cloud model [ATT 78, KAR 92]:

$$\sigma^0_{veg} = \frac{\sigma_v cos\theta}{2\,\kappa_e}\left(1 - \frac{1}{L^2}\right) \qquad [3.13]$$

where σ_v expresses the backscatter coefficient of volume, κ_e is the extinction coefficient, and L is the transmission coefficient of the vegetation canopy. For a layer of height h and a radar beam at an incidence angle θ, $L = e^{\frac{\tau}{cos\theta}}$, where $\tau = \kappa_e\,.h$ is the optical depth of an element of vegetation volume, depending on its extinction coefficient κ_e.

Magagi and Kerr [MAG 97] estimate the parameters σ_v and κ_e (as well as the soil moisture) by inversion of the signal. Frison *et al.* [FRI 98] and Jarlan *et al.* [JAR 02] estimate the paramaters σ_v and κ_e in a physical manner, based on a radiative transfer model which integrates the response of an elementary diffusor (representing a herbaceous leaf by an ellipsoid) across the whole covered area [KAR 92]. Simulating the interaction of an electromagnetic wave with the surface therefore requires several parameters to depict the scene (superficial soil moisture, aerial biomass, vegetation height) as well as their temporal evolution. In "direct" mode (as opposed to the inverse mode which aims to invert the signal in order to retrieve the biophysical characteristics of the surface) these are obtained using a dynamic model of the vegetation growth in Sahelian regions [MOU 95b] over several sites where *in situ* measurements necessary for its calibration are available. Figure 3.10 shows the results of a comparison between the backscatter coefficient measured by the satellite ERS for five seasons of growth on a site in the Sahelian zone in northern Mali and the results of the simulation. We are able to note that the strong seasonality of the signal, well reproduced in the simulation, is linked to the change from the dry season to the rainy season: bare and dry soils during the dry season are characterized by a low backscatter coefficient almost constant around –19 dB; the arrival of the monsoon at the beginning of summer is accompanied by a marked increase ofthe signal linked to the humidification of the soil and the growth of vegetation, mainly herbaceous in this regions. Finally, the end of the rainy season at the end of summer is linked to a drastic decrease of the backscatter coefficient as a result of the soil drying and of the senescence of the vegetation.

Unlike the direct problem which aims to reproduce the signal observed, the inverse problem aims to estimate certain characteristics of the observed surface (aerial mass of vegetation, vegetation fraction cover, leaf area index, soil moisture, soil roughness) using one or several observations of the backscatter coefficient. Magagi and Kerr [MAG 97] and Woodhouse and Hoekman [WOO 00] simulatenously retrieve information on vegetation (the equivalent water content of the vegetation in one cases, the vegetation cover fraction in the other) and on soil (superficial moisture and roughness for both studies). Owing to a lack of *in situ* data, the results were only partially validated. In order to better constraint the inverse problem, the surface soil moisture was computed *a priori* using a simple soil hydric budget model forced by satellite precipitation fields in [JAR 03]. This soil moisture time series is used as an input of the radiative transfer model; the roughness is determined using observations of the dry season given that the soil is bare and dry. Finally, inversion is focused on the characteristics of vegetation only.

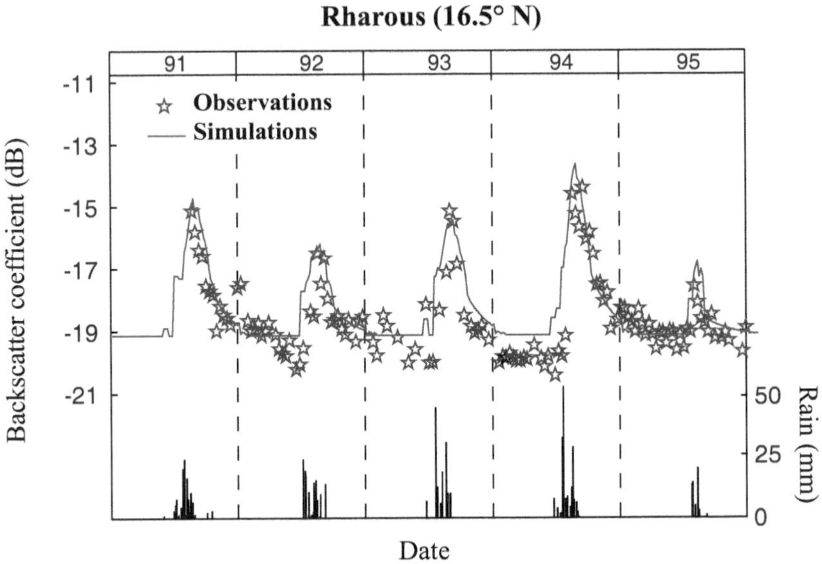

Figure 3.10. *Time series of the backscatter coefficient from the ERS scatterometer normalized at 45° of incidence observed and simulated using Karam's radiative transfer model on a site in the Sahelian region near Gourma-Rharous (Mali) for five seasons of growth (1991–1995) [JAR 02]*

For illustration purposes, Figure 3.11 shows maps of herbaceous mass in the Sahelian band in West Africa for four years. A clear distinction is made along the north–south gradient of the herbaceous mass between the desert in the north and the savannahs in the south. Moreover, the vegetation production for years 1994 and 1999 characterized by above normal rainfall condition, is markedly higher than in 1995, a very dry year, especially in the east of the region. The black line depicts the 200 kg of dry matter per hectare isoline and can be interpreted as the transition between the Sahara and the Sahel. The continuous acquisition of scatterometer data since 1992 by ERS-1 and ERS-2 and then ASCAT provides a unique opportunity to study the fluctuations of the northern limits of the Sahel, and thus to contribute to the current debate around the regreening or the desertification of this region.

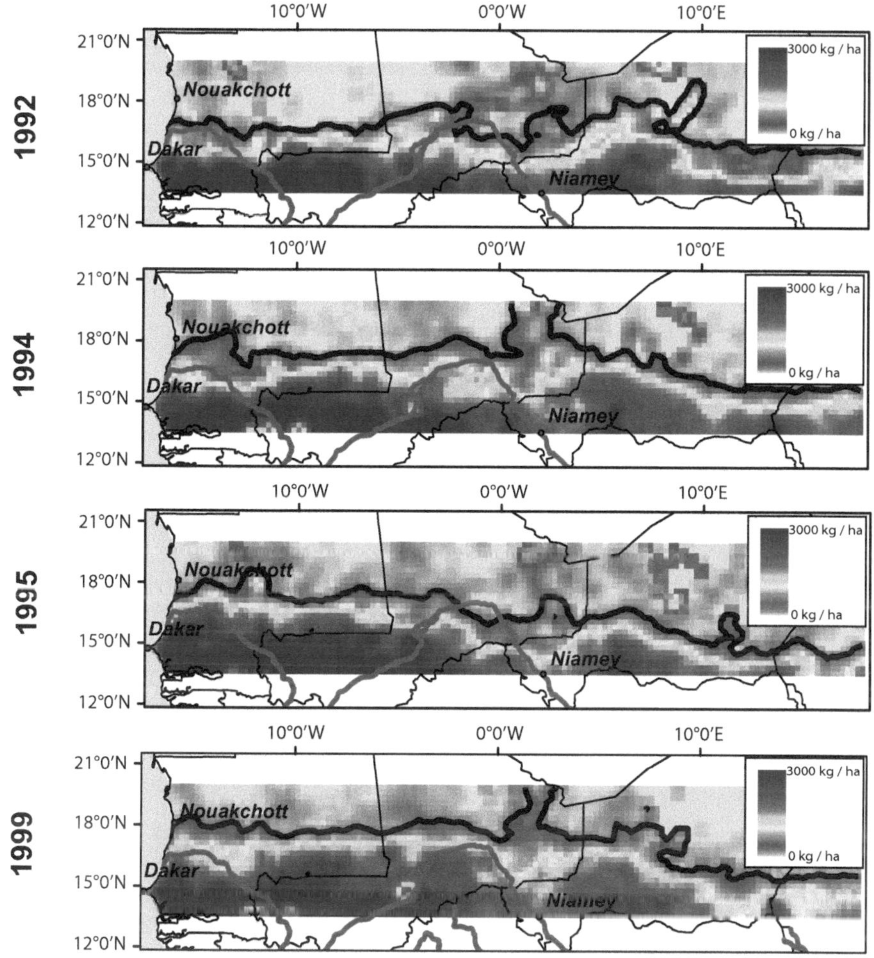

Figure 3.11. *Maps showing the aerial herbaceous mass derived from ERS Scatterometer data in the Sahelian region for four contrasting years following the approach of [JAR 03]. The color level ranges from 0 tonnes/ha (beige) to 3 tonnes/ha (dark green). For a color version of this figure, see www.iste.co.uk/baghdadi/4.zip*

3.4.2.2. *Detection of the vegetation hydric stress*

According to the physics of radar signals, the main parameter characteristics of the vegetation to which radar is sensitive is the equivalent water of aerial mass. While this adds an additional unknown to the inversion process, an original application of scatterometer data is for the detection of the hydric stress of the plant, given that its water content will decrease when stress is present. Nonetheless, there have been few studies relating to this subject. Steele-Dunne *et al.* [STE 12] analyzed the backscattering coefficient difference between data from a descending orbit (at 10 am) and from an ascending orbit (at 10 pm) of the ERS satellite, based on the work of Friesen [FRI 08]. They were able to prove that there was a systematic difference in favor of data acquired in descending orbit and that this difference tends to be maximal at the end of the rainy season and minimal at the end of the dry season. Differences of 1 dB between ascending and descending orbits were also highlighted for equatorial forests by Frison and Mougin [FRI 96a]. Plants, and more specifically trees, lose water by transpiration during the day and recharge their storage of water by root extraction at night. This explains the higher backscatter value for the descending orbit. Moreover, when the rains stop, the ground dries and there is a significant decrease in transpiration and root extraction. Thus, the water content of the vegetation is identical in the morning and in the evening. This could prove to be a promising application for the detection of hydric stress, in other words for estimating the water content of the root zone using the water content of the vegetation. We can also imagine how these applications could be used in order to map risks of wildfire; the water content of the vegetation being a key variable in this context.

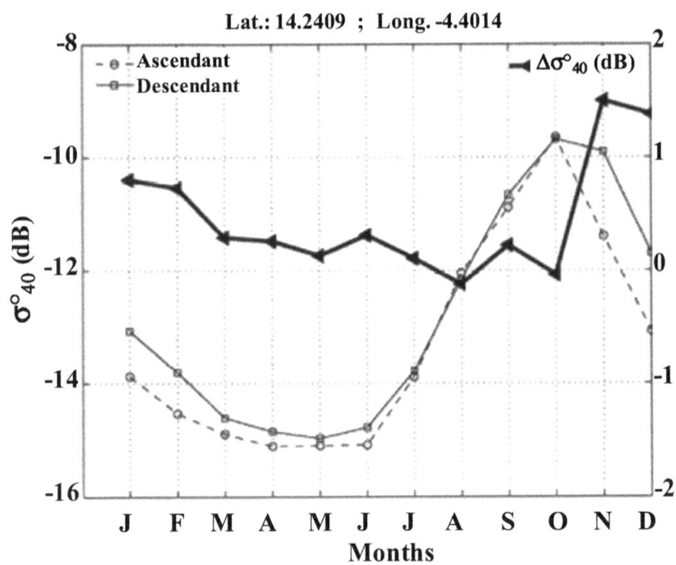

Figure 3.12. *Climatology (1991–2008) of the monthly backscatter coefficient for the ascending and descending orbit obtained by the scatterometer ERS for a pixel in West Africa. The black curve represents the difference between the two orbits*

Based on the same principles, Frolking *et al.* [FRO 11] were also able to demonstrate the potential of scatterometer data for detecting droughts in humid tropical forests.

3.4.2.3. *Other uses*

Radar data, and in particular scatterometer data, shows a sensitivity to the water content and to the structure of the vegetation canopy. This data therefore contains different and complementary information than observations acquired in the visible and near-infrared part of the electromagnetic spectrum which is traditionally used in order to map terrestrial ecosystems. Scatterometer data has thus been used for: (1) mapping the vegetation type [FRI 96a, MOU 95a, PRI 01b], in particular demonstrating the potential of the ERS scatterometer to distinguish between the different types of vegetation compared to the AVHRR and SSMI sensors; (2) monitoring regional phenology – Frolking *et al.* [FRO 06] analyzed the potential of QuikSCAT for detecting the main phenological stages of several types of vegetation in North America, most notably showing the added value of scatterometer data compared to visible and near-infrared data (LAI MODIS) for detecting bud-break in deciduous trees using the heightened sensitivity of the Ku-band to the increase in water content of woody parts which precedes leaves emergence; (3) monitoring the deforestation of equatorial forests in South America, Africa and Asia [FRO 12]. Although this has yet to be fully adapted in terms of resolution for mapping deforestation, these authors have been able to demonstrate a significant drop in backscatter coefficient of 1 to 2 dB in certain regions having experienced large-scale deforestation.

3.5. Conclusions and prospects

A scatterometer is a radar sensor that was initially developed in order to monitor oceanic winds. It is characterized by a revisit time of a few days but this strong repetitivity is obtained at the expense of spatial resolution (> 10 km). Although the first series of data has only been available to the public since the launch of the satellite ERS in 1991, scatterometer data has since been the object of increasing interest among the scientific community as the range of applications explored in this chapter demonstrate. In particular, it is the sensitivity of this data to variations in the dielectric properties of continental surfaces which makes it highly complementary with data obtained in the Vis/PIR domains that have been used since the early 1970s. Essentially, variations in dielectric permittivity are directly linked to variations in water content which opens the way to monitoring the water content of vegetation and therefore to detecting hydric stress and drought from space and to directly estimating the superficial moisture of soil. For this last point, the level of maturity of the (many) studies undertaken over more than 30 years has made it possible to develop an operational product of the superficial moisture of soil based on scatterometer data which responds to the high expectations among the meteorological community for forecasting weather and climate. As for the monitoring of vegetation, studies are still in the research phase but it would seem to be only a matter of time before we see its full potential.

Programmes introduced by space agencies have meant that global observation has been possible in the C-band since 1991 and in the Ku-band since 1996. These long-time series represent a hugely valuable dataset for analyzing the evolution of continental surfaces over both short and long terms. In addition, the use of this scatterometer data together with data from passive sensors in order to improve soil humidity products and the quantification of their uncertainties is an exciting prospect.

3.6. Key points

– Unlike Vis/PIR remote sensing, from which the first images for observing the Earth date back to the early 1970s, global observations in radar carried out by scatterometers are more recent and have enabled continuous C-band observation since 1991 and continuous Ku-band observation since 1999.

– This data is directly linked to the surface soil water content and to the quantity of aerial biomass. Given that they are almost not, sensitive to the atmosphere, quantitative estimations of these parameters should be possible.

– Radars' sensitivity to the superficial moisture of soil has been a major contributing factor in the use of this data for monitoring land surface as part of various applications in meteorology and hydrology.

– For monitoring vegetation, the wavelengths used mean that they are particularly suited to monitoring arid and semi-arid zones.

– A number of more exploratory applications using the data from scatterometers have been carried out. Although these studies are not quite as developed, many, including the mapping of floodplains and methods for detecting the hydric stress of vegetation, have been brought about thanks to the uninterrupted presence of space programmes which enable data to be obtained over long periods of time.

3.7. Bibliography

[ALB 08] Albergel C., Rüdiger C., Pellarin T. *et al.*, "From near-surface to root-zone soil moisture using an exponential filter: an assessment of the method based on in-situ observations and model simulations", *Hydrology and Earth System Sciences Discussions*, vol. 5, no. 3, pp. 1603–1640, 2008.

[ALB 09] Albergel C., Ruediger C., Carrer D. *et al.*, "An evaluation of ASCAT surface soil moisture products with *in situ* observations in southwestern France", *Hydrology and Earth System Sciences*, vol. 13, no. 2, pp. 115–124, 2009.

[ALB 12] Albergel C., De Rosnay P., Gruhier C. *et al.*, "Evaluation of remotely sensed and modelled soil moisture products using global ground-based in situ observations", *Remote Sensing of Environment*, vol. 118, pp. 215–226, 2012.

[AMR 12] Amri R., Zribi M., Lili-Chabaane Z. *et al.*, "Analysis of C-Band Scatterometer Moisture Estimations Derived over a Semiarid Region", *IEEE Transactions on Geoscience and Remote Sensing*, vol. 50, no. 7, pp. 2630–2638, 2012.

[ATT 78] Attema E.P.W., Ulaby F.W., "Vegetation model as a water cloud", *Radio Science*, vol. 13, no. 2, pp. 357–364, 1978.

[BAR 07] Bartsch A., Kidd R.A., Wagner W. *et al.*, "Temporal and Spatial Variability of the Beginning and End of Daily Spring Freeze/thaw Cycles Derived from Scatterometer Data", *Remote Sensing of Environment*, vol. 106, no. 3, pp. 360–374, 2007.

[BAR 14] Barbu A.L., Calvet J.C., Mahfouf J.F. *et al.*, "Integrating ASCAT Surface Soil Moisture and GEOV1 Leaf Area Index into the SURFEX Modelling Platform: A Land Data Assimilation Application over France", *Hydrology and Earth System Sciences*, vol. 18, no. 1, pp. 173–192, 2014.

[BRO 09] Brocca L., Melone F., Moramarco T. *et al.*, "Antecedent Wetness Conditions Based on ERS Scatterometer Data", *Journal of Hydrology*, vol. 364, nos. 1–2, pp. 73–87, 2009.

[CHO 15] CHO E., CHOI M., WAGNER W., "An Assessment of Remotely Sensed Surface and Root Zone Soil Moisture through Active and Passive Sensors in Northeast Asia", *Remote Sensing of Environment*, vol. 160, pp. 166–179, 2015.

[DHA 11] DHARSSI I., BOVIS K.J., MACPHERSON B. *et al.*, "Operational assimilation of ASCAT surface soil wetness at the Met Office", *Hydrology and Earth System Sciences*, vol. 15, pp. 2729–2746.

[DER 12] DE ROSNAY P., DRUSCH M., VASILJEVIC D. *et al.*, "A simplified Extended Kalman Filter for the global operational soil moisture analysis at ECMWF", *Quarterly Journal of the Royal Meteorological Society*, vol. 139, no. 674, pp. 1199–1213, 2012.

[DOU 99] DOUVILLE H., BAZILE E., CAILLE P. *et al.*, "Global Soil Wetness Project: Forecast and Assimilation Experiments Performed at Meteo-France", *Journal of the Meteorological Society of Japan*, vol. 77, no. 1B, pp. 305–16, 1999.

[DRA 11] DRAPER C., MAHFOUF J.F.,CALVET J.C. *et al.*, "Assimilation of ASCAT near-Surface Soil Moisture into the SIM Hydrological Model over France", *Hydrology and Earth System Sciences*, vol. 15, no. 12, pp. 3829–3841, 2011.

[FAR 15] FARHADI L., REICHLE R.H., DE LANNOY G.J.M. *et al.*, "Assimilation of Freeze–Thaw Observations into the NASA Catchment Land Surface Model", *Journal of Hydrometeorology*, vol. 16, no. 2, pp. 730–743, 2015.

[FER 16] FERRO-FAMIL L., POTTIER, E., "Synthetic Aperture Radar Imagery", in BAGHDADI N., ZRIBI M. (eds), *Microwave Remote Sensing of Land Surfaces*, ISTE Press Ltd, London and Elsevier Ltd, Oxford, 2016.

[FRI 96a] FRISON P.L., MOUGIN E., "Use of ERS-1 Wind scatterometer data over land surfaces", *IEEE Transactions on Geoscience and Remote Sensing*, vol. 34, no. 2, pp. 550–560, 1996.

[FRI 96b] FRISON P.L., MOUGIN E., "Monitoring global vegetation dynamics with ERS-I wind scatterometer data", *International Journal of Remote Sensing*, vol. 17, no. 16, pp. 3201–3218, 1996.

[FRI 98] FRISON P.L., MOUGIN E., HIERNAUX P., "Observations and Interpretation of Seasonal ERS-1 Wind Scatterometer Data over Northern Sahel (Mali)", *Remote Sensing of Environment*, vol. 63, no. 3, pp. 233–242, 1998.

[FRI 08] FRIESEN J.C., "Regional vegetation water effects on satellite soil moisture estimations for West Africa", PhD Thesis, Zentrum für Entwicklungsforschung ZEF, Bonn, Germany, 2008.

[FRI 13] FRISON P.L., PAILLOU PH., SAYAH N. *et al.*, "Spatio-temporal monitoring of evaporitic processes using multi-resolution C band radar remote sensing data: example over the Chott el Djerid", *Canadian Journal of Remote Sensing*, vol. 39, no. 2, pp. 127–137, 2013.

[FRO 06] FROLKING S., MILLIMAN T., MCDONALD K. *et al.*, "Evaluation of the Sea Winds Scatterometer for Regional Monitoring of Vegetation Phenology", *Journal of Geophysical Research: Atmospheres*, vol. 111, no. 17, pp. 1–14, 2006.

[FRO 11] FROLKING S., MILLIMAN T., PALACE M. *et al.*, "Tropical Forest Backscatter Anomaly Evident in SeaWinds Scatterometer Morning Overpass Data during 2005 Drought in Amazonia", *Remote Sensing of Environment*, vol. 115, no. 3, pp. 897–907, 2001.

[FRO 12] FROLKING S., HAGEN S., MILLIMAN T. *et al.*, "Detection of Large-Scale Forest Canopy Change in Pan-Tropical Humid Forests 2000-2009 with the Seawinds Ku-Band Scatterometer", *IEEE Transactions on Geoscience and Remote Sensing*, vol. 50, no. 7, pp. 2603–2617, 2012.

[FUN 94] FUNG A.K., *Microwave Scattering and Emission Models and Applications*, Artech House, 1994.

[GRU 09] GRUHIER C., DE ROSNAY P., HASENAUER S. *et al.*, "Soil Moisture Active and Passive Microwave Products: Intercomparison and Evaluation over a Sahelian Site", *Hydrology and Earth System Sciences Discussions*, vol. 6, no. 4, pp. 5303–5339, 2009.

[JAR 03] JARLAN L., MAZZEGA P., MOUGIN E. *et al.*, "Mapping of Sahelian Vegetation Parameters from ERS Scatterometer Data with an Evolution Strategies Algorithm", *Remote Sensing of Environment*, vol. 87, no. 1, pp. 72–84, 2003.

[JAR 02] JARLAN L., MOUGIN E., FRISON P.L. *et al.*, "Analysis of ERS Wind Scatterometer Time Series over Sahel (Mali)", *Remote Sensing of Environment*, vol. 81, pp. 404–415, 2002.

[JIM 09] JIMÉNEZ C., PRIGENT C., AIRES P., "Toward an Estimation of Global Land Surface Heat Fluxes from Multisatellite Observations", *Journal of Geophysical Research*, vol. 114, no. D6, p. D06305, 2009.

[KAR 92] KARAM M.A., FUNG A.K., LANG R.H. *et al.*, "A Microwave Scattering Model for Layered Vegetation", *IEEE Transactions on Geoscience and Remote Sensing*, vol. 30, no. 4, pp. 767–784, 1992.

[KIM 01] KIMBALL J., MCDONALD K.C., RUNNING S.W. *et al.*, "Satellite Radar Remote Sensing of Seasonal Growing Seasons for Boreal and Subalpine Evergreen Forests", *Remote Sensing of Environment*, vol. 90, no. 2, pp. 243–258, 2004.

[KIM 04] KIMBALL J., MCDONALD K.C., RUNNING S.W. *et al.*, "Satellite Radar Remote Sensing of Seasonal Growing Seasons for Boreal and Subalpine Evergreen Forests", *Remote Sensing of Environment*, vol. 90, no. 2, pp. 243–258, 2004.

[KEN 89a] KENNETT G., LI F.K., "Seasat over land scatterometer data, part I: Global overview of the Ku-band backscatter coefficients", *IEEE Transactions on Geoscience and Remote Sensing*, vol. 27, no. 5, pp. 592–605, 1989.

[KEN 89b] KENNETT R.G., LI F.K., "Seasat over land scatterometer data, part II: Selection of extended area land target sites for the calibration of spacebornescatterometers", *IEEE Transactions on Geoscience and Remote Sensing*, vol. 27, no. 5, pp. 779–788, 1989.

[LEC 98] LECOMTE P., "The ERS Scatterometer instrument and the on-Ground processing of its Data", *Proceedings of ESA-EUMETSAT Workshop Emerging Scatterometer Applications – From Research to Operations*, Nordwijk, Holland, 5–7 October, 1998.

[LER 14] LEROUX D.J., KERR Y., AL BITAR A. *et al.*, "Comparison Between SMOS, VUA, ASCAT, and ECMWF Soil Moisture Products Over Four Watersheds in U.S.", *IEEE Transactions on Geoscience and Remote Sensing*, vol. 52, no. 3, pp. 1562–1571, 2014.

[MAG 97] MAGAGI R., KERR Y., "Retrieval of Soil Moisture and Vegetation Characteristics by Use of ERS-1 Wind Scatterometer over Arid and Semi-Arid Areas", *Journal of Hydrology*, vol. 189, nos. 1-4, pp. 361–384, 1997.

[MAN 14] MANFREDA S., BROCCA L., MORAMARCO T. *et al.*, "A physically based approach for the estimation of root-zone soil moisture from surface measurements", *Hydrology and Earth System Sciences*, vol. 18, pp. 1199–1212, 2014.

[MAS 13] MASSONNET D., SOUYRIS J.C., *Imaging with Synthetic Aperture Radar*, EPFL Press, 2013.

[MES 97] MESSEH M., QUEGAN S., "The use of ERS-1 scatterometer for global vegetation monitoring", *Proceedings of EARSelSymp. Future Trends Remote Sensing*, Lyngby, Denmark, June 17–20, 1997.

[MOU 95a] MOUGIN E., LOPES A., FRISON P.L. *et al.*, "Preliminary analysis of ERS-1 wind scatterometer data over land surfaces", *International Journal of Remote Sensing*, vol. 16, no. 2, pp. 391–398, 1995.

[MOU 95b] MOUGIN E., LO SEEN D., RAMBAL S. *et al.*, "A Regional Sahelian Grassland Model To Be Coupled with Multispectral Satellite Data. I: Model Description and Validation", *Remote Sensing of Environment*, vol. 52, pp. 181–193, 1995.

[NAE 12] NAEIMI V., PAULIK C., BARTSCH A. *et al.*, "ASCAT Surface State Flag (SSF): Extracting Information on Surface Freeze/Thaw Conditions From Backscatter Data Using Empirical Threshold-Analysis Algorithm", *IEEE Transactions on Geoscience and Remote Sensing*, vol. 50, no. 7, pp. 2566–2582, 2012.

[OH 92] OH U., SARABANDI K., ULABY F.T., "An empirical model and an inversion technique for radar scattering from bare soil surfaces", *IEEE Transactions on Geoscience and Remote Sensing*, vol. 30, no. 2, pp. 370–381, 1992.

[PAP 10] PAPA F., PRIGENT C., AIRES F. *et al.*, "Interannual Variability of Surface Water Extent at the Global Scale, 1993–2004", *Journal of Geophysical Research*, vol. 115, no. D12, p. D12111, 2010.

[PAR 08] PARIS ANGUELA T., ZRIBI M., HASENAUER S. *et al.*, "Analysis of Surface and Root-Zone Soil Moisture Dynamics with ERS Scatterometer and the Hydrometeorological Model SAFRAN-ISBA-MODCOU at Grand Morin Watershed (France)", *Hydrology and Earth System Sciences Discussions*, vol. 5, no. 4, pp. 1903–1926, 2008.

[PEL 06] PELLARIN T., CALVET J.C., WAGNER W., "Evaluation of ERS Scatterometer Soil Moisture Products over a Half-Degree Region in Southwestern France", *Geophysical Research Letters*, vol. 33, no. 17, p. L17401, 2006.

[PRI 01a] PRIGENT C., MATTHEWS E., AIRES F. *et al.*, "Remote Sensing of Global Wetland Dynamics with Multiple Satellite Data Sets", *Geophysical Research Letters*, vol. 28, no. 24, pp. 4631–4634, 2001.

[PRI 01b] PRIGENT C., AIRES F., ROSSOW W. *et al.*, "Joint Characterization of Vegetation by Satellite Observations from Visible to Microwave Wavelengths: A Sensitivity Analysis", *Journal of Geophysical Research*, vol. 106, no. D18, p. 20665, 2001.

[PRI 05] PRIGENT C., TEGEN I., AIRES F. *et al.*, "Estimation of the Aerodynamic Roughness Length in Arid and Semi-Arid Regions over the Globe with the ERS Scatterometer", *Journal of Geophysical Research*, vol. 110, no. D9, p. D09205, 2005.

[SCI 08] SCIPAL K., DRUSCH M., WAGNER W., "Assimilation of a ERS Scatterometer Derived Soil Moisture Index in the ECMWF Numerical Weather Prediction System", *Advances in Water Resources*, vol. 31, no. 8, pp. 1101–1112, 2008.

[SEE 14] SEELYE M., *An Introduction to Ocean Remote Sensing*, Cambridge University Press, 2014.

[SHU 82] SHUKLA J., MINTZ Y., "Influence of Land-Surface Evapotranspiration on the Earth Climate", *Science*, vol. 215, no. 7, pp. 1498–1501, 1982.

[SPE 97] SPENCER M.W., WU C., LONG D.G., "Tradeoffs in the Design of a Spaceborne Scanning Pencil beam Scatterometer: Application to SeaWinds", *IEEE Transactions on Geoscience and Remote Sensing*, vol. 35, no. 1, pp. 115–126, 1997.

[SPE 01] SPENCER M.W., A methodology for the design of spaceborne pencil-beam scatterometer systems, PhD Thesis, Brigham Young University, 1997.

[STE 05] STEPHEN H., LONG D.G., "Microwave backscatter modeling of Erg surfaces in the Sahara Desert", *IEEE Transactions on Geoscience and Remote Sensing*, vol. 43, no. 2, pp. 238–247, 2005.

[STE 12] STEELE-DUNNE S.C., FRIESEN J., VAN DE GIESEN N., "Using Diurnal Variation in Backscatter to Detect Vegetation Water Stress", *IEEE Transactions on Geoscience and Remote Sensing*, vol. 50, no. 7, pp. 2618–2629, 2012.

[TRA 11] TRAMBLAY Y., BOUVIER C., AYRAL P.A. *et al.*, "Impact of Rainfall Spatial Distribution on Rainfall-Runoff Modelling Efficiency and Initial Soil Moisture Conditions Estimation", *Natural Hazards and Earth System Science*, vol. 11, no. 1, pp. 157–170, 2011.

[ULA 86] ULABY F.T., MOORE R.K., FUNG A.K., *Microwave Remote Sensing: Active and Passive*, Addison-Wesley, Reading, 1982.

[ULA 88] ULABY F.T., SARABANDI K., MC DONALD K. *et al.*, "Michigan microwave canopy scattering model (MIMICS)", Tech. Report 022486-T-1, University of Michigan, Ann Arbor, 1988.

[WAG 99a] WAGNER W., LEMOINE G., BORGEAUD M. *et al.*, "A study of vegetation cover effect on ERS scatterometer data", *IEEE Transactions on Geoscience and Remote Sensing*, vol. 37, no. 2, pp. 938–948, 1999.

[WAG 99b] WAGNER W., LEMOINE G., ROTT H., "A method for estimating soil moisture from ERS scatterometer and soil data", *Remote Sensing of Environment*, vol. 70, no. 2, pp. 191–207, 1999.

[WAG 13] WAGNER W. *et al.*, "The ASCAT soil moisture product: a review of its specification, validation results, and emerging applications", *Meteorologische Zeitschrift*, vol. 2, no. 1, pp. 5–33, 2013.

[WIS 00] WISMANN V., "Monitoring of Seasonal Thawing in Siberia with ERS Scatterometer Data", *IEEE Transactions on Geoscience and Remote Sensing*, vol. 38, no. 4, pp. 1804–1809, 2000.

[WOO 00] WOODHOUSE I.H., HOEKMAN D.H., "Determining Land-Surface Parameters from the ERS Wind Scatterometer", *IEEE Transactions on Geoscience and Remote Sensing*, vol. 38, no. 1, pp. 126–140, 2000.

[ZIN 04] ZINE S., JARLAN L., FRISON P.L. *et al.*, "Land surface parameter monitoring with ERS scatterometer data over the Sahel: a comparison between agro-pastoral and pastoral areas", *Remote Sensing of Environment*, vol. 96, nos. 3–4, pp. 438–452, 2004.

[ZRI 08] ZRIBI M., ANDRÉ C., DECHARME B., "A Method for Soil Moisture Estimation in Western Africa Based on the ERS Scatterometer", *IEEE Transactions on Geoscience and Remote Sensing*, vol. 46, no. 2, pp. 438–448, 2008.

[ZWI 15] ZWIEBACK S., PAULIK C., WAGNER W., "Frozen Soil Detection Based on Advanced Scatterometer Observations and Air Temperature Data as Part of Soil Moisture Retrieval", *Remote Sensing*, vol. 7, no. 3, pp. 3206–3231, 2015.

4

Optical Remote Sensing of Snow Cover

4.1. Introduction: the importance of snow cover

Snow, which covers up to 53% of the northern hemisphere's land mass and 44% of the Earth, is of crucial importance for our climate [AMS 08]. Snow cover modifies the energy balance of the continental surface because it reflects a significant fraction of the incident solar radiation. The presence of snow also affects surface humidity and temperature at the lower boundary of the atmosphere, which in turn influences the climate dynamics [BAR 89, CAS 16]. This "climatic" role played by snow cover has a significant impact on boreal and polar regions (Siberia, Greenland, Antarctica, etc.). Remote sensing is a suitable tool to use for studying the snowpack properties for such vast and insufficiently monitored areas.

Snow also represents, in certain regions, an essential source of fresh water. It is estimated that a sixth of the world's population depends on snowmelt for their fresh water supply. In a number of semi-arid and Mediterranean regions, the spring thaw in upstream drainage basins in particular makes it possible to irrigate cultivated land downstream just as vegetation is beginning to grow. In the mountains, snowmelt is also used for hydropower. In mountain areas, *in situ* measurements cannot depict the high natural variability of the snowpack due to topographical conditions that can be observed using satellite data.

Chapter written by Marie DUMONT and Simon GASCOIN.

Knowledge of the characteristics of snow cover is also essential for predicting natural hazards such as avalanches or floods in mountainous regions. Finally, in many mountain regions, snow is an economic resource for winter tourism.

Snow cover is one of components of the Earth system which is most vulnerable to climate change. Rising temperatures have a direct impact on the spread, the quantity and the physical properties of snow cover, which then have significant consequences on how ecosystems function, on the hydrology of catchments impacted by snow cover and the very functioning of the climate itself. To take an example, Stewart *et al.* [STE 05] were able to observe that, in catchments located in the western states of USA and Canada, there has been a trend in the onset of the spring peak flow due to earlier snowmelt since the middle of the 20th century. The consequence is a reduction of the river discharge in the summer, when pressure on water resources is at its highest.

The second section of this chapter will deal with the optical properties of snow. These properties enable us to understand which snowpack parameters are accessible through optical remote sensing. The third section is focused on the latter. The use of optical snow remote sensing products is then illustrated by an application to hydrology. The chapter concludes by considering what possible developments we can expect in optical remote sensing in the coming years.

4.2. Optical properties of snow

Snow on the ground differs from most of the Earth surfaces by its high reflectance or albedo in the visible and near-infrared wavelengths (0.350 to 1 μm), where snow albedo is close to 1 (see Figure 4.1): in other words, almost all the incident radiation is reflected by snow. For longer wavelengths (typically those above 1 μm that are often referred to as mid-infrared or "MIR"), the snow's reflectance decreases significantly. In these wavelengths, snow is even less reflective than certain types of vegetation. Most of the incident radiation in these wavelengths is absorbed in the snowpack. These unique spectral characteristics are used in remote sensing in order to distinguish between snow and other types of surfaces (section 4.3.1). Snow optical properties also imply that the penetration depth of sunlight in the snowpack varies with wavelength. In the visible spectrum, light can penetrate into the snowpack up to half a meter, while in the infrared wavelength, the signal is quickly absorbed under the surface and the reflectance only reflects the properties of the top few millimeters of snow cover.

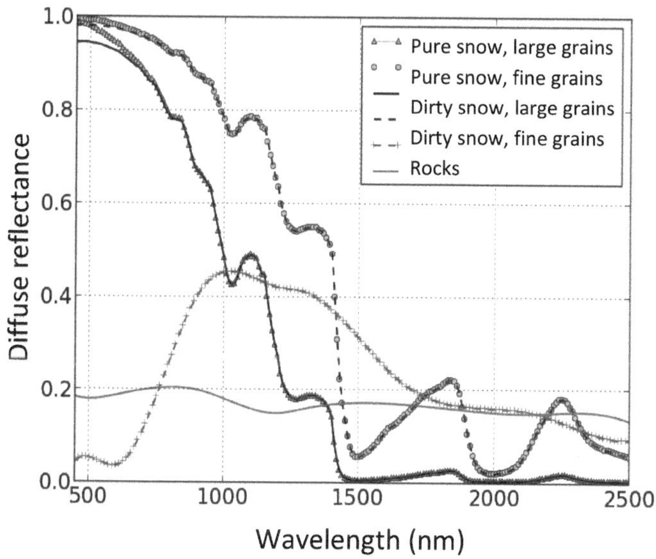

Figure 4.1. *Examples of diffuse reflectance spectra (diffuse isotropic incident radiation) for different types of snow and soil*

As shown in Figure 4.1, the optical properties of snow change according to its microstructure (size and shape of snow grains) and the quantity of light absorbing impurities such as mineral dust or soot.

In near-infrared wavelengths snow reflectance is mainly affected by its microstructure. Snow microstructure can be described in the first approximation with the snow density and the ratio between the total surface of ice crystals and their mass, which is known as the surface specific area (SSA). The SSA is inversely proportional to the optical radius which is often used to calculate reflectance. The smaller the ice crystals (and, therefore, the higher the SSA) the greater the reflectance. Essentially, for solar wavelengths, the scattering is proportional to the surface while absorption is proportional to the volume. This dependence implies that the bigger the crystals, the lower the reflectance and therefore the higher the quantity of energy absorbed by the snow cover. Finally, this last property has the effect of amplifying snow metamorphism: in other words, it accelerates grain growth. In turn, SSA continues to decrease and the albedo decreases. This effect is known as snow–albedo feedback (or ice–albedo feedback) [FLA11] and it is highly significant for climate feedbacks.

In visible wavelengths, snow reflectance is also sensitive to the presence of light-absorbing impurities such as soot or mineral dust, because of the high transparency of ice. The greater the quantity of impurities, the lower the reflectance. Additional feedbacks can then take place. Essentially, the lower the SSA, the greater the impact of impurities on reflectance. As a consequence, the presence of impurities will have the effect of reinforcing the snow–albedo feedback [DOH 13].

Furthermore, reflectance is sensitive to the angular characteristics of incident radiation. Essentially, the closer the incident radiation to the vertical, the weaker the reflectance will be. A photon that hits perpendicular to the snow cover will find it more difficult to exit the snowpack than a photon which only grazes the surface [WAR 82].

The refractive index of liquid water is very similar to the refractive index of ice, but there is a slight spectral shift between the indices of the two states of water. The reflectance of dry snow is therefore very similar to that of wet snow, other than that snow grains become larger in wet snow, causing a drop in reflectance. This is particularly noticeable in mid-infrared. By using hyperspectral data, it is possible to identify the presence of water in snow in liquid form through the slight spectral discrepancy between the two indices [GRE 06].

In addition, the surface roughness of snow (the formation of structures because of wind, for example) will modify snow albedo. In general, increased surface roughness will cause the albedo to decrease.

To conclude, snow is a complex material whose optical properties depend on the characteristics of the incident radiation, on its physical properties and on the quantity of light-absorbing impurities that it contains.

4.3. Properties of snow cover observable by optical remote sensing

4.3.1. *The presence of snow*

4.3.1.1. *History*

Optical remote sensing of snow-covered surfaces dates back to November 1966 when the first map showing the extent of snow cover was plotted by NOAA, using satellite images taken by the NASA satellite ESSA-3 (Environmental Science Services Administration) [MAT 81]. The spatial resolution of the images, to the order of 4 km, was already very high, but the images only had a single spectral band (500–750 nm). From this date onwards, NOAA produced a weekly map of the spread of snow cover across the northern hemisphere (Figure 4.2). After 1972, NOAA used 1 km resolution multispectral data from the VHRR and then the AVHRR (Advanced Very High Resolution Radiometer) on board NOAA Polar Orbiting Environmental Satellites (POES). From 1975 onward, NOAA began integrating high frequency images from geostationary satellites, then in 1997 they began using data from microwave sensors (SSM/I) to improve their ability to detect snow coverage in instances of excess cloud cover or poor solar illumination [RAM 98]. Up until 1999, NOAA maps showing the snow cover area in the northern hemisphere were drawn manually on paper by an analyst each week before being digitized into a grid with a resolution close to 190 km, depending on the latitude. Interactive software was later introduced to make it easier to combine multi-source data. These snow coverage maps, made available in the short-term, are still in operational use for the input of atmospheric models in weather forecasting. This currently represents the longest environmental time series produced using satellite observations [RAM 98].

4.3.1.2. *Available products and sensors*

There is now an array of optical sensors which make it possible to calculate snow cover using a variety of spatial and temporal resolutions (Table 4.1). NOAA, NASA and the ESA have made freely available products which allow users to produce snow cover maps without having to carry out any processing. The most widely used among these products is without doubt the MOD10A1 that is distributed by the National Snow and

Ice Data Center (NSIDC) [HAL 02]. Using Terra/MODIS data, this product indicates the presence or absence of snow and the percentage of snow in a pixel at a resolution of 500 m. This has been possible since the satellite Terra was put into orbit in 2000. The algorithm of MOD10A1 gives highly satisfactory results across various geographical contexts [HAL 07]. Given that MODIS is a passive optical sensor, it can only be used to measure snow coverage when there are no clouds. However, the daily repetitivity of MODIS observations enables frequent observation without clouds. Where there is no cloud cover, omissions (the non-detection of snow) are generally caused by (1) the geometric distortion of MODIS images at the edge of the swath (the bow tie effect) or (2) obstruction caused by the canopy in forests. In addition, the edges of areas of snow cover are often mistaken for clouds. This reduces the information available but it does not provide false information on snow coverage (section 4.4.1). Currently, this product offers the best compromise in terms of spatial and temporal resolution, length of observation and ease of use.

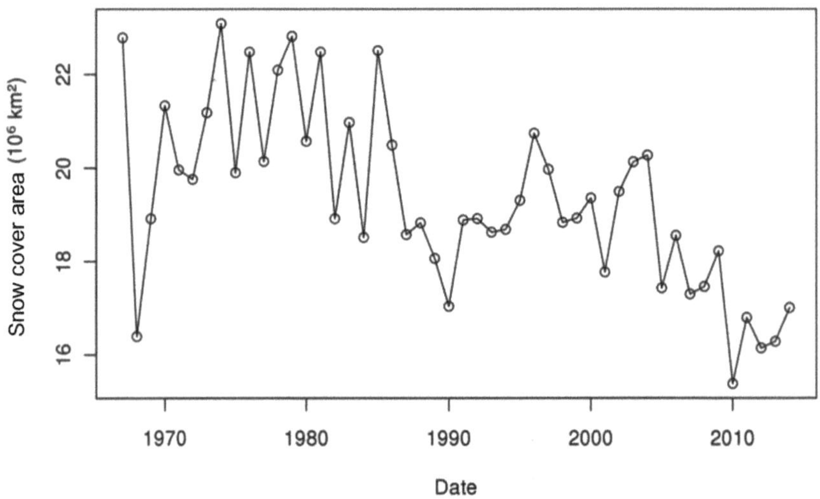

Figure 4.2. *The evolution of the average level of snow coverage in the northern hemisphere for the month of May (1967–2014) established using NOAA snow-coverage maps (Source: Rutgers University Global Snow Laboratory, [BRO 11])*

Table 4.1 does not list the sensors on board the European satellites ERS and ENVISAT, which stopped operating in 2011 and 2012. Nevertheless, data collected by the ATSR-2 (ESR-2) and AATSR (ENVISAT) in the visible spectrum–MIR domain made it possible to generate a daily map of snow coverage in the northern hemisphere with a spatial resolution of 1 km (Globsnow-2, [SOL 10]). Lastly, it is still possible to determine the total snow coverage using sensors which do not have spectral bands in MIR, but it becomes more difficult to distinguish between snow and clouds. One possible solution is to use supervised classification [MAR 15]. We would invite the reader to consult Frei *et al.* [FRE 12] and Dietz *et al.* [DIE 11] for a more extensive review of the products and algorithms available for the optical remote sensing of snow cover, as well as for other wavelengths. In addition, the use of active microwaves for monitoring snow cover is outlined in Chapter 5 of this book.

Sensor	Platform (s)	Spatial resolution (m)	Year launched	Field of vision (km)	Orbital cycle (days)	Revisit time (days)	Snow product available
TM/ETM+/OLI	Landsat 5 8	30	1982	185	16	16 at the most	Yes
MSI	Sentinel-2A (B)	10–60	2015	290	10	10 (5)	In preparation
MODIS	Aqua/Terra	250–1000	2000	2330	16	1 (global)	Yes (NSIDC)
VIIRS	Suomi NPP	400–750	2012	3040	16	1 (global)	Expected (NSIDC)
VEGETATION	SPOT/ PROBA-V	1000 (100 PROBA-V)	1998	2250	26	1 (global)	No
AVHRR	NOAA POES	1000	1981	2900	24	1 (global)	Yes (combined with other sources)

Table 4.1. *A selection of radiometric imagers on polar orbit satellites (we have excluded geostationary satellites) which (i) are adapted to detecting snow (i.e. those whose spectral bands extend from the visible spectrum to mid-infrared), (ii) are operational as of 2015, (iii) produce data that is free to access, and (iv) make systematic observations (not on programming). The resolution indicated is what can be obtained for detecting snow. Sentinel 3, the ESA's optical mission which has not been launched at the time of going to press, will provide data relevant for the study of snow coverage. For Sentinel-2, the revisit time of 5 days will be reached with the launch of Sentinel 2B, expected in late 2016*

4.3.1.3. *Methods*

Because of the specific reflectance of snow, methods used to detect its presence are relatively efficient and these are briefly outlined in the following section. The method of detection used by the Normalized Difference Snow Index (NDSIT, [DOZ 89]) is based on the fact that the reflectance of snow is higher in the visible spectrum and lower in the MIR (1-3 μm), which makes it possible to distinguish between snow and clouds, whose reflectance remains high in MIR [HAG 14]. The NDSI was defined using Landsat as:

$$NDSI = \frac{\rho_V - \rho_M}{\rho_V + \rho_M} \qquad [4.1]$$

where ρ_V and ρ_M represent the reflectance of the surface in green (band 2 of the Landsat TM sensor) and in MIR (band 5 of the TM). It is then possible to indentify snow-covered pixels using a threshold on the NDSI. By way of an example, the algorithm of MOD10A1 uses a minimal NDSI of 0.4 to denote a pixel as having sufficient snow coverage to be classed as such (Figure 4.3). Zhu and Woodcock [ZHU 12] proposed a threshold of 0.15 for Landsat and Sentinel-2. Moreover, this threshold can be adapted to the type of vegetal coverage: thus for MODIS products, the NDSI threshold changes in forested areas in a way that is inversely proportional to the Normalized Difference Vegetation Index (NDVI) [KLE 98]. Nevertheless, snow coverage in forests is still underestimated, especially in cases where higher sensor view angles increase obstruction caused by the canopy [XIN 12]. Additional tests on absolute values of reflectance are often necessary to exclude lakes or coniferous forests which, like snow, have a reflectance which is proportionally higher in the green than the MIR.

The NDSI can also be used to give an indication of the percentage of snow coverage in a pixel. In the case of MOD10, the percentage of snow is calculated using a linear function of NDSI [SAL 04]. This relationship was established empirically by making a comparison against higher resolution maps of snow coverage (Landsat) and it will therefore not necessarily be valid everywhere with the same precision or with a sensor other than MODIS [CHO 10].

More advanced models make it possible to calculate the percentage of snow coverage in a pixel, which, for the most part, are methods of spectral unmixing. These methods rely on a library of spectral references of different types of surfaces (snow which is old or fresh, dirty or clean, ice, rocks, vegetation) and this makes it possible to describe the spectrum of a pixel as a (generally) linear combination of spectral references. The coefficients of this linear combination, termed abundance, are what characterize the percentage of each type of surface within the pixel. These methods are a more costly means of calculation than NDSI methods, but they do provide a more accurate diagnosis [PAI 03, SIR 09].

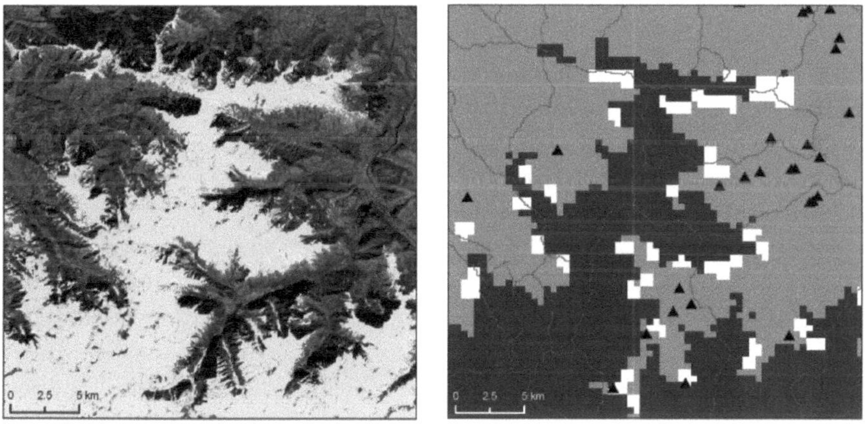

Figure 4.3. *Two different views of the same snow cover (Vicdessos valley, Ariége Pyrénées, 19th March 2014, coordinates at the center of the images: 1°27'30"E, 42°49'32"N). Left: an RVB composition of reflectance from Landsat 8 in MIR/red/green bands. Here, surface reflectance corrected for atmospheric effects and the effects of slopes and produced by the Theia center (http://www.theia-land.fr/) have been used. Snow appears as turquoise because its reflectance is low in MIR and high in red and green. The yellow contours indicate the spread of snow detected by thresholding on the NDSI > 0.15. Right: the spread of snow coverage according to MOD10A1 (in blue). Pixels at the edges of areas of snow coverage can sometimes be confused with clouds (in white). The black triangles indicate hydraulic plants used for the production of electricity. The spatial resolution of MODIS is sometimes insufficient for hydrological uses in mountain areas where catchments of interest for water management are small. For a color version of this figure, see www.iste.co.uk/baghdadi/4.zip*

4.3.2. *Albedo and optical radius, light-absorbing impurities*

Aside from the snow-covered area, reflectance measured by satellite allows the retrieval of some of the physical and chemical properties of the snow itself (section 4.2). Two main parameters can be generally estimated using satellite imaging: the albedo, which is the bihemispheric reflectance integrated across the solar spectrum, and the optical radius, the radius of spheres with the same SSA as the snow.

The albedo is used to estimate radiative and surface energy balance or to evaluate detailed snow models. There are two main types of methods for retrieving snow albedo from multi-spectral imagers such as MODIS or VIIRS. The first method, and the simplest, consists in establishing a parameterization between satellite spectral reflectances and *in situ* measured albedo. It is then possible to apply this linear or quadratic relationship to a larger spatial or temporal area [KNA 99]. With this simple method it is not possible to take into account the spectral variations of snow's optical properties with the characteristics of the surface. The second method consists in calculating the physical parameters of snow cover using spectral data and extrapolating the values of the bands across the whole spectrum using a radiative transfer model for snow cover [DUM 12]. This last method is used with MOD10 products distributed by the National Snow and Ice Data Centre to calculate snow albedo [KLE 02]. Finally, some products, such as MCD43 which is based on MODIS, use several images obtained at different solar angles and viewpoints to reconstruct angular data and to take into account angular variations in reflectance when calculating snow albedo [SCH 10].

Optical radius is an example of another variable that is calculated using satellite images, either for calculating the albedo, for monitoring the physical properties of the snowpack or for evaluating snowpack simulations. Processing algorithms are often based on the use of an infrared band and either one or several bands in the visible spectrum, with the infrared band carrying the signature of the optical radius and the visible band(s) which make it possible to disentangle the influence of the atmosphere on the satellite data. The algorithms differ in terms of the relationship linking optical radius to reflectance based on different types of radiative transfer models [MAR 13, NEG 11]. An example of an optical radius map obtained using MODIS imagery is shown in Figure 4.4.

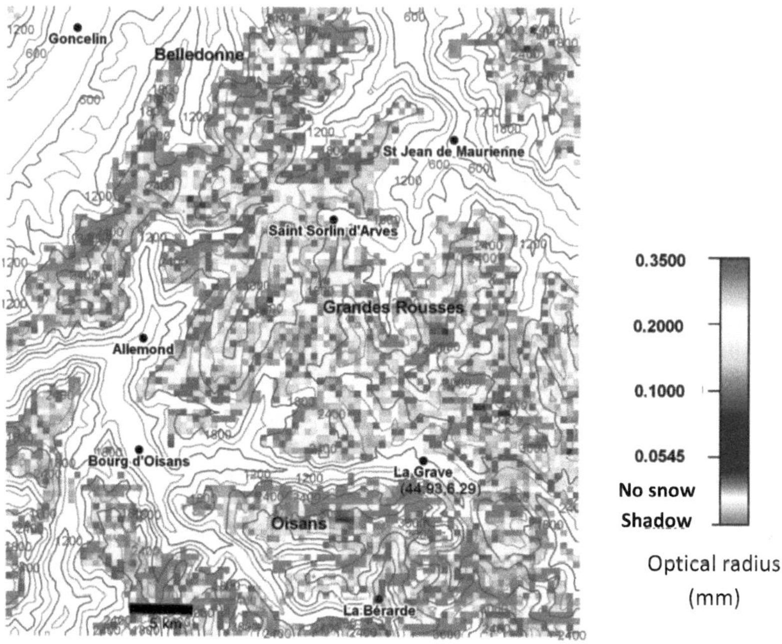

Figure 4.4. *Snow optical radius (0.0 to 0.35 mm), retrieved over the French Alps using MODIS data obtained from the 21st March 2009 at 10:40 UTC [MAR 13]. For a color version of this figure, see www.iste.co.uk/baghdadi/4.zip*

As explained in section 4.2, snow reflectance also depends on the type and the quantity of the impurities contained in the snowpack. Retrieving impurities content using satellite reflectance is a difficult exercise, given:

– the accuracy of satellite reflectance;

– the quality of the atmospheric correction;

– the uncertainties on refractive indices of different types of impurities [WAR 13].

As far as this subject is concerned, it is therefore preferable to restrict ourselves to impact or trend studies without seeking to retrieve accurate impurity content [PAI 12].

In summary, satellite reflectance allows us to establish the albedo and the optical radius of snow cover. It also provides us with information on how clean the snow is. It is important to note that the accuracy of satellite retrieval of snow properties is limited by the atmospheric correction, the angular distribution of light reflected by snow and the snow roughness.

4.4. The use of data produced from snow-covered surfaces in hydrology

4.4.1. *What can be done about clouds?*

The main disadvantage of data produced from snow-covered surfaces using optical remote sensing is the obstruction caused by clouds. How are we able to calculate the snow coverage of a catchment if part of its surface is hidden by clouds? Over the course of a year, the average percentage of pixels hidden by clouds is somewhere in the region of 20% in the High Atlas Mountains, 50% in the Pyrenees [GAS 15], 60% in the Austrian Alps [PAR 08] and 70% in British Columbia [TON 09].

In some casese, it is possible to exploit only the clear parts of images, for example in order to validate, calibrate or assimilate data using distributed models of snow cover where the resolution is close to that of the images. For hydrological models which operate at the level of drainage basins, it is possible to choose clear images to extract the snow cover for these particular dates. Nevertheless, this considerably reduces the data available to the extent that a preliminary interpolation of snow cover is often carried out. Even a partial interpolation of the hidden pixels has made it possible to improve the performance of the data assimilation for a continental surface model [HAL 10] and for a hydrological model [THI 13]. Most importantly, interpolating all the data gaps is a necessary stage in order to produce climatologies of snow cover or to calculate the interannual trends of snow cover area (section 4.4.2).

Again, there are a number of gap-filling algorithms, but these are based on two key ideas: using data (1) in the temporal vicinity and (2) in the spatial vicinity. In the first instance, interpolation makes use of the fact that the extent of snow cover changes less quickly than that of cloud cover. For example, if a "cloud" pixel is denoted as "snow" the day before and the day after, it is safe to reclassify this pixel as "snow" (Figure 4.5). This method is the most effective, especially given that it can function with a moving time window over several days without introducing a significant number of errors [PAR 08, GAF 09, GAO 11]. In the second instance, there are several possible strategies. For example, hidden pixels can be reclassified by a majority filter (Figure 4.5) but with this method it is not possible to eliminate a large number of gaps. A more effective strategy for use in mountainous regions consists of using topographical information, which could involve determining the lower limit of snow coverage using the clear section of the image in order to then reclassify the pixels depending on their altitude.

These methods can be applied to "binary" snow maps (snow/no snow) but temporal and spatial interpolation can be generalized to snow maps using percentages (where the value of a pixel is its percentage of snow coverage). What is important is to apply criteria that is consistent with the natural variability of snow cover in the region being studied.

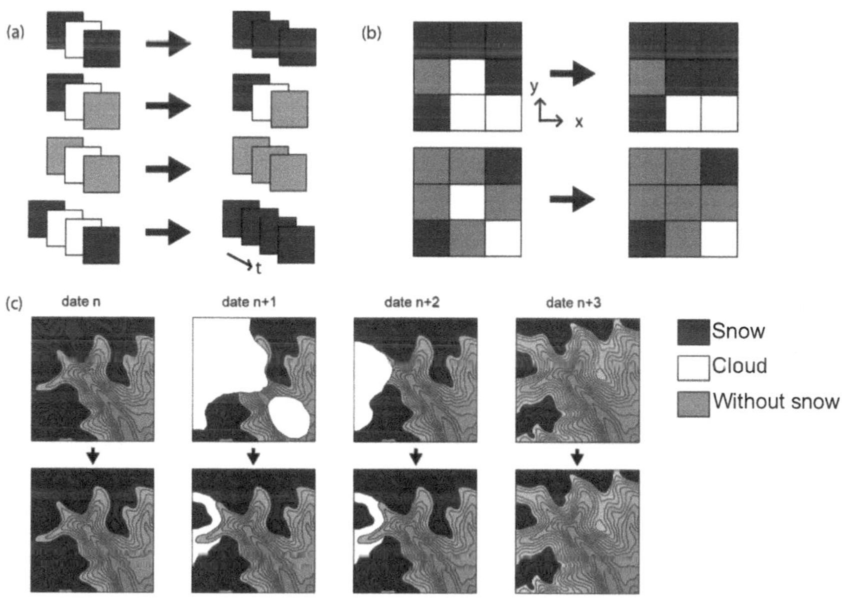

Figure 4.5. *Two methods for interpolating pixels hidden by clouds; a) by temporal combination; b) by a majority filter; c) schematic of a temporal combination*

4.4.2. *Generating hydro-climatic indicators*

A number of relevant hydro-climatic indicators can be generated using the simple fact of the presence or absence of snow: most often, these include the snow cover duration (or number of snow days), the date on which snow appears or disappears, the snow-covered surface of a given zone (such as a catchment area), the altitude of the snow line and so on. These indicators can be studied over time (trend analysis) or in space. By way of an example, the number of snow days per pixel was calculated for the Pyrenees using data produced from MODIS for different elevation bands and different classes of slope aspect (Figure 4.6). This tells us that above 1600 m snow is present at least 50% of the time between December and April. There is also a clear increase in the number of snow days as altitude increases between intervals of 800–1600 m and 1600–2400 m, an observation that is consistent with the average altitude of the 0°C isotherm in the Pyrenees, which is estimated to be between 1600–1700 m between November and April [LOP 04]. The length of the snow season also varies according to exposure: on average, the snow cover duration is longer for pixels in north-facing slopes, since they receive less solar energy.

Figure 4.6. *The snow cover duration in the Pyrenees by elevation band (left) and for different levels of slope exposure above 800 m (right) [GAS 15]. For a color version of this figure, see www.iste.co.uk/baghdadi/4.zip*

The retro-ablation method makes it possible to generate an even more relevant indicator in hydrology using snow cover maps. For each pixel it is possible to determine the date on which the snow disappeared, then, using a model, to simulate the quantity of snow that would have melted by going back to the date of peak snow accumulation (for example, the 1st April) [CLI 98, RAL 12]. It is thus possible to obtain a map of the snow water equivalent which was theoretically present before the beginning of the ablation season. The accuracy of the result is still restricted by the quality of atmospheric forcing used and uncertainties linked to the model itself.

4.4.3. *Hydrological modeling*

Data from snow-covered surfaces (snow cover area, SCA) allows us to advance further in hydrology if used in conjunction with an appropriate model. SCA data makes it possible to validate a model of snow cover in environments where observations often default (Figure 4.7). The snowmelt runoff model (SRM) uses SCA satellite data directly in forcing in order to calculate melt using the air temperature [MAR 86]. Several agencies in charge of water management in mountainous regions use this model in order to forecast water flow [DEW 08]. Lastly, by assimilating SCA data it is possible to improve simulations of snow cover in land surface models [ROD 04, CLA 06]. However, performance gains obtained using SCA data for the simulation of stream flows are still relatively low, whether in calibration [PAR 08] or assimilation [CLA 06, THI 13]. This is due to the fact that SCA is still an incomplete form of data in a hydrological sense, which means that it is unable to compensate for the various other uncertainties which arise in mountain catchments at high latitude. Nevertheless, snow cover is the most widely available form of data from remote sensing on a global level for sufficiently long periods and at increasingly high spatio-temporal resolutions thanks to the latest sensors (PROBA-V, VIIRS) and those still to come (Sentinel-2 and Sentinel-3). Currently, there are no satellites capable of directly determining the snow water equivalent (SWE) at a sufficient resolution for mountain regions [LET 15]. How to make best use of SCA data for hydrological modeling is an important issue and it is currently the subject of a number of studies. This link is established through the declining curve equation of the SWE in accordance with the SCA on a grid of the snow depletion curve (SDC) model. Several possible functions have been put forward, ranging from a binary conversion where the percentage of snow coverage is equal to 0 or 1 depending on the simulated value of SWE, to a probabilistic distribution model within a grid [LIS 04]. These methods depend on the spatial resolution of satellite observations and on the model used, because the distribution of snow across a given area is linked to that area's topography. The diversity of terrain results in a hysteresis in the SWE–SCA relationship at the scale of the computation unit (Figure 4.8). The inclination might be to think that these questions will be simplified with the arrival of high spatio-temporal resolution data such as that expected from Sentinel-2.

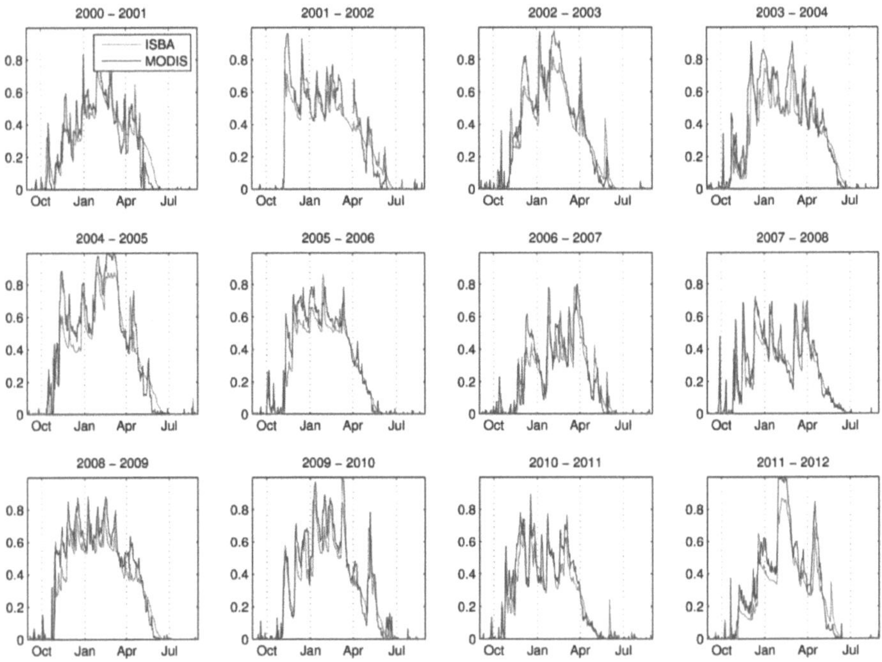

Figure 4.7. *The evolution of the snow-covered surface of the Ariège à Foix basin (1357 km², Pyrenees, France) simulated by the ISBA surface model and taken from MYD10A1 and MOD10A1. The period covered by the MODIS observations is now long enough to be able to evaluate the capacity of hydrological models to represent the interannual variability of snow coverage [MAR 16]*

4.5. Possibilities

As this chapter concludes, we are now able to understand why optical remote sensing by satellite provides us with key information on the properties and evolution of snow cover. Rapid progress made in spatial technology will undoubtedly allow us to move beyond current limits in optical remote sensing and thus to improve the way in which snow cover is monitored. The Sentinel-2 and Sentinel-3 missions in particular will very soon allow us to move past the compromise between temporal resolution (MODIS, VGT, AVHRR) and spatial resolution (Landsat, SPOT) for monitoring snow-covered surfaces and their optical properties. Very high resolution sensors, such as Pléaides, provide a new option for determining the snow height in mountains by satellite by using stereoscopy [MAR 16]. On the other hand, the reprocessing of old data such as AVHRR makes it possible to look back at how snow cover has evolved over sufficiently long periods to study the effects of climate change [CHO 13, HUS 14]. Snow coverage is considered to be one of the thirteen "essential climate variables" which require systematic monitoring to support the work carried out by the Intergovernmental Panel on Climate Change (IPCC). As a consequence, it is highly likely that space agencies will continue to work together for the monitoring of snow cover through new products and new missions.

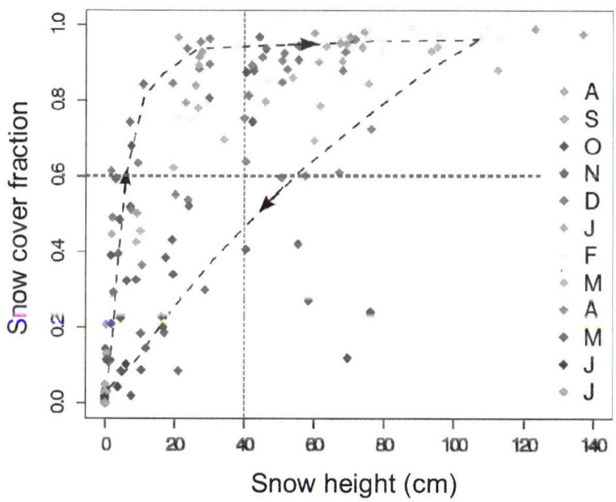

Figure 4.8. *The fraction of the snow-covered area against the height of snow in the Guilat Montdauphin catchment (723 km², French Alps). Months are indicated by color. The same height of snow tends to cover a larger area during the accumulation period (winter) than during the ablation season (spring) [MAG 14]. For a color version of this figure, see www.iste.co.uk/baghdadi/4.zip*

Finally, improvements in the description of the snow cover in land surface, climate and hydrological models enables us to foresee exciting possibilities in terms of the assimilation of optical remote sensing data for better simulation and better forecasting of snow cover evolution. Data from optical remote sensing, in contrast to in situ measurements, effectively makes it possible to portray the spatial variability of snow cover.

4.6. Key points

– Snow has unique optical properties (very high reflectivity in visible wavelengths) which distinguishes it from any other kind of terrestrial surface.

– Optical sensors make it possible to detect (1) the presence of snow on the ground, and (2) several surface properties of the snow cover such as presence of light-absorbing impurities and microstructural characteristics.

– Continuous and reliable observations through the remote sensing of snow coverage have been available for sufficiently long periods for hydrological and climatological applications. However, these observations are incomplete because they do not describe all the physical properties of snow cover and therefore must be used in combination with snow cover models or *in situ* observations in order to properly benefit from it.

4.7. Bibliography

[ARM 08] ARMSTRONG R.L., BRUN E. (eds), *Snow and Climate: Physical Processes, Surface Energy Exchange and Modeling*, Cambridge University Press, 2008.

[BAR 89] BARNETT T.P., DÜMENIL L., SCHLESE U. *et al.*, "The effect of Eurasian snow cover on regional and global climate variations", *Journal of the Atmospheric Sciences,* vol. 46, pp. 661–686, 1989.

[BAR 05] BARNETT T.P., ADAM J.C., LETTENMAIER D.P., "Potential impacts of a warming climate on water availability in snow-dominated regions", *Nature*, vol. 438, pp. 303–309, 2005.

[BRO 11] BROWN R.D., ROBINSON D.A., "Northern Hemisphere spring snow cover variability and change over 1922–2010 including an assessment of uncertainty", *The Cryosphere*, vol. 5, pp. 219–229, 2011.

[CAS 16] CASSOU C., CATTIAUX J., "Disruption of the European climate seasonal clock in a warming world", *Nature Climate Change*, available at: www.nature.com/nclimate/journal/vaop/ncurrent/full/nclimate296.htm, 2016.

[CHO 10] CHOKMANI K., DEVER K., BERNIER M. *et al.*, "Adaptation of the SNOWMAP algorithm for snow mapping over eastern Canada using Landsat-TM imagery", *Hydrological Sciences Journal*, vol. 55, pp. 649–660, 2010.

[CHO 13] CHOKMANI K., BERNIER M., ROYER A., "A merging algorithm for regional snow mapping over Eastern Canada from AVHRR and SSM/I data", *Remote Sensing*, vol. 5, pp. 5463–5487, 2013.

[CLA 06] CLARK M.P., SLATER A.G., BARRETT A.P. *et al.*, "Assimilation of snow covered area information into hydrologic and land-surface models", *Advances in Water Resources*, vol. 29, pp. 1209–1221, 2006.

[CLI 98] CLINE D.W., BALES R.C., DOZIER J., "Estimating the spatial distribution of snow in mountain basins using remote sensing and energy balance modeling", *Water Resour. Res.*, vol. 34, pp. 1275–128, 1998.

[DEW 08] DEWALLE D.R., RANGO A., *Principles of Snow Hydrology*, Cambridge University Press, 2008.

[DIE 12] DIETZ A.J., KUENZER C., GESSNER U. *et al.*, "Remote sensing of snow – a review of available methods", *International Journal of Remote Sensing*, vol. 33, pp. 4094–4134, 2012.

[DOH 13] DOHERTY S.J., GRENFELL T.C., FORSSTRÖM S. *et al.*, "Observed vertical redistribution of black carbon and other insoluble light-absorbing particles in melting snow", *J. Geophys. Res. Atmos.*, vol. 118, pp. 5553–5569. 2013.

[DOZ 89] DOZIER J., "Spectral signature of alpine snow cover from the Landsat thematic mapper", *Remote Sensing of Environment*, vol. 28, pp. 9–22, 1989.

[DUM 12] DUMONT M., GARDELLE J., SIRGUEY P. *et al.*, "Linking glacier annual mass balance and glacier albedo retrieved from MODIS data", *The Cryosphere*, vol. 6, pp. 1527–1539, 2012.

[FLA 11] FLANNER M.G., SHELL K.M., BARLAGE M. *et al.*, "Radiative forcing and albedo feedback from the Northern Hemisphere cryosphere between 1979 and 2008", *Nature Geosci.*, vol. 4, pp. 151–155, 2011.

[FRE 12] FREI A., TEDESCO M., LEE S. *et al.*, "A review of global satellite-derived snow products", *Advances in Space Research, Oceanography, Cryosphere and Freshwater Flux to the Ocean*, vol. 50, pp. 1007–1029, 2012.

[GAF 09] GAFUROV A., BÁRDOSSY A., "Cloud removal methodology from MODIS snow cover product", *Hydrol. Earth Syst. Sci.*, vol. 13, pp. 1361–1373, 2009.

[GAO 11] GAO Y., LU N., YAO T., "Evaluation of a cloud-gap-filled MODIS daily snow cover product over the Pacific Northwest USA", *Journal of Hydrology*, vol. 404, pp. 157–165, 2011.

[GAS 15] GASCOIN S., HAGOLLE O., HUC M. *et al.,* "A snow cover climatology for the Pyrenees from MODIS snow products", *Hydrol. Earth Syst. Sci.*, vol. 19, pp. 2337–2351, 2015.

[GRE 06] GREEN R.O., PAINTER T.H., ROBERTS D.A. *et al.*, "Measuring the expressed abundance of the three phases of water with an imaging spectrometer over melting snow", *Water Resour. Res.*, vol. 42, p. W10402, 2006.

[HAL 02] HALL D.K., RIGGS G.A., SALOMONSON V.V. *et al.*, "MODIS snow-cover products", *Remote Sensing of Environment*, vol. 83, pp. 181–194, 2002.

[HAL 07] HALL D.K., RIGGS G.A., "Accuracy assessment of the MODIS snow products", *Hydrol. Process.*, vol. 21, pp. 1534–1547, 2007.

[HAL 10] HALL D.K., RIGGS G.A., FOSTER J.L. *et al.*, "Development and evaluation of a cloud-gap-filled MODIS daily snow-cover product", *Remote Sensing of Environment,* vol. 114, pp. 496–503, 2010.

[HUS 14] HÜSLER F., JONAS T., RIFFLER M. *et al.*, "A satellite-based snow cover climatology (1985–2011) for the European Alps derived from AVHRR data", *The Cryosphere*, vol. 8, pp. 73–90, 2014.

[KLE 98] KLEIN A.G., HALL D.K., RIGGS G.A., "Improving snow cover mapping in forests through the use of a canopy reflectance model", *Hydrol. Process.*, vol. 12, pp. 1723–1744, 1998.

[KNA 99] KNAP W.H., REIJMER C.H., OERLEMANS J., "Narrowband to broadband conversion of Landsat TM glacier albedos", *International Journal of Remote Sensing*, vol. 20, pp. 2091–2110, 1999.

[LET 15] LETTENMAIE, D.P., ALSDORF D., DOZIER J. *et al.*, "Inroads of remote sensing into hydrologic science during the WRR era", *Water Resour. Res.*, vol. 51, pp. 7309–7342, 2015.

[LIS 04] LISTON G.E., "Representing subgrid snow cover heterogeneities in regional and global models", *J. Climate*, vol. 17, pp. 1381–1397, 2004.

[LOP 04] LÓPEZ-MORENO J.I., GARCÍA-RUIZ J.M., "Influence of snow accumulation and snowmelt on streamflow in the central Spanish Pyrenees", *Hydrological Sciences Journal*, vol. 49, pp. 787–802, 2004.

[MAR 86] MARTINEC J., RANGO A., “Parameter values for snowmelt runoff modeling”, *Journal of Hydrology*, vol. 84, pp. 197–219, 1986.

[MAR 13] MARY A., DUMONT, M., DEDIEU, J.P. *et al.*, “Intercomparison of retrieval algorithms for the specific surface area of snow from near-infrared satellite data in mountainous terrain, and comparison with the output of a semi-distributed snowpack model”, *The Cryosphere*, vol. 7, pp. 741–761, 2013.

[MAR 15] MARCHANE A., JARLAN L., HANICH L. *et al.*, “Assessment of daily MODIS snow cover products to monitor snow cover dynamics over the Moroccan Atlas mountain range”, *Remote Sensing of Environment*, vol. 160, pp. 72–86, 2015.

[MAR 16a] MARTI R., GASCOIN S., BERTHIER E. *et al.*, “Mapping snow depth in open alpine terrain from stereo satellite imagery”, *The Cryosphere Discuss.*, under review, 2016.

[MAR 16b] MARTIN E., GASCOIN S., GRUSSON Y. *et al.*, “On the use of hydrological models and satellite data to study the water budget of river basins affected by human activities: examples from the Garonne basin of France”, *Surveys of Geophysics*, vol. 37, no. 2, pp. 223–247, 2016.

[MAT 81] MATSON M., WIESNET D.R., “New data base for climate studies”, *Nature*, vol. 289, pp. 451–456, 1981.

[NEG 11] NEGI H.S., KOKHANOVSKY A., “Retrieval of snow grain size and albedo of Western Himalayan snow cover using satellite data”, *The Cryosphere*, vol. 5, pp. 831–847, 2011.

[PAI 03] PAINTER T.H., DOZIER J., ROBERTS D.A. *et al.*, “Retrieval of subpixel snow-covered area and grain size from imaging spectrometer data”, *Remote Sensing of Environment*, vol. 85, pp. 64–77, 2003.

[PAI 12] PAINTER T.H., BRYANT A.C., SKILES S.M., “Radiative forcing by light absorbing impurities in snow from MODIS surface reflectance data”, *Geophys. Res. Lett.*, vol. 39, p. L17502, 2012.

[PAR 08a] PARAJKA J., BLÖSCHL G., “Spatio-temporal combination of MODIS images – potential for snow cover mapping”, *Water Resour. Res.*, vol. 44, p. W03406, 2008.

[PAR 08b] PARAJKA J., BLÖSCHL G., “The value of MODIS snow cover data in validating and calibrating conceptual hydrologic models”, *Journal of Hydrology*, vol. 358, pp. 240–258, 2008.

[RAL 12] RALEIGH M.S., LUNDQUIST J.D., "Comparing and combining SWE estimates from the SNOW-17 model using PRISM and SWE reconstruction", *Water Resour. Res.*, vol. 48, p. W01506, 2012.

[RAM 98] RAMSAY B.H., "The interactive multisensor snow and ice mapping system", *Hydrological Processes*, vol. 12, pp. 1537–1546, 1998.

[ROD 04] RODELL M., HOUSER P.R., "Updating a land surface model with MODIS-derived snow cover", *J. Hydrometeor*, vol. 5, pp. 1064–1075, 2004.

[SAL 04] SALOMONSON V.V., APPEL I., "Estimating fractional snow cover from MODIS using the normalized difference snow index", *Remote Sensing of Environment*, vol. 89, pp. 351–360, 2004.

[SCH 11] SCHAAF C.B., WANG Z., STRAHLER A.H., "Commentary on Wang and Zender—MODIS snow albedo bias at high solar zenith angles relative to theory and to in situ observations in Greenland", *Remote Sensing of Environment*, vol. 115, pp. 1296–1300, 2011.

[SIR 09] SIRGUEY P., MATHIEU R., ARNAUD Y., "Subpixel monitoring of the seasonal snow cover with MODIS at 250 m spatial resolution in the Southern Alps of New Zealand: Methodology and accuracy assessment", *Remote Sensing of Environment*, vol. 113, pp. 160–181, 2009.

[SOL 10] SOLBERG R., WANGENSTEEN B., METSÄMÄKI S. *et al.*, GlobSnow snow extent product guide product version 1.2, (ESRIN contract 21703/08/I-EC), European Space Agency, 2010.

[STE 05] STEWART I.T., CAYAN D.R., DETTINGER M.D., "Changes toward earlier streamflow timing across western North America", *J. Climate*, vol. 18, pp. 1136–1155. 2005.

[THI 13] THIREL G., SALAMON P., BUREK P. *et al.*, "Assimilation of MODIS snow cover area data in a distributed hydrological model using the particle filter", *Remote Sensing*, vol. 5, pp. 5825–5850, 2013.

[TON 09] TONG J., DÉRY S.J., JACKSON P.L., "Topographic control of snow distribution in an alpine watershed of western Canada inferred from spatially-filtered MODIS snow products", *Hydrol. Earth Syst. Sci.*, vol. 13, pp. 319–326, 2009.

[WAR 82] WARREN S.G., "Optical properties of snow", *Rev. Geophys.*, vol. 20, pp. 67–89, 1982.

[WAR 13] WARREN S.G., "Can black carbon in snow be detected by remote sensing", *J. Geophys. Res. Atmos.*, vol. 118, pp. 779–786, 2013.

[XIN 12] XIN Q., WOODCOCK C.E., LIU J. *et al.*, "View angle effects on MODIS snow mapping in forests", *Remote Sensing of Environment*, vol. 118, pp. 50–59, 2012.

[ZHU 12] ZHU Z., WOODCOCK C.E., "Object-based cloud and cloud shadow detection in Landsat imagery", *Remote Sensing of Environment*, vol. 118, pp. 83–94, 2012.

5

Snow Characterization Using Radar Imaging

5.1. Introduction

Aside from its impact on human activity in various regions around the world, snow also plays an important role in a range of environmental processes. In climatology, snow cover has the effect of lowering temperatures due to its high albedo which reflects between 40 and 95% of incident solar radiation [GRA 81] and it can therefore be used as a barometer for climate change and climate variations. When snow cover melts, the quantity of water generated will have a significant impact on a region's hydrological regime and, in fact, depending on the region, this can represent up to 40% of peak rises in water level. Snow cover also has an isolating effect on the soil below as it influences when it will freeze. In temperate or northern regions, a lack of snow in winter can affect agricultural productivity for certain cultures and can also even cause plants to die off. In the subarctic, the presence of thicker snow cover helps keep the soil warmer and helps accelerate the thawing of permafrost in the summer.

Mapping snow cover and estimating its characteristics have become essential tools, not only for studying climate change but for managing hydroelectric dams and forecasting rises in water level in the spring. The

Chapter written by Monique BERNIER, Jean-Pierre DEDIEU and Yannick DUGUAY spatial distribution of snow cover is one parameter that has been studied for a number of years using optical satellites (see Chapter 4). For the measurement of quantitative parameters such as the height or the snow–water equivalent (SWE) it is necessary to turn to microwave sensors. Essentially, the most frequently used wavelengths, ranging from 3 cm (X-band) to 25 cm (L-band) make it possible not only to cover clouds without causing too much disruption but also to penetrate snow cover, enabling the acquisition of data. Passive microwave radiometers (microwave frequency) measure the radiation emitted by the terrestrial surface. This radiation remains very weak, which means that the energy emitted must be measured over a wider surface area in order to obtain a significant signal, which reduces the spatial resolution. Thus, satellite remote sensors such as SSM/I and AMSR-E feature spatial resolutions in the region of tens of kilometers. These remote sensors make it possible to evaluate and map parameters such as the SWE and snow thickness, but the data collected is mostly used for studies on a global or regional level.

For studies on a more detailed scale, where it is possible to make connections to the hydrology of small drainage basins, to the freezing of soil in agriculture or to the thawing of permafrost, it is therefore necessary to turn towards active remote sensors, or synthetic aperture radar (SAR). Its capacity to image terrestrial surfaces regardless of light or cloud cover conditions makes it a very reliable tool, particularly for regions where cloud cover poses an issue for optical remote sensors. In addition, the range of wavelengths used means it is possible to penetrate through the snow cover, thus allowing the acquisition of data on its vertical structure.

5.2. Radar interaction and snow cover

5.2.1. *Physical characterization of snow*

Snow is a stratified environment made of air and ice crystals whose composition and structure vary through time and space. A certain quantity of liquid water can also be present in the snow cover when temperatures are above freezing point. The structure and composition of snow cover can also be very complex and can help determine how electromagnetic waves interact in the environment.

5.2.1.1. *The transformation of snow cover*

Consecutive instances of snowfall throughout winter leads to snow accumulating and the formation of snow cover. The atmospheric conditions in which snow grains form will determine their height and their shape when they are deposited [PRU 78]. Different strata of snow grains then go through various transformations over the winter season following two main mechanisms of transformation: mechanical transformation or thermodynamic transformation. Mechanical transformation is the first to affect snow grains once they have been deposited. This can be in the form of the erosion of particles when they are being moved by the wind through saltation. The weight of the layers which follow leads to the transformation of crystals by compression. These destructive mechanisms increase the density of snow due to the way in which particles accumulate.

Thermodynamic transformation is more complex and evolves up until the snow cover has completely melted. This involves a mass flow between the solid and gas states (plus the liquid state in instances where the snow cover is wet) and is driven by a need for thermodynamic equilibrium. These mechanisms differ depending on whether or not the snow cover is dry or damp: in other words, whether or not there is liquid present in the snow cover.

5.2.1.1.1. Dry snow coverage

The transformation of dry snow is carried out following two processes. The first is produced within each individual crystal and results in the formation of curves. These zones where the radiation curve is weak and positive undergo a process of sublimation, where the emerging water vapor is placed in zones with a weak and negative radiation curve. This phenomenon is relatively slow and tends to create rounded snow grains, which reduces the tension of the vapor at the surface and stimulates thermodynamic equilibrium [COL 83].

The second mechanism is produced by the influence of a temperature gradient within the snow cover. The isolating properties of snow mean that, during winter, the temperature of the soil remains relatively high in relation to the air, which leads to the establishment of a temperature gradient. The sublimation of snow particles in the warmest sections of the snow cover frees the water vapor, which builds up on the coldest particles located below. Snow crystals will tend to take a form that is somewhat elongated vertically. The higher the temperature gradient, the quicker this phenomenon takes place. When the temperature gradient reaches a critical threshold between 10 and 20° C/m, crystals start to form in the shape of angular grains, also called flat face grains. If the gradient exceeds the critical threshold (> 20°C/m) we are able to observe the appearance of deep frost comprised of hollow, pyramid-shaped grains [COL 83].

5.2.1.1.2. Wet snow cover

During a period of thaw or during spells of winter rain, liquid water lodges in the crevices of the snow cover. It is possible to distinguish between two regimes of wet snow depending on the liquid water content of the snow cover: the pendular regime and the funicular regime. The pendular regime is characterized by a liquid water content of less than 7%. Water found in the pores is fragmented and is lodged in the concavities of the grains or around the points of contact between the grains, by capillarity. In these conditions, snow crystals tend to form a mass. In the funicular regime, snow cover is considered as having reached saturation point. Water occupies an uninterrupted space in the pores and the capillary forces become negligible [COL 83].

5.2.1.2. *Descriptive parameters of snow cover*

The processes outlined in the previous section lead to the formation of complex structures in snow cover which will have a significant influence on the electromagnetic signature of the snow viewed by the radar sensor. In order to model the electromagnetic response of the snow cover, it is necessary to define certain parameters which will make it possible to describe these structures. Different types of parameters make it possible to describe snow cover depending on the spatial scale involved. The size and the shape of particles reflects the snow cover at the level of masses or individual grains. The density, the snow thickness, the SWE and the liquid water content enable us to describe the vertical profile of the snow cover at a given point. Snow cover is generally divided into consistent layers, each with different characteristics, and this makes it possible to devise a more precise model of the snow cover. Figure 5.1(a) shows a profile of snow cover obtained using a hole dug in the snow. In this photograph, only the ice crust (darker shade) at the center and towards the bottom of the image are distinguished from the layers of snow.

5.2.1.2.1. Size and shape of particles

Even within a consistent layer of snow, it is sometimes possible to observe a large degree of variability in the size and shape of grains. For anyone looking to familiarize themselves with the way in which grains of snow are described and named, we would direct them towards the international classification for seasonal snow [FIE 09]. The measurement of the size of a grain of snow is generally considered to be the size in millimeters across its widest point [FIE 09]. The classical method for evaluating the size of grains from a homogeneous layer of snow is carried out using a gridded surface and a magnifying glass. Grains of snow from the layer under observation are deposited on the gridded surface and the average size of snow grains is determined by visual observation. New methods involving digital image processing have been developed in order to compensate for the subjective nature of classical methods of measurement [GAY 02, ING 12].

a) b)

Figure 5.1. *An example of a snow profile a) and a photograph of snow grains deposited on a millimetric sheet for classification b)*

The shape of particles is generally determined by a visual and subjective evaluation according to the transformation mechanism to which this shape of grain corresponds. A good knowledge of transformation mechanisms and the environmental conditions in which snow cover develops is necessary for an accurate assessment. Given the wide diversity of shapes found [FIE 09], the size of a snow grain at its widest point is sometimes not representative of its electromagnetic properties. Figure 5.1(b) shows snow grains roughly a millimeter in size where the size and shape varieties give a good indication of the diversity of snow grains found in one single layer of snow. The majority of scatter models use spherical particles to represent snow grains and the use of the size of a snow grain at its widest point means that the size of particles for modeling is often overestimated. Mätzler [MAT 02] suggests using parameters such as the spherical surface *s*, defined as the ratio between the total surface of the air/ice interface per unit of snow mass (in $m^2\ kg^{-1}$) or by unit of ice volume (in $m^2\ m^{-3}$). Studies led by Gallet *et al.* [GAL 09] made it possible to develop a relatively simple and precise method which is easy to carry out on the ground. This method directly links snow's reflectance in mid-infrared wavelengths (1310 and 1550 nm) to the specific surface and makes it possible to establish a vertical profile of *s* using a portable sensor. Another method put forward by Ingvander *et al.* [ING 12] would enable users to link the size of grains measured to their specific surface using a directed object classification of photographs of snow cover which take into consideration the height, shape and reflectance of snow grains.

5.2.1.2.2. Thickness, density and snow–water equivalent

The thickness of an area covered by snow (d) is generally measured in centimeters and measures the vertical distance between the air–snow and snow–soil interface. On uneven terrain, it becomes necessary to indicate if the measurement was taken in relation to the vertical or in relation to the perpendicular, as well as the inclination and the orientation of the slope. The density of the snow (p) is a measurement of mass per unit of volume given in $kg{\cdot}m^{-3}$ or in $g{\cdot}cm^{-3}$. Alternatively, density can be defined as the relation between the volumetric mass of snow m_n over the volumetric mass of the water m_e:

$$\rho = \frac{m_n}{m_e} = \frac{v_e m_e + v_g m_g}{m_e} \quad [5.1]$$

where m_g is the volumetric mass of ice, and v_e and v_g represent the volumetric percentage of liquid water and ice, respectively. The volumetric mass of water and that of ice are defined as:

$$m_e = 1000kg \cdot m^{-3} \; ; \; m_g = 917kg \cdot m^{-3} \quad [5.2]$$

Equation [5.1] represents the general case of wet snow cover; for dry snow, the numerator on the right is expressed as $v_g \, m_g$. Density is measured by weighing a predefined volume of snow for each layer of the snow cover. The density and the type of grains are generally the two parameters used in order to define a homogeneous layer of snow.

The snow–water equivalent (SWE) is a measurement grouping together the average depth and density of the snow cover. It represents the quantity of liquid water obtained once the cover has completely melted:

$$SWE = d \cdot \rho \quad [5.3]$$

where d is in meters and ρ is in kg·m^{-3}. Although the resulting units are given in kg·m^{-2}, SWE is ordinarily expressed in millimeters, or the equivalent height of liquid water.

5.2.1.2.3. Liquid water content

The quantity of liquid water present in an area covered by snow has an effect not only on the transformation mechanisms but also on its dielectric properties. The liquid water content (LWC) is expressed in percentage form and it represents a mass or volumetric ratio:

$$LWC = \frac{mass\ of\ liquid\ water}{total\ mass\ of\ snow} = \frac{volume\ of\ liquid\ water}{total\ volume\ of\ snow} \quad [5.4]$$

5.2.1.2.4. Thermal resistance of snow cover

The thermal resistance of snow (R) is a measurement of the isolating capacities of a snow-covered area. This parameter has a direct impact on the cycle of freezing and thawing of permafrost [GOO 82] and is a function of the thermal conductivity (k) and the thickness (d) of the coverage for dry snow. Thermal conductivity is the key factor and several studies have been able to establish a clear relationship between the density of snow and its thermal conductivity. Sturm *et al.* [STU 97] puts forward the relation:

$$k(\rho) = \begin{cases} 0.138 - 1.01\rho + 3.233\rho^2 & for \quad 0.156 \leq \rho \leq 0.6 \\ 0.23 - 0.234\rho & for \quad \rho \leq 0.156 \end{cases} \qquad [5.5]$$

where ρ is in g·cm^{-3} and k is in W·m^{-1}·K^{-1}. The thermal conductivity of a homogeneous layer of snow i can be connected to its thermal resistance R_i (in $K{\cdot}m^2{\cdot}W^{-1}$) and the total resistance of the volume of snow using:

$$R_i = \frac{d_i}{k_i} \qquad [5.6]$$

$$R = \sum R_i \qquad [5.7]$$

where d_i is the thickness of each layer in meters. By generalizing [5.6] in order to take into account the average density of the snow cover, it is possible to establish a relation between R and the SWE:

$$SWE = R \cdot k(\rho) \qquad [5.8]$$

where $k(\rho)$ indicates the thermal conductivity given in [5.5]. Relationship [5.8] was used by Bernier and Fortin [BER 98] in order to estimate SWE using SAR data.

5.2.2. *Electromagnetic modeling of radar signals*

In order to gain a better understanding and carry out a better analysis on data generated by SAR sensors for snow-covered areas, it is necessary to turn to physical models depicting the phenomenon under observation. Essentially, given the complexity of the interactions and the various parameters which must be considered, a straightforward empirical study makes the comparison between the images generated and the characteristics of snow cover a very complex undertaking. Modeling therefore becomes an important tool for evaluating the behavior of an electromagnetic wave when snow is present. This next section will spend some time considering the effects of snow cover on an electromagnetic wave. Although soil will tend to play an important role in the total backscatter of a target covered in snow, the modeling of this element is outlined in Chapter 1.

5.2.2.1. *Direct modeling*

The radar backscatter of a snow-covered area is a function of various parameters connected to the characteristics of the sensor being used, to the parameters of the snow cover as well as to the characteristics of the soil underneath. In the presence of a heterogeneous media like snow, the total signal measured is a combination of scatters emanating from the air–snow interface, from the volume of snow and from the snow–soil interface. One of the most commonly used methods for modeling the contribution of these different elements is the method of radiative transfer. This is based on the principle of the conservation of energy through a volume element. In addition to the characteristics of the snow cover (section 5.2.1), the characteristics of the sensor and of the soil underneath must also be taken into account. In order to model backscatter using radiative transfer, there are three sets of parameters which come under consideration [SHI 00a]:

– parameters linked to the sensor; frequency, polarization and the incidence angle;

– parameters linked to the characteristics of the soil; the dielectric constant, the standard deviation of the heights, the length of correlation and the function of the associated correlation (see Chapter 1);

– parameters linked to snow cover; depth, density, stratification, the size and the variation in size of its particles, the liquid water content, the spatial distribution of particles and their cohesion [DIN 94].

5.2.2.1.1. Parameters linked to the sensor

Generally, the expression of the radiative transfer for a snow-covered surface is represented by four main elements:

– the soil surface scattering;

– the snow volume scattering;

– the surface–volume interaction between the snow cover and the soil;

– the surface scattering of the air–snow interface (Figure 5.2).

The relative importance of each of these depends on the characteristics of the snow cover, the soil and the sensor used.

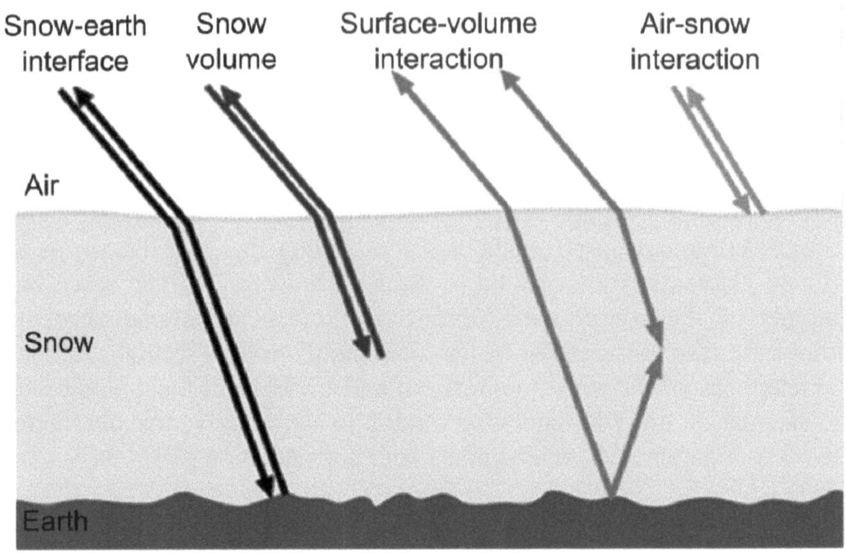

Figure 5.2. *The principle scattering mechanisms from a snow-covered area (figure adapted from [FUN 94])*

The parameters of the sensor have a significant impact on the behavior of an electromagnetic wave when observing a snow-covered area. Frequency will have a direct effect on the scattering and the extinction of the signal by the snow cover. As a general rule, the higher the frequency, the closer the wavelength gets to the height of the snow particles, which in turn increases the scatter and the extinction of the volume of snow. The incidence angle is another important element to consider because it will affect the wave's path through the snow. The higher the incidence angle, the greater the volume of snow covering the wave, allowing it to generate more scattering and more extinction of the signal from the snow cover [FUN 94, WEN 90].

Given that snow particles are generally modeled using spherical shapes and that a sphere generates no backscatter in cross polarization (HV and VH), parallel polarization (HH and VV) is preferred for studying snow cover. Nonetheless, studies undertaken by Kendra *et al.* [KEN 98] and Yueh *et al.* [YUE 09] were able to demonstrate that the backscatter of snow cover in cross polarization is significant and can be in correlation with the height of snow under certain conditions. There are two main phenomena which can generate cross polarization in snow cover: multiple scattering between grains inside the snow cover and the presence of non-spherical particles. The first mechanism will dominate in the case of an environment with a relatively high radar albedo, or when particles which are larger in relation to the wavelength are present. In instances of weaker radar albedo, the influence of particle shape will dominate [DU 10].

5.2.2.1.2. Parameters linked to snow cover

The conventional formulation of radiative transfer considers independent scatters: in other words, the fields diffused by a particle will not affect the scattering of surrounding particles [FUN 94]. The scatter of particles is therefore modeled according to Rayleigh scattering. This approximation is valid if the size of the particles at their widest point is less than $\lambda/2\pi$ which is generally the case for snow particles in the microwave range. In scatter models, a spherical shape tends to be used for snow particles in order to simplify calculations and make inversion easier, given that particles can be represented by one single factor, radiation.

Formulation of the radiative transfer equation using Rayleigh scattering has been successfully used in order to map dry snow [MAR 05]. There are some limits to this method however, and it does not make it possible to properly link the behavior of a wave to the physical characteristics of a covering of snow [KEN 98]. To begin with, to consider the independent scatter of particles implies that the intensity of the scatter from a given volume is equal to the sum of the different measurements of intensity diffused by each of the particles. Studies have nevertheless been able to demonstrate that the relationship between particle density and the behavior of a signal is not linear [WEN 90]. The reduction and the intensity of the bi-static scatter increase with density before reaching a maximum, after which they will decrease. We can therefore state that the hypothesis of independent scatter is invalid, because the fields diffused by the particles influence neighboring particles. Furthermore, Rayleigh scattering requires a random distribution of particles in the volume being observed in order to justify the inconsistent addition of intensities which is not necessarily the case for a volume of snow.

The most commonly used method for addressing these issues is the use of the dense media radiative transfer (DMRT) [WEN 90]. This model is a product of quasi-crystalline approximation (QCA) and quasi-crystalline approximation with coherent potential (QCA-CP) introduced by [TSA 80]. These theories take into account the coherent interactions between the diffused fields of particles and make it possible to correct certain errors from conventional radiative transfer. One advantage of DMRT is that it addresses the criteria of energy conservation and can therefore be integrated into the formulation of radiative transfer [WEN 90]. The model introduces the concept of the effective dielectric constant ε_{eff} of a dense environment which is calculated using quasi-crystalline approximation. For wet snow, the interstitial presence of liquid water must be taken into account when estimating ε_{eff}. Different methods exist for calculating the effective dielectric constant from such an environment including the strong fluctuation theory (SFT) [STO 84, TED 06]. This method can also be used in order to estimate the effective dielectric constant (ε_{eff}) for dry snow cover. Figure 5.3 shows a comparison between the response of the Rayleigh model and the response of SFT in the L-band and in the C-band for a snow-covered surface. We are able to observe that the Rayleigh model underestimates volume scattering in relation to the SFT.

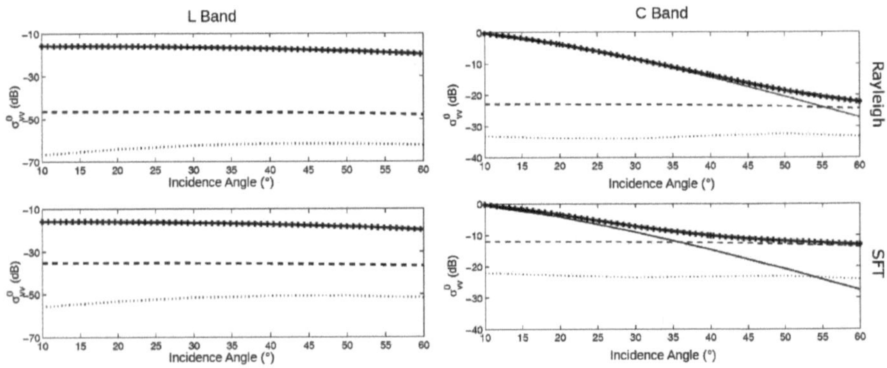

Figure 5.3. *Backscattering from a dry snow covered surface as a function of the incidence angle with the mechanisms of soil scattering (—), volume scattering (--), surface-volume interaction (...) and total scattering (+) according to Rayleigh backscatter and SFT (taken from [LON 08])*

The DMRT method makes it possible to compensate for the shortcomings of Rayleigh backscatter, but it does not make it possible to model the multiple scatters between grains within snow cover in order to estimate the contribution of cross polarization. Essentially, when spheres are used to model snow particles, it is necessary to resolve the radiative transfer equation for solutions of a greater magnitude in order to generate cross polarization backscatter. By using the iterative method, it is possible to find the second order solution using the first order solution. Nevertheless, the higher the order of the equation, the more complex the equation becomes. Thus, if the first order solution has four terms, its second order solution will then have sixteen terms [TSA 07]. Digital methods have therefore been put forward in order to find an exact solution for the radiative transfer equation. One of these, the matrix doubling method introduced by Ulaby *et al.* [ULA 86], puts forward a relatively intuitive and effective calculation method when the environment has a relatively high optical thickness.

The modeling of a volume made up of non-spherical particles is a widely-known problem [TSA 81], but its use has generally been limited to studying vegetation. However, snow particles are not necessarily spherical. In subarctic or northern regions, deep frost can form a significant proportion of the snow cover [STU 95]. These relatively large, non-spherical particles can have a non-negligible effect on the signal, but to model them is complicated because it requires the addition of several parameters. For example, the model put forward by Tsang *et al.* [TSA 81] not only requires the parameters of the height of the different axes of an ellipsoidal particle, but other parameters linked to the orientation of particles as well as the distribution functions of these parameters. Du *et al.* [DU 10] put forward a simplified approach using fixed-parameter spheroids in the relationship between the large axis and the small axis in order to describe the shape of particles.

Electromagnetic modeling has also been used in order to integrate the parameters obtained from SAR images, such as the backscatter coefficient, in a physical model of the snow cover metamorphism (SURFEX-Crocus) which makes it possible to monitor the seasonal evolution of snow cover's characteristics [LON 09].

5.2.2.2. *Inverse modeling*

Inverse modeling aims to obtain an environment's physical characteristics through the use of radar measurements. The problem of inversion for a snow-covered surface is very complicated given the number of physical parameters which need to be identified.

The number of variables to evaluate demands a huge number of independent radar measurements in order to solve the problem, and for several years this limited the development of inversion models. The radiative transfer equation is also complicated to invert because of the integration terms included in the formulation.

Shi and Dozier [SHI 00a, SHI 00b] were the first to formulate a complete inversion model which makes it possible to retrieve the parameters of snow and soil. The series of independent measurements necessary for inversion were generated using multi polarization, multi frequency data in the L-band ($\lambda \approx 24$ cm), C-band ($\lambda \approx 5$ cm) and X-band ($\lambda \approx 3$ cm) from the Jet Propulsion Laboratory's (JPL) AirSAR (Airborne Synthetic Aperture Radar) platform. The model is configured in accordance with the wide range of simulations of the first order solution of radiative transfer. By varying the input parameters of the radiative transfer model, it is possible to generate a bank of simulated backscatter values. The frequency and polarization combinations that are most sensitive to the sought-after parameters are then chosen and multivariate regressions are carried out in order to configure the model. The shape of the radiative transfer model used in this study uses the DMRT in order to model snow cover while the modeling of soil scatter is carried out using the integral equation model (IEM) [FUN 94]. Initially, the soil's characteristics are evaluated using L-band data. Given that the size of snow grains is extremely small compared to the wavelength, the snow cover's scatter and extinction become negligible at the L-band and the only clear effect is the refraction of the wave when it penetrates the snow cover. Refraction, which affects the incidence angle and the wavelength, is directly linked to the snow's dielectric constant. This can be correlated with the density using empirical or semi-empirical equations [MAT 96]. We can see therefore that the model determines a relationship between the co-polarized channels HH and VV (parallel) which makes it possible to minimize the influence of the parameters of the soil and to maximize the effects the incidence angle and the wavelength have on the signal. After the new parameters linked to the incident wave (the incidence angle and the wavelength) have been evaluated, it is then possible to evaluate the soil's roughness and the dielectric constant using the refracted incident wave [SHI 00b]. The depth of the snow cover and the size of the grains are then evaluated using C-band and X-band data. The characteristics of the soil as well as the snow's dielectric constant that are evaluated using the L-band are input back into the radiative transfer equation and the unknown quantities linked to the snow's characteristics are estimated using the co-polarized signal in the C-band and X-band [SHI 00a].

Although the model developed by Shi and Dozier makes it possible to evaluate the set of characteristics of a snow-covered surface, it does have certain limitations. To begin with, the precision of the model in terms of measuring the height of snow is relatively poor. The mean squared error obtained is 34 cm, for an average snow height of 190 cm with a standard deviation of 82 cm and there is a tendency to overestimate the snow's height. This could prove problematic in subarctic regions where the height of snow is generally quite low. From the point of view of modeling snow particles, DMRT is resolved by using spherical particles and the multiple scattering of particles is not taken into account, which restricts the model to the co-polarized signal. Furthermore, although satellite sensors operating in the L-, C- and X-band are currently in use, it can be complicated to coordinate simultaneous acquisitions for the three sensors and in the short term this could limit the model's practical use. In order to arrive at a better understanding and a better analysis of the data generated by SAR sensors for snow-covered surfaces, it becomes necessary to turn to physical models which describe the phenomenon under observation.

5.2.3. *The polarimetric study of snow*

Radar polarimetry is the study of the effect of an object or a natural environment on the polarization state of an incident wave. Polarimetry aims to interpret the signal received by a polarimetric radar sensor in order to obtain information on the physical structure of the object being observed. This makes it possible to extract the portion of the signal resulting from the various elementary scatter mechanisms such as surface diffusion, volume diffusion and double rebound diffusion (see [FER 16a and FER 16b]). Within the framework of studying snow, the mechanisms which are of interest to us are the volume diffusion generated by the snow cover which makes it possible to obtain a better characterization of the vertical structure of the surface in the case of dry snow, plus the surface diffusion, which is the dominant mechanism when wet snow is present.

By using polarimetry, the aim is to extract the portion of the SAR signal on which the physical parameters of the snow cover has an effect. In order to facilitate the extraction of this data, we turn to the use of polarimetric decompositions which can be divided into two main categories: decompositions based on scatter models (section 5.2.3.1) and those based on the extraction of specific values and specific vectors of the coherence matrix. These two forms of decompositions have both advantages and disadvantages with respect to the study of snow, and these will be discussed in the following sections.

5.2.3.1. *Decompositions founded on scattering models*

The following polarimetric decompositions are based on simplified scattering models in order to extract the relative powers of each scattering mechanism. The chief mechanisms to be considered will be surface scattering, volume scattering and the interaction term. The decomposition developed by Freeman and Durden [FRE 98] was one of the first to employ this type of model. This decomposition takes into account the total power diffused (P_t) and considers it as a combination of three scattering mechanisms:

$$P_t = P_s + P_d + P_v \qquad [5.9]$$

where P_s corresponds to the surface scatter, P_d corresponds to the double reflection and P_v corresponds to the volume scatter. The surface scatter is obtained using a first order Bragg scatter model based on the model of low disturbance [FUN 94], and can be used with soil with low roughness. The interaction term is modeled using a local dihedral reflector whose horizontal and vertical elements (soil and tree trunk) possess different dielectric constants. As for volume scatter, it is modeled using a cloud of randomly-angled cylindrical dipoles which represent the small branches of trees and shrubs. This general model was developed in order to characterize different types of soil activity and the activity of forested environments. This method of decomposition was tested in the framework of a study looking at snow cover in forested areas in Colorado and Wyoming in the USA [TRU 09]. Results for the signals in the C-band showed that the snow contributes to the volume scatter and that the double reflection signal is weak because of the ice crusts contained in the snow cover. However, the model used by Freeman and Durden [FRE 98] suggests certain limitations. Essentially, the method is only applicable if targets responding to the criteria of reflective symmetry are present, and the scattering model is relatively fixed and quite inflexible. Thus, other researchers looked towards the decomposition method developed by Yamaguchi *et al.* [YAM 05], which avoids the issue of a target's reflective symmetry by adding an additional scattering mechanism, and uses a more flexible volume scattering model. Tests were carried out using L-band images from mountainous regions of Japan [PAR 13] and in the Himalayas [SIN 14] in order to determine the snow cover's contribution to the total backscatter. The results show that in the L-band, the soil surface scattering remains the dominant mechanism when snow is present. For a slightly wet stratified snow-cover on top of rice paddies [PAR 13], the surface scattering will remain unchanged when snow is present and the volume and double reflection scatterings will be smaller in winter. This indicates a reduction of the effect of the surface's roughness and rice crops on the total signal. For wet snow [SIN 14], the element of surface scattering will be the dominant factor given the weak penetration of the signal in snow cover.

Decompositions based on scatter models have the advantage of being relatively easy to interpret because of the simplicity of representing the signal's different elements. However, they are rarely used for studying snow cover because the models used in order to represent the signal's different elements have not been adapted. These models which represent the volume scattering take into consideration a set of more or less randomly angled dielectric dipoles while snow cover is generally modeled using a set of spheres. In this context, Arii *et al.* [ARI 11] were able to develop a decomposition method which uses a more flexible scatter model that is able to adapt to different types of scatterometers, but it has yet to be tested for studying snow at the time of going to press. Thus, decomposition methods that are better adapted to the specific characteristics of snow cover have yet to be developed.

5.2.3.2. *Decompositions founded on the specific values and vectors of the coherence matrix*

Cloude [CLO 85] was the first to introduce this type of decomposition. The extraction of the specific values (λ_k) and specific vectors ($\overrightarrow{v_k}$) of the coherence matrix makes it possible to uniquely decompose a distributed target. Each specific value represents the power of a scatter mechanism whose nature is defined by the specific vector associated with this specific value. Given that the three specific vectors are orthogonal to one another, each one makes it possible to define an independent mechanism. Using these principles as a base, Cloude and Pottier [CLO 97] were able to define three parameters which make it possible to describe the nature of the scatter mechanisms under observation as well as the relative importance in the backscattered signal: entropy (H), anisotropy (A) and the alpha angle (α). Entropy, which comes from the relative difference between the specific values, expresses the random nature of scattering mechanisms which make up the target. Non-existent entropy indicates the presence of one single scatter mechanism and that this is therefore a deterministic target. An entropy with a value of 1 indicates that the three specific values are equal and that the scattering mechanisms are completely random. Anisotropy is compatible with entropy and expresses the relative importance of secondary mechanisms linked to λ_2 and λ_3. Non-existent anisotropy tells us that the two secondary mechanisms are equal while an anisotropy of 1 means that the target includes only two dominant mechanisms linked to λ_1 and λ_2. The angle α is connected to the nature of the scattering mechanism observed, it varies between 0 and $\pi/2$ and is generally used in combination with the entropy in order to interpret the various scattering mechanisms present for a distributed target. When entropy is zero, the angle α is directly linked to an orthodox diffuser. When a distributed target is present, entropy generally becomes non-zero and an average is taken of all the angles α linked to each of the specific vectors, defined as $\bar{\alpha}$, in order to represent the dominant scattering mechanism. The values of $\bar{\alpha} < \pi/4$ are interpreted as the surface scatter. When $\bar{\alpha} = 0$ the Geometric Optics (GO) model is used and when $\bar{\alpha} > 0$ the Bragg or Physical Optics (PC) model is used. When $\bar{\alpha} \approx \pi/4$, the volume scattering from the angled cylindrical dipoles is taken to be the dominant mechanism, while values of $\bar{\alpha} > \pi/4$ represent the double bounce such as the interaction terms of soil/tree trunk or the ground/building. The parameters resulting from this decomposition were tested for different environmental conditions and different frequencies, generally by analyzing those decomposition methods based on scattering models. The results are roughly the same in terms of interpreting the snow effects on various scattering mechanisms. However, the parameters resulting from decompositions based on specific values and specific vectors have tended to provide better results in terms of the classification of snow-covered areas (section 5.3.1). Essentially, they provide a degree of flexibility and are able to represent a wider range of conditions than decompositions based on predefined scattering models.

Although this type of decomposition, as well as the interpretation of parameters H and α which result from it, is widely used, there are a few ambiguities which create problems when studying snow cover. As was mentioned previously (section 5.2.2) snow cover can be modeled using an environment comprised of spheres or spheroids. However, one of the orthodox diffusers which corresponds to $\alpha = 0$ means that for an environment comprised of spheres, $\alpha = 0$ for the simple scatter of particles and the values of α and H increase when multiple scatters within the snow cover are present [CLO 97].

Given that the significance of multiple scatters decreases for low radar albedos [DU 10], which is the case in C-band, it can become difficult to distinguish the snow volume scattering from the soil surface scattering using only H and α, if the model being used underestimates multiple scatters. In general, for the C and X bands, the soil surface scattering is the dominant mechanism for dry snow and incidence angle lower than 35° [MAR 05, DED 14]. For incidence angle greater than 35°, the volume scattering and the double rebound scattering are the dominant mechanisms, given that the snow particles are the main diffusers, especially in cross-polarization HV or VH. In the C-band, the variations of the alpha angle and the entropy for a dry snow cover will be higher than those for bare soil, irrespective of the incidence angle [MAR 05]. In L-band, the surface scattering is the dominant mechanism for all incidence angle: the soil surface scattering under the dry snow cover and the surface scattering at the snow/air interface for a wet snow cover. At L-band (26 cm), the soil surface scattering prevails over the volume scattering from the ice particles which are comfortably smaller than the wavelength (except for damp snow). When considering the total scattering of a snow-covered surface (surface +volume +double rebound) in L- and C- band, an increase in the parameters of Entropy and the angle α with the incidence angle can be noted, particularly for incidence angle between 30° and 40°.

For illustration and interpretative purposes (Figure 5.4), the average values of entropy (H) and the alpha angle (α) taken from a RADARSAT-2 (C-band) image taken in the French Alps in April 2011 were organized using the framework H/α, split up into eight distinct zones by [CLO 97].

Lastly, in order to optimize the polarimetric contrast and to reproduce the height of the snow cover [DED 14], other scattering mechanisms were also considered [BES 15] and compatible polarimetric descriptors were used in combination with H and α. These include simple reflection (SERD) and multiple reflection (DERD) parameters [ALL 06]. At the time of going to press, these new approaches are currently being validated over long temporal series.

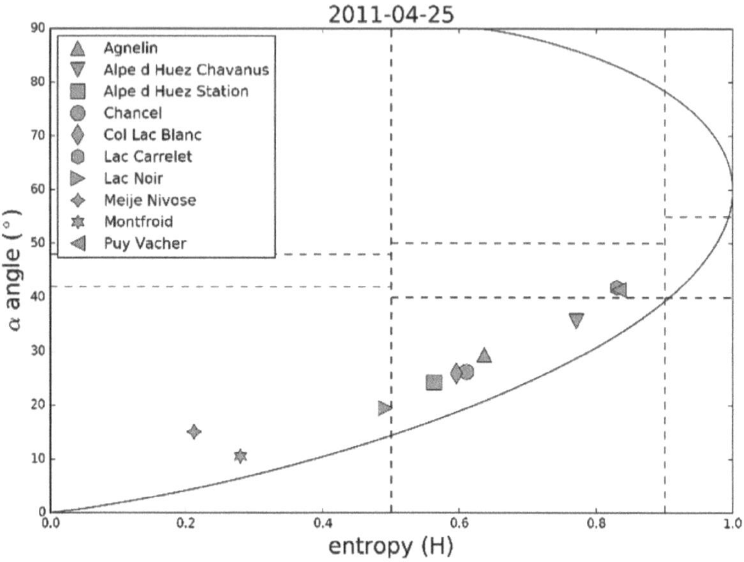

Figure 5.4. *Example of average values of entropy (H) and alpha angle obtained from a RADARSAT-2 image taken in the French Alps in April 2011. These values are set using the H/α plane which is split up into eight regions each representing separate scattering mechanisms [CLO 97]. Thus, for the sites on the bottom left, the entropy is low and this corresponds to homogeneous snow, that is cold and dry. On the contrary, the further we move to the right, the more we notice an increase in entropy and this expresses an increasing level of moisture in the snow cover, with multiple scattering interaction [ROB 15]*

5.3. Mapping snow cover

5.3.1. *Mapping the extent of snow coverage*

The first studies showing the potential of SAR imaging for mapping snow extent were undertaken using C-band data from the ERS-1 satellite which was launched in 1991 by the European Space Agency (ESA). The main motivation for using radar data is that cloud cover can last several days during the spring thaw, which is a critical period for the management of dams and the forecasting of rising water levels for a given drainage basin. Guneriussen *et al.* [GUN 96] and Baghdadi *et al.* [BAG 97] were able to demonstrate the potential of this sensor for mapping the percentage of a surface covered by wet snow. Using ERS images, [BAG 97] would go on to show that the introduction of a simple thresholding method of the difference between radar signals obtained with wet snow ($\sigma_{wet\ snow}$) and radar signals obtained with dry snow ($\sigma_{dry\ snow}$) or with no snow would provide a far more effective way of mapping wet snow:

$$Wet\ snow\ if: \sigma^0_{wetsnow} - \sigma^0_{drysnow} \leq -3dB \qquad [5.10]$$

Figure 5.5 shows a RADARSAT-2 image of the Matapedia region of Quebec (Canada) taken on 25th April 2015 after a snowfall. Agricultural fields and open areas covered by damp snow appear black (weaker backscatter) while areas where snow has melted appear gray and are hard to distinguish from the forested areas which make up most of the region.

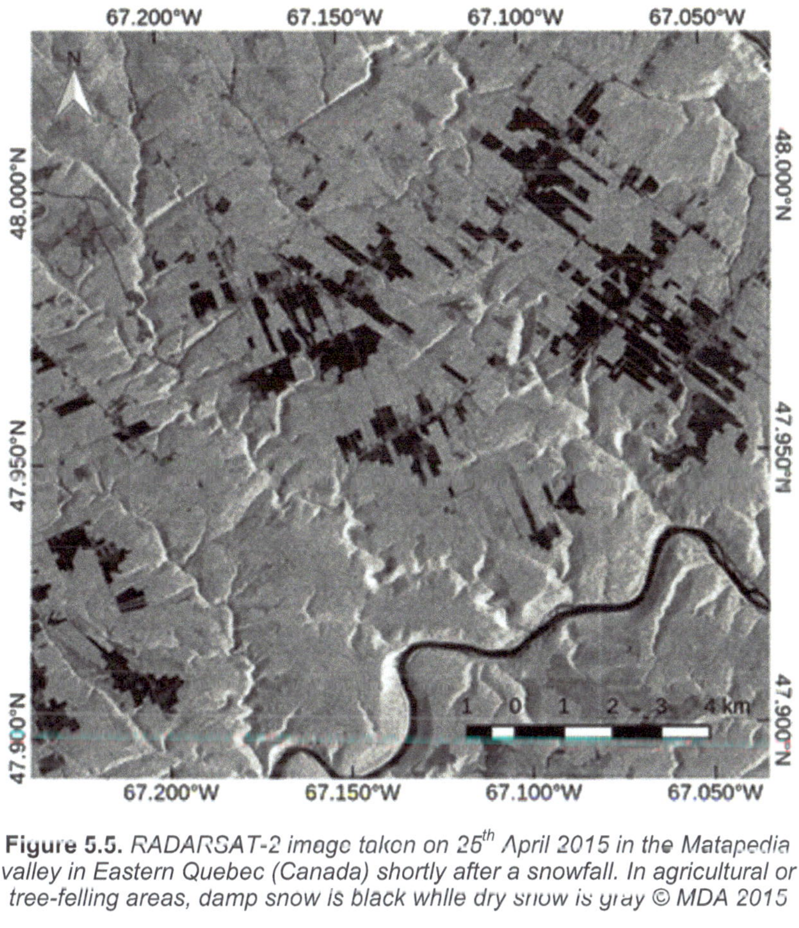

Figure 5.5. *RADARSAT-2 image taken on 25th April 2015 in the Matapedia valley in Eastern Quebec (Canada) shortly after a snowfall. In agricultural or tree-felling areas, damp snow is black while dry snow is gray © MDA 2015*

With the arrival of polarimetric sensors, new methods for mapping snow cover were developed which take advantage of the capabilities of these sensors (section 5.2.3.2). Methods of classification which use polarimetric decompositions based on specific values and specific vectors made it possible to improve the precision with which wet snow could be mapped. Thus, Singh *et al.* [SIN 14] were able to classify the extent of wet snow using only the specific values of the coherence matrix. Two thresholds were used in order to classify the extent of snow in the Himalayas. The first threshold is applied to the polarization percentage that was determined using the ratio between the third specific value (λ_3) and the total power of the signal. The second threshold was applied to the value of (λ_3) in order to exclude areas of distortion linked to the topography. For their part, Park *et al.* [PAR 13] used the temporal ratios of the entropy and the cross-polarized backscatter (σ^0_{HV}) in order to map the extent of snow coverage in Japan. The additional information provided by the polarimetric data among other elements made it possible to develop a method for mapping the extent of dry snow cover [MAR 06].

5.3.2. *Mapping snow–water equivalent*

The quantitative evaluation of the parameters of snow cover, such as the snow–water equivalent, present serious challenges. The backscatter coming from the volume of dry snow cover in the L-band (λ=26 cm) will be zero (transparent coverage), very low in the C-band (λ=5 cm) and even zero for snow coverage less than 20 cm thick, present in the X-band (λ=3cm) and higher in Ku-band (λ=1.5cm) but this is not currently used due to the absence of a SAR spatial sensor in this frequency range. However, a method developed by Bernier and Fortin [BER 98] makes it possible to indirectly obtain SWE data using radar data in HH or VV polarization in the C-band for farmland or taiga (open coniferous forests). It is called an indirect method because the signal measured does not come directly from the volume scattering of the snow cover but instead uses the snow cover's thermal resistance and its effect on the soil beneath which will either become colder or less cold. When the snow is dry, the signal that is recorded is effectively coming from the soil below. The radar is therefore sensitive to changes in the dielectric properties of the soil (wetness, freezing).

In the EQeau approach, four types of data are necessary in order to produce a map showing the snow–water equivalent [BER 99]:

– a reference SAR image in the C-band, ideally obtained towards the end of autumn, when the soil is frozen at the surface but when there is little snow present. Soil freezing restricts variations in the signal relating to the soil's water content since the water content in the pores of the soil is in the form of ice and the soil then has the characteristics of a dry soil;

– a SAR image taken in winter, obtained on the date for which you are looking to obtain the snow–water equivalent. This image should be acquired following the same technical parameters (acquisition mode, incidence angle, polarization) as the reference image. The combination of these two images makes it possible to reduce the effects linked to topography and vegetation (shrubs, isolated trees) and to maximize the ability to detect changes linked to the characteristics of the snow cover;

– the average density of the snow cover. This makes it possible to calculate the snow–water equivalent using the snow cover's thermal resistance (section 5.2.1.2) estimated from the radar images. This density can be obtained by carrying out *in situ* measurements by mathematical modeling of the snow cover or by using historical data;

– the fourth piece of data is the land use. This makes it possible to spatially distribute the density values of the snow cover that are available. The density of the snow cover will vary depending on whether or not the given area is a closed or open forest, in a moorland or on burnt land. The land use map also makes it possible to mask the lakes and rivers. Essentially, the EQeau algorithm can only be used on terrestrial areas. Soil activity can be obtained using the classification of optical images where the spatial resolution is similar to that of the SAR data used.

The EQeau approach involves establishing a relationship between the signal recorded by the radar (backscatter coefficient) and the thermal resistance of the snow cover, in order to estimate the water equivalent. In winter, the backscatter signal of the snow cover is largely coming from the soil beneath. As is illustrated in Figure 5.6, in cold weather, the thinner and denser the snow cover (low thermal resistance) the lower the temperature of the first centimeters of soil (below freezing point). This leads to a drop in the dielectric properties of the soil and as a consequence leads to a drop in backscatter in winter.

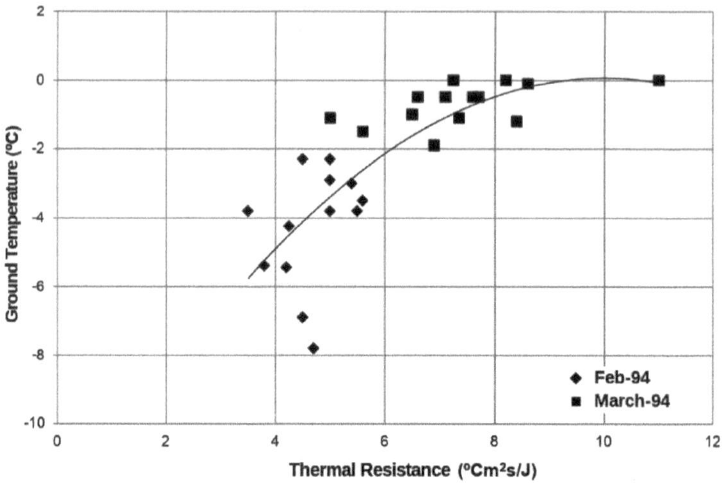

Figure 5.6. *Relationship between the soil temperature and the thermal resistance of the snow cover. An increase in the thermal resistance of the snow cover will reduce the heat lost by the soil beneath which will stay warmer. Measurements carried out in the Saint Lawrence Valley in Quebec (Canada)*

The EQeau model includes two equations. The first makes it possible to estimate the thermal resistance (R_{es}) of the snow cover using the relationship between the backscattering coefficient of a snow-covered surface (using an image taken in winter σ_w^0) divided by the backscattering coefficient of soil uncovered by snow (using an image taken in autumn σ_a^0):

$$R_{est} = m\left(\frac{\sigma_w^0}{\sigma_a^0}\right) + b \qquad [5.11]$$

where m and b are, respectively, the slope and the ordinate at the root of the linear relationship between the thermal resistance and the backscatter relationship σ_w^0/σ_a^0. This equation is determined empirically using the linear relationship between the thermal resistance and the backscattering relationship at test sites (Figure 5.7).

The second equation calculates the snow–water equivalent (SWE) using the estimated thermal resistance (R_{est}) of the snow cover and the typical density of the area:

$$SWE_{est} = k(\rho)R_{est} \qquad [5.12]$$

where k is the thermal conductivity defined by Raudkivi [RAU 79] as:

$$k(\rho) = A\rho^2 + B\rho + C \quad [5.13]$$

where A = 2.83056×10^{-6}, B = -9.09947×10^{-5} and C = 0.0319739

ρ : snow density (kg/m^3)

The relationship between the SWE and the thermal density of the snow is linear, but the regression slope $k(\rho)$ varies depending on the snow density.

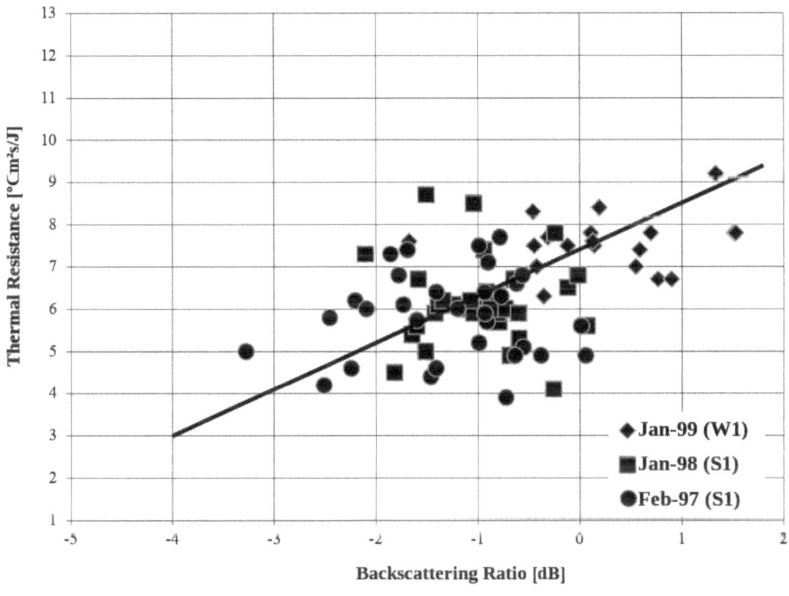

Figure 5.7. *Relationship between the backscatter relation (C-band, HH polarization) and the thermal resistance of the snow cover using data from RADARSAT-1 images (W1, wide mode and S1 standard mode) in the La Grande River basin in Quebec (Canada)*

The EQeau model uses a relationship between two SAR images [5.11]. This relationship can also be applied to mountainous regions [COR 05] and makes it possible to exclude radiometric bias due to the topography (section 5.3.3.1) and to reduce to a minimum the effect of the soil's roughness. The study carried out by Bernier *et al.* [BER 02] showed that the local slope effects (aside from shadow and layover) can be offset with a backscatter relation by using original images when the images were obtained in the same relative orbit (24-hour cycle for RADARSAT-1). However, radiometric calibration [ULA 96] is necessary in order to study the effect of the geographical or environmental parameters or to map the extent of snow cover, for a given SAR image (section 5.3.3.1).

An uncertainty analysis was also carried out on the EQeau model [CHO 06] using the parameters established by means of a series of RADARSAT-1 images (HH polarization) ScanSAR Narrow (2000–2004) obtained over the taiga (open boreal forest) in Canada. The SWE measured varied between 80 and 490 mm. The analysis proved that estimating the SWE is sensitive to fluctuations in the snow's average density (ρ) and to a lesser extent to fluctuations in the backscatter relation (σ_w^0/σ_a^0, [5.11]). Density is therefore the main source of errors in the EQeau model due to the heightened sensitivity of this parameter and the high evaluation uncertainty (Table 5.1). The model is more sensitive for backscatter relation values below -3 dB. Even in optimal conditions (soil that is already frozen in the reference image and dry snow in the winter image), SWE values are obtained with a high level of uncertainty (relative error of between 38% and 50%). Given that these error levels are not acceptable for those responsible for the management of hydroelectric facilities and that it was not possible to reduce them for boreal forests [CHO 06], the EQeau model has not been adopted as an operational approach.

Parameter	Parameter value	Uncertainty values	
		Absolute values	Relative values
m	1.1	0.0963	8.7 %
b	6.4	0.1193	1.9 %
σ_w^0/σ_a^0	-5 dB – 5 dB	0.98 dB	33 % – 98 %*
ρ	80 kg/m^3 – 500 kg/m^3	39 kg/m^3	13 % – 19.5 %**

Table 5.1. *Uncertainty analysis of EQeau model parameters. (*) corresponds to the backscatter relation values σ_w^0/σ_a^0 between –3 and 1 dB; (**) corresponds to the values of ρ between 200 and 300 kg/m^3*

However, the EQeau approach was tested in northern China using two ENVISAT-ASAR images (VV polarization), one from the 14th March 2008 and the other from the 9th September 2009 [SUN 15]. The study site is a basin in a mountainous region covered by prairies and rocky outcrops. The average depth of the snow coverage was 30 cm and the snow water equivalent was 53 mm. The relationship between the thermal resistance and the backscattering coefficient ratio [5.11] was established using half of the data gathered on the ground. The other half was used in order to validate the results. The SWE was estimated using the estimated thermal resistance (R_{est}) and the average density [5.12]. The RMS error was 7.4 mm (or 6.2%). The resulting SWE map showed that the quantity of snow was higher on the southern basin than the northern basin and that the thickest accumulations were to be found in valleys. The estimated SWE values were compared to the values measured by a passive microwave airborne radiometer and it was found that the two approaches produce the same spatial distribution depending on the topography.

Concurrently, studies undertaken by Shi and Dozier [SHI 00a, 00b] showed the potential of multi polarized (HH and VV) and multi frequency (bands L, C and X) in order to evaluate snow water equivalent. This research was carried out using data from the Spaceborne Imaging Radar-C/X-band Synthetic Aperture Radar (SIR-C/X-SAR) sensors onboard the space shuttle Endeavour. These sensors were the focus of two missions in 1994. Despite the relatively interesting results (RMS error of 42 kg/m^{-3} for the density and 34 cm for the snow depth), it is not worth considering using this method in the short or long term because of the difficulty of putting a series of multi frequency sensors on a satellite or organizing the acquisition of data from three different sensors. Likewise, several studies have been able to demonstrate the suitability of using double frequency (X and Ku) data in the framework of the $CoReH_2O$ program [ROT 10, ROT 14], but a similar mission in the Ku-band has not yet been put into orbit. However, recent results using multi polarized data with a single sensor in the X-band (TerraSAR X) indicate some potential by using the parameter of the phase difference between the HH and VV polarizations [LEI 14]. With the higher revisit frequency allowed by the TanDEM-X mission (two satellites), it is possible to monitor the temporal evolution of snow cover more precisely and thus to improve the detection of changes which occur in the depths of the snow cover.

5.3.3. *Specific environments*

5.3.3.1. *Mountainous regions*

Due to the water budget resulting from snow and glaciers melting, rivers originating in mountain ranges are responsible for roughly 40% of the world's water supply (farming, drinking water, energy). Spatial remote sensing used in order to monitor snow cover in mountains is mainly dedicated to the optical spectrum (mapping) because the spatial resolution of passive microwaves is too rough (25 km) to allow the retrieval of the snow's physical properties (height, water equivalent), given the strong variations of the signal caused by the local topography. "All-weather" active radar systems are an alternative, but require necessary preprocessing steps. This will be outlined in the following section.

5.3.3.1.1. Constraints linked to radar acquisition geometry

A simple look at a radar image of a mountainous region gives an immediate indication of the patterns in which variations in the signal intensity (or amplitude) are implicitly connected to the changes in the topography. The influence of the mountainous area on the backscattered signal is expressed by the presence of a local incidence angle (angle between the direction of the antenna-ground view angle and the direction of the slope of the ground in relation to the normal). Given that these principles have already been presented by [FER 16a], we will not describe them in this section. For reference, Figure 5.8 presents a simplified model showing the principal elements of the reflectivity of a mountain surface. What renders Alpine regions different is that apart from the roughness of the illuminated surface, the three elements of a mountainous area (altitude, slope, orientation) have an impact on the target and must be integrated into the processing procedures in order to calibrate the raw signal [SHI 94, DED 12].

Firstly, it is essential to ensure that the SAR images mostly provided by Agencies are set in the original acquisition geometry (slant range), in order to be able to calibrate the effects of the topography, radiometrically as well as geometrically speaking. This rectification relies on the use of a digital elevation model (DEM) where the spatial resolution in x, y and the mean squared error in z (elevation) values must be lower than or equal to the size of the radar pixel, in order to obtain a proper calibration of the signal in alpine regions [SMA 11]. Improvements in the spatial resolution of active radar sensors that have been observed over recent years (with pixel < 10 m) offer sophisticated digital elevation models (interferometry, airborne LiDAR) in order to characterize with a better accuracy and at a detailed scale the target's physical properties. For example, the world DEM database from the German Aerospace Center's (DLR) TanDEM-TerraSAR-X mission has provided worldwide DEMs with a resolution of 12 m since 2015.

The look angle of the radar sensor is another factor which impacts on the quality of results obtained when monitoring snow-covered surfaces. Essentially, a low incidence (look angle between 20° and 25°) such as those in the first generation sensors ERS-1/2 or JERS-1/2 will increase in the image the areas of geometric distortion this is not workable, including layover and foreshortening. Conversely, high view angles, up to 50°, will lead to more areas in the radar shadow, also not workable. Since the 1980s, literature has agreed that images obtained high in the mountains should be at a medium look angle of 40° in order to balance these two opposing constraints [ROT 84, NAG 00]. This configuration thus minimizes the areas which are not usable in order to optimize the mapping of snow in mountainous regions or the estimation of its physical properties (wetness, water equivalent).

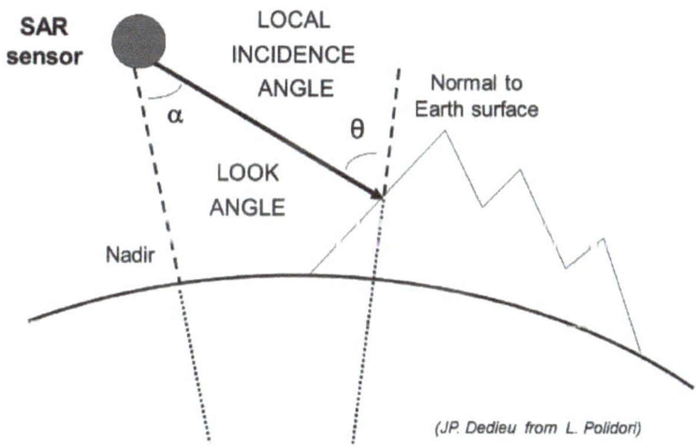

Figure 5.8. *Simplified model of radar geometry acquisition for mountainous regions*

5.3.3.1.2. Terrain effect correction

Two types of terrain effects can be observed in radar images: distortion of the target's initial ground geometry and disturbance in its radiometry (brightness). Taking in account that the two phenomena happen simultaneously, we will separate them here for a better understanding.

Geometry

As a reminder, geometric distortion mainly occurs in the slant range, perpendicular to the orbital axis, rarely in azimuth distance. Despite this complexity of radar imaging compared to the optical mode, different protocols for geometric calibration have been produced and proven since the 1990s [ULA 96, VAN 93]. These are now implemented into commercial or freeware software (PCI, ERDAS, ENVI, Nest/ESA, Orfeo-Toolbox/CNES, etc.) and have been adapted for various imaging sensors (ERS, ENVISAT, RADARSAT, TerraSAR-X, COSMO-SkyMed, ALOS, Sentinel-1, etc.). It is important to note that making these products available in so few years has proved a huge challenge to accomplish.

Given that the topology of geometric radar is different to documents set in geographic coordinates, the aim is to move from a raw slant range image to a ground range image. This is done using all the geometric acquisition parameters of the image (metadata) and a suitable DEM. Figure 5.9 provides an example of how a RADARSAT-2 image can be used for studying snow in the French Alps. The points not seen by the incident wave and which are therefore unusable (shadow, layover, foreshortening) are then put together beneath a mask.

The more consistent the mountainous area is, the larger the unusable areas will be, even when the optimal look angle of 40° is used. Thus, most of mapping applications for snow studies will combine a descending track (morning) with an ascending track (evening) in order to obtain a complete coverage of the snow-cover in Alpine regions, essentially for wet snow mapping [NAG 00].

Radiometry

In mountainous regions, the position of the surface-target in relation to the sensor is directly linked to three terrain components: elevation, slope and aspect. The local incidence angle (Figure 5.8) has to be calculated in order to correct the raw radiometric values and to be able to obtain consistent values similar to acquisition conditions for flat surfaces, either (i) for the backscattering coefficients (dB) of images in intensity mode, or (ii) for the statistical descriptors of images in polarimetric mode. In order to retrieve the projected local incidence angle (see Figure 5.8) we need to calculate for each pixel of the relevant DEM a map of slope gradients (0 to 90°) and aspects (0 to 360°) .

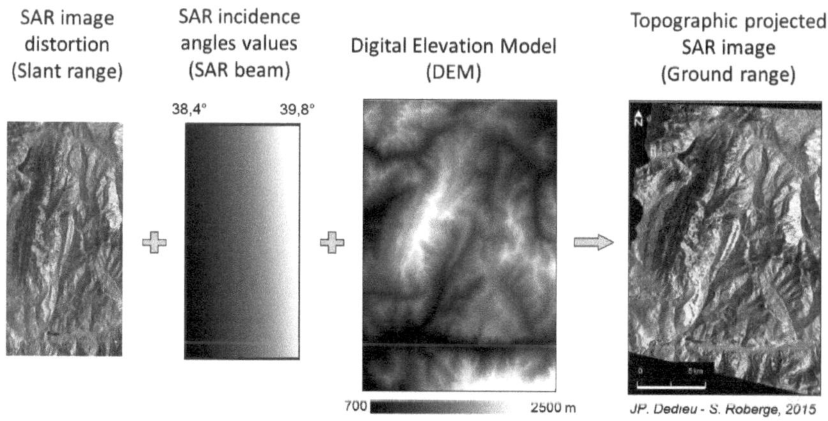

Figure 5.9. *Steps of geometric calibration for a radar image in a mountainous region using a digital model of the terrain. The calibrated image is a RADARSAT-2 image taken on May 27th 2014, in France (Grandes Rousses, French Alps, lat. 45° 09' N, long. 06° 10' E). © RSI-MDA, DEM 25m © IGN-France. In this color image of HH-2VH-VV polarizations, wet snow appears blue. For a color version of this figure, see www.iste.co.uk/baghdadi/4.zip*

The backscattering coefficient normalized by the topography can be given as (γ^0):

$$\gamma^0 = \sigma^0 \frac{A_{flat}}{A_{slope}} \left(\frac{\cos\,\theta_{ref}}{\cos\,\theta_{loc}}\right) \qquad [5.14]$$

where:

σ^0: radar backscatter coefficient ;

A_{flat} : size of the pixel of the radar image for theoretically flat terrain;

A_{slope} : real size of the pixel of the radar image for mountainous terrain;

θ_{ref}: radar incidence angle at centre of the image;

θ_{loc}: local incidence angle.

Then, for each pixel of the radar image, the radar acquisition geometry values (look angle, azimuth and radar-target distance), are corrected with these two output maps calculated from the DEM (slope, aspect) by trigonometric relation and bilinear interpolation. This method has been applied for 15 years and is put forward by “open source” toolboxes (i.e. ESA Toolbox). Another more sophisticated, up-to-date methodology on the other hand consists of reprojecting the DEM values into the original topology of the radar acquisition, using cubic interpolation. The aim is to integrate the local variations from the “distorted” DEM into the Doppler range geometry of the radar [SMA 11]. This enhanced protocol improves the performance of the previous method for radiometric terrain correction in order to compare different tracks or sensor acquisitions. Lastly, the speckle effect leads to a random granular “noise”, and damages the backscattering coefficient calculation, causing difficulties for image interpretation. Given its multiplying geometric effect, speckle will have a greater impact on slopes illuminated by the signal than those not facing it [POL 97]. Thus, it is important in rugged terrain to correct it using conventional algorithmic boxcar windows (Lee, Gamma) following the previous geometric and radiometric steps.

5.3.3.2. *Forested areas*

A large part of the total snow-covered area can be found in boreal regions. Boreal forests present a significant challenge because, depending on the density of the forest, conifers hide the snow cover which completely diffuses the radar signal. For open forest, the canopy makes an additional contribution to the volume scatter (because of the branches) or the double rebound scatter (tree-soil). Magagi *et al.* [MAG 02a] were able to model the effect of the forest canopy of black spruces on the backscattered signal in C-band. The contribution of the canopy on the signal remains low for an incidence angle of 20°, even for a mature forest where the density varies from 0.10 to 0.24 per meter squared (43% of coverage). The contribution of the soil or that of the snow cover (dry or wet) on the total signal in winter decreases as the incidence angle increases between 20° and 50°, while the contribution of the vegetation and the tree-soil interaction increases [MAG 02b]. Where the surface is rough, wet snow is harder to distinguish than with wet snow where the surface is smooth. These authors recommend incidence angles lower than 35° when mapping snow in forested areas.

Beginning towards the end of the 1990s, Finnish researchers devised methods which make it possible to minimize the impact of forest cover on evaluating the characteristics of snow cover. In the forested region of Savukoski in Finland, Koskinen *et al.* [KOS 10] were able to observe that the backscatter coefficient is higher during the winter when the ground is covered by dry snow. A drop in backscatter can be observed when the snow becomes wet (illustrated in Figure 5.5), due to the signal being absorbed by the snow (increased dielectric loss). There is a further increase in forest backscatter when the snow begins to disappear and certain areas will become uncovered depending on how exposed they are to sunlight. Lastly, the backscatter is maximal when the snow has completely melted and the soil is saturated. In order to map the snow cover area (SCA) using one single polarization (VV or HH), the optimal threshold in boreal forests is -0.4 dB [LUO 09a]. This threshold varies depending on the reference image being used (dry snow). An optimal threshold lower than that observed for high mountainous regions (-3 dB, [5.10]) is mainly due to the forest cover which will reduce the snow's signal, depending on density. From one year to the next, it is possible to use the same reference image. However, it is important that each image is taken in the same conditions (method of acquisition and incidence angle).

In order to improve the classification of the SCA, Luojus *et al.* [LUO 09b] and [KOS 10] were able to validate the so-called TKK method (Helsinki University of Technology), using a form of linear interpolation which uses two reference images. The first of these is an image without any snow cover and the second is an image with 100% wet snow coverage.

$$\mathrm{SCA} = 100\% \ \frac{\sigma^0_{surf} - \sigma^0_{soil,ref}}{\sigma^0_{snow,ref} - \sigma^0_{soil,ref}} \qquad [5.15]$$

where σ^0_{surf} is the surface backscatter observed or estimated (value modeled for the forest), $\sigma^0_{soil,ref}$ is the reference backscatter of the soil without snow, and $\sigma^0_{snow,ref}$ is the reference backscatter for soil completely covered by wet snow.

The contribution of the forest canopy to the total backscatter signal (σ^0_{canopy}) is a source of errors when estimating the SCA. However, as this contribution is independent of the quantity of snow on the ground, the effect of the canopy can be reduced by using a compensation method based on a semi-empirical backscattering model [LUO 09b]. The model estimates the backscatter signal of the forest for a given incidence angle by using the density of the forest (in meters cubed per hectare) as a variable input [PUL 99, PUL 01]. This contribution (σ^0_{canopy}) is subtracted from the total backscatter coefficient (σ^0_{total}) resulting in a calibrated value of surface backscatter (σ^0_{surf}) in order to estimate the percentage of snow coverage using equation [5.15]. This interpolation is based on the hypothesis that the observed or estimated backscatter coefficient (σ^0_{surf}) is a linear combination of the backscatter of a surface covered in wet snow and the backscatter of a non-snow covered surface.

The SCA calculation can be carried out pixel by pixel. However, because of the speckle effect, it is often preferable to estimate the SCA by groups of pixels representing forested areas of consistent density. The compensation of the effect of the forest makes it possible to obtain higher percentages of snow cover area in forests than in clear areas (tree-felling locations or farmland) which correspond to the natural process of snowmelt [PUL 01].

5.3.3.3. *Subarctic environments*

Arctic and subarctic environments have undergone significant changes over the course of the past few decades. Warming observed in these regions leads to various changes in terms of ecosystems and affects the natural environment as much as the human populations of these regions. Snow cover plays an important role in these environments and mapping it using SAR sensors poses particular challenges. The accumulation of snow is influenced by a variety of factors including the direction of dominant winds, the topography and the presence of shrub vegetation. As a consequence, snow will tend to accumulate in thickets of shrubs in these environments, which makes it more difficult to isolate the signal backscattered by the snow cover. We are able to add the volume scattering caused by the snow to the volume scattering caused by the branches of shrubs which can be found within the snow cover [DUG 12, DUG 15]. Figure 5.10 shows the effect of snow in the presence of shrub vegetation on polarimetric images in the C-band. The illustrations on the left-hand side are the pseudo channels of entropy and the angle α taken from a RADARSAT-2 image on 22nd March 2010 with a dry snow cover. The illustrations on the right-hand side are the pseudo channels taken from a RADARSAT-2 image taken on 9th May 2010 shortly after almost all of the snow cover had melted. The black lines on the images roughly correspond to the visual field of two photographs taken on the ground on the same days. The entropy and the angle α are lower in winter when dry snow is present, which implies that snow reduces the volume scattering created by the shrubbery.

Figure 5.10. *Effect of snow cover on the entropy and the angle α in a subarctic environment. The column on the left shows the pseudo channels taken from a RADARSAT-2 image (©MDA 2010) taken on 22nd March 2010 with dry snow present and the column on the right shows pseudo channels taken from a RADARSAT-2 image (©MDA 2010) taken on 9th May 2010 shortly after almost all of the snow cover had melted. The black lines on the image roughly correspond to the visual field of two photographs taken on the same days. For a color version of this figure, see www.iste.co.uk/baghdadi/4.zip*

5.4. Current users and future prospects

The mapping and temporal monitoring of snow are of great interest for energy production companies (water supply), national meteorological services (climate) and scientific research institutions. Several European research programs listed the needs of users (i.e. EnviSnow, CryoLand), and the results indicate that the regular mapping of snow cover makes up 80% of the priority attempts formulated by users, instead of other and more sophisticated products such as SWE or snow albedo. Despite the constraints of cloud cover, products derived from optical imagery (AVHRR, MODIS) are favored in pre-operational processes for public and industrial users: longer archive, proven protocols, daily repetitivity of acquisition, and free access to the databases. There are many examples: Canada (Hydro Quebec, BC Hydro), France (EDF), Switzerland (Météo), Finland, USA, Austria; sometimes used in combination with passive microwaves (GlobSnow/ESA database). This choice is reinforced by the arrival of the ESA's new optical missions, Sentinel-2 (2A launched in June 2015 and 2B planned for mid-2016) and Sentinel-3 (3A launched in February 2016). Active radar is currently still in the domain of research and development, applied for targeted operations. However, significant advances have been made in Norway (NORUT) and Austria (ENVEO) where SAR imaging is used with industrial operators (hydroelectricity) in order to compensate for the lack of optical data due to cloud cover in the mountains during spring snowmelt. Lastly, one compromise is to use the multi-sensor approach by generating fusion images [SOL 08] between optics (reflectance) and SAR (backscattering) or to focus on the snow cover fraction (SCF) at the sub-pixel level (percentage) [MAL 02].

The current implementation of new space programs with SAR sensors will strengthen users' interest towards this domain: high spatial resolution, high temporal relativity but also accessibility to free data, like the European Copernicus Program with the Sentinel-1 satellites (C-band, 10 m, daily registration at high latitudes, five days at a latitude of 45°). Sentinel-1A has been in orbit since April 2014 and its twin 1B since April 2016. The RADARSAT Constellation Mission (RCM) is also a promising program following the current Radarsat-2 mission (C-band, 3 satellites, compact polarimetry). All these new SAR sensors and missions will make it possible to improve the operational use of radar images as assimilation data for hydrological and climate modeling.

5.5. Key points

– On a radar image, dry snow cover is difficult to distinguish from surfaces without snow, unless at the lowest wavelengths (λ< 2 cm). Indeed, the backscattering signal coming from the volume of dry snow cover is zero at the L-band (λ=26 cm), very low at the C-band (λ=5cm), existent at the X-band (λ=3cm) and significant at the Ku-band (λ=1.5cm).

– However, there is a method which makes it possible to directly obtain data on the water equivalent of snow cover in C-band and in HH or VV polarization in farmland or in taiga (open coniferous forest, with a coverage of between 25% and 40%). The additional information provided by C-band polarimetric data also made it possible to develop a method for mapping the extent of dry snow cover in Alpine regions. Furthermore, recent results using multi polarized data in X-band indicate the potential for mapping snow height using the phase difference between HH and VV polarization. Likewise, several studies have shown the relevance of using double frequency data (X and Ku) in the framework of the program $CoReH_2O$, but there has yet to be a mission put into orbit operating in the Ku-band.

– Wet snow is relatively easy to distinguish from bare surfaces. Simple algorithms make it possible to map the percentage of snow cover during snowmelt, a critical period for the management of dams and forecasting rising water levels. This is applicable to mountainous regions as well as boreal forests when optical images are not available. Finnish researchers have devised approaches which make it possible to minimize the impact of forest cover on the evaluation of snow cover characteristics at the C-band. The precision with which wet snow can be mapped has also been improved using classification methods based on polarimetric decompositions.

– Despite the complexity of radar imaging compared to optical imaging, processing algorithms have become routinely used in commercial or free-to-access software and have become accessible for all users. Accessibility to free radar data, high temporal repetitivity and high spatial resolution will serve to increase the interest of water supply administrators towards this media over the next decade.

5.6. Bibliography

[ALL 06] ALLAIN S., FERRO-FAMIL L., POTTIER E., "A polarimetric classification fompolsar data using SERD/DERD parameters", *EUSAR 2006 – 6th European Conference on Synthetic Aperture Radar*, Dresden, Germany, May 16–18, 2006.

[ARI 11] ARII M., VAN ZYL J.J., KIM Y., "Adaptive Model-Based Decomposition of Polarimetric SAR Covariance Matrices", *IEEE Transactions on Geoscience and Remote Sensing*, vol. 49, pp. 1104–1113, 2011.

[BAG 97] BAGHDADI N., GAUTHIER Y., BERNIER M., "Capability of multitemporal ERS-1 SAR data for wet-snow mapping", *Remote Sensing of Environment*, vol. 60, pp. 174–186, 1997.

[BER 98] BERNIER M., FORTIN J.P., "The potential of times series of C-band SAR data to monitor dry and shallow snow cover", *IEEE Transactions on Geoscience and Remote Sensing*, vol. 36, pp. 226–243, 1998.

[BER 99] BERNIER M., FORTIN J.P., GAUTHIER Y. *et al.*, "Estimation of the water equivalent of a snowpack using spaceborne radar images", *Revue des sciences de l'eau*, vol. 12, pp. 407–423, 1999.

[BER 02] BERNIER M., GAUTHIER Y., BRIAND P. *et al.*, "Radiometric correction of RADARSAT-1 images for mapping the snow water equivalent (SWE) in a mountainous environment", *IEEE International Geoscience and Remote Sensing Symposium* (IGARSS'02), vol. 1, pp. 227–230, 2002.

[BES 15] BESIC N., VASILE G., DEDIEU J.P. *et al.*, Stochastic Approach in Wet Snow Detection Using Multitemporal SAR Data", *IEEE Geoscience and Remote Sensing Letters*, vol. 12, pp. 244–248, 2015.

[CHO 06] CHOKMANI K., BERNIER M., GAUTHIER Y., "Uncertainty Analysis of EQeau, a Remote Sensing Based Model for Snow Water Equivalent Estimation", *International Journal of Remote Sensing*, vol. 27, pp. 4337–4346, 2006.

[CLO 85] CLOUDE S.R., "Target decomposition theorems in radar scattering", *Electronics Letters*, vol. 21, pp. 22–24, 1985.

[CLO 97] CLOUDE S.R., POTTIER E., "An entropy based classification scheme for land applications of polarimetric SAR", *IEEE Transactions on Geoscience and Remote Sensing*, vol. 35, pp. 68–78, 1997.

[COL 83] COLBECK S.C., "Theory of Metamorphism of Dry Snow", *J. Geophys. Res.*, vol. 88, pp. 5475–5482, 1983.

[COR 05] CORBANE C., SOMMA J., BERNIER M. *et al.,* "Estimation of water equivalent of the snow cover in Lebanese mountains by means of RADARSAT-1 images", *Hydrological Sciences Journal,* vol. 50, pp. 355–370, 2005.

[DED 12] DEDIEU J.P., DE FARIAS G.B., CASTAINGS T. *et al.*, "Interpretation of a RADARSAT-2 fully polarimetric time-series for snow cover studies in an Alpine context – first results", *Canadian Journal of Remote Sensing*, vol. 38, pp. 336–351, 2012.

[DED 14] DEDIEU J.P., BESIC N., VASILE G. *et al.*, "Dry snow analysis in alpine regions using RADARSAT-2 full polarimetry data. Comparison with *in situ* measurements", *IEEE International Geoscience and Remote Sensing Symposium* (IGARSS′14), July 13–18, Quebec, Canada, pp. 3658–3661, 2014.

[DIN 94] DING K.H., ZURK L.M., TSANG L., "Pair Distribution-Functions and Attenuation Rates For Sticky Particles In Dense Media", *Journal of Electromagnetic Waves and Applications*, vol. 8, pp. 1585–1604, 1994.

[DU 10] DU J., SHI J., ROTT H., "Comparison between a multi-scattering and multi-layer snow scattering model and its parameterized snow backscattering model", *Remote Sensing of Environment*, vol. 114, pp. 1089–1098, 2010.

[DUG 12] DUGUAY Y., BERNIER M., "The use of RADARSAT-2 and TerraSAR-X data for the evaluation of snow characteristics in subarctic regions", *IEEE International Geoscience and Remote Sensing Symposium (IGARSS′12)*, pp. 3556–3559, 2012.

[DUG 15] DUGUAY Y., BERNIER M., LÉVESQUE E. *et al.*, "Potential of C- and X-Band SAR for Shrub Growth Monitoring in Subarctic Environments", *Remote Sensing*, vol. 7, no. 7, pp. 9410–9430, 2015.

[FER 16a] FERRO-FAMIL L., POTTIER E., "Synthetic Aperture Radar Imagery", in BAGHDADI N., ZRIBI M. (eds), *Microwave Remote Sensing of Land Surfaces*, ISTE Ltd, London and Elsevier Ltd, Oxford, 2016.

[FER 16b] FERRO-FAMIL L., POTTIER E., "Coherent Diversity Mode SAR Imaging: SAR Polarimetry, Interferometry and Tomography", in BAGHDADI N., ZRIBI M. (eds), *Microwave Remote Sensing of Land Surfaces*, ISTE Ltd, London and Elsevier Ltd, Oxford, 2016.

[FIE 09] FIERZ C., ARMSTRONG R.L., DURAND Y. *et al.*, The international classification for seasonal snow on the ground, UNESCO, Paris, 2009.

[FRE 98] FREEMAN A., DURDEN S.L., "A three component scattering model for polarimetric SAR data", *IEEE Transactions on Geoscience and Remote Sensing*, vol. 36, pp. 963–973, 1998.

[FUN 94] FUNG A.K., *Microwave Scattering and emission models and their applications*, Artech House, Norwood, MA, 1994.

[GAL 09] GALLET J.C., DOMINE F., ZENDER C.S. *et al.*, "Measurement of the specific surface area of snow using infrared reflectance in an integrating sphere at 1310 and 1550 nm", *The Cryosphere*, vol. 3, pp. 167–182, 2009.

[GAY 02] GAY M., FILY M., GENTHON C. *et al.*, "Snow grain-size measurements in Antarctica", *Journal Glaciology*, vol. 48, pp. 527–535. 2002.

[GOO 82] GOODRICH L.E., "The influence of snow cover on the ground thermal regime", *Canadian Geotechnical Journal*, vol. 19, pp. 421–432, 1982.

[GRA 81] GRAY D.M., MALE D.H., *Handbook of Snow*, Pergamon Press, Toronto, 1981.

[GUN 96] GUNERIUSSEN T., JOHNSEN H., SAND K., "DEM corrected ERS-1 SAR data for snow monitoring", *International Journal of Remote Sensing*, vol. 17, pp. 181–195, 1996.

[ING 12] INGVANDER S., JOHANSSON C., JANSSON P. *et al.*, "Comparison of digital and manual methods of snow particle size estimation", *Hydrology Research*, vol. 43, pp. 192–202, 2012.

[KEN 98] KENDRA J.R., SARABANDI K., ULABY F.T., "Radar measurements of snow: Experiment and analysis", *IEEE Transactions on Geoscience and Remote Sensing*, vol. 36, pp. 864–879, 1998.

[KOS 10] KOSKINEN J.T., PULLIAINEN J.T., LUOJUS K.P. *et al.*, "Monitoring of snow-cover properties during the spring melting period in forested areas", *IEEE Transactions on Geoscience and Remote Sensing*, vol. 48, pp. 50–58, 2010.

[LEI 14] LEINSS S., PARRELLA G., HAJNSEK I., "Snow height determination by polarimetric phase differences in X-band SAR data", *IEEE Journal of Selected Topics in Applied Earth Observations and Remote Sensing,* vol. 7, pp. 3794–3810, 2014.

[LON 08] LONGÉPÉ N., Apport de l'imagerie SAR satellitaire en bande L et C pour la caractérisation du couvert neigeux, PhD Thesis, University of Rennes 1, 2008.

[LON 09] LONGÉPÉ N., ALLAIN S., FERRO-FAMIL L. *et al.*, "Snowpack characterization in mountainous regions using C-band SAR data and a meteorological model", *IEEE Transactions on Geoscience and Remote Sensing*, vol. 47, pp. 406–418. 2009.

[LUO 09a] LUOJUS K.P., PULLIAINEN J.T., BLASCO CUTRONA A. *et al.*, "Comparison of SAR-based snow-covered area estimation methods for the boreal forest zone", *IEEE Transactions on Geoscience and Remote Sensing Letters*, vol. 6, pp. 403–407, 2009.

[LUO 09b] LUOJUS K.P., PULLIAINEN J.T., METSAMAKI S.J. *et al.*, "Enhanced SAR-based snow-covered area estimation method for boreal forest zone", *IEEE Transactions on Geoscience and Remote Sensing Letters*, vol. 47, pp. 922–935, 2009.

[MAG 02a] MAGAGI R., BERNIER M., BOUCHARD M.C., "Use of ground observations to simulate the seasonal changes in the backscattering coefficient of the subarctic forest", *IEEE Transactions on Geoscience and Remote Sensing*, vol. 40, pp. 281–297, 2002.

[MAG 02b] MAGAGI R., BERNIER M., UNG C.H., "Quantitative analysis of RADARSAT SAR data over a sparse forest canopy", *IEEE Transactions on Geoscience and Remote Sensing*, vol. 40, pp. 1301–1313, 2002.

[MAL 02] MALNES E., GUNERIUSSEN T., "Mapping of snow covered area with Radarsat in Norway", *IEEE International Geoscience and Remote Sensing Symposium* (IGARSS'02), pp. 683–685, 2002.

[MAR 05] MARTINI A., Télédétection d'un couvert neigeux en milieux alpins à partir de données SAR polarimétriques multi-fréquentielles et multi-temporelles, PhD Thesis, University of Rennes 1, p. 178, 2005.

[MAR 06] MARTINI A., FERRO-FAMIL L., POTTIER E. *et al.*, "Dry snow discrimination in alpine areas from multi-frequency and multi-temporal SAR data", *IEEE Proceedings-Radar Sonar and Navigation,* vol. 153, pp. 271–278, 2006.

[MAT 96] MÄTZLER C., "Microwave permittivity of dry snow", *IEEE Transactions on Geoscience and Remote Sensing*, vol. 34, pp. 573–581, 1996.

[MAT 02] MÄTZLER C., "Relation between grain-size and correlation length of snow", *Journal of Glaciology*, vol. 48, pp. 461–466, 2002.

[NAG 00] NAGLER T., ROTT H., "Retrieval of wet snow by means of multitemporal SAR data", *IEEE Transactions on Geoscience and Remote Sensing,* vol. 38, no. 2, pp. 754–765, 2000.

[PAR 13] PARK S.E., YAMAGUCHI Y., SINGH G. *et al.*, "Polarimetric SAR response of snow-covered area observed by multi-temporal ALOS PALSAR fully polarimetric mode", *IEEE Transactions on Geoscience and Remote Sensing*, pp. 1–12, 2013.

[POL 97] POLIDORI L., *Cartographie Radar*, Taylor & Francis, 1997.

[PRU 78] PRUPPACHER H.R., KLETT J.D., *Microphysics of Clouds and Precipitation, Atmospheric and Oceanographic Sciences Library*, Springer Netherlands, Dordrecht, 1978.

[PUL 01] PULLIAINEN J., KOSKINEN J., HALLIKAINEN M., "Compensation of forest canopy effects in the estimation of snow covered area from SAR data", *IEEE International Geoscience and Remote Sensing Symposium* (IGARSS'01), Sydney, Australia, Jul. 9–13, pp. 813–815, 2001.

[PUL 99] PULLIAINEN J., KURVONEN L., HALLIKAINEN M., "Multitemporal behavior of L- and C-band SAR observations of boreal forest", *IEEE Transactions on Geoscience and Remote Sensing*, vol. 37, no. 2, pp. 927–937, 1999.

[RAU 79] RAUDKIVI A.J., *Hydrology: An Advanced Introduction to Hydrological Processes and Modelling*, Pergamon Press, 1979.

[ROT 84] ROTT H., "The analysis of backscattering properties from SAR data of mountain regions", *IEEE Journal of Oceanic Engineering*, vol. 9, no. 5, pp. 347–355, 1984.

[ROT 10] ROTT H., YUEH S.H., CLINE D.W. *et al.*, "Cold Regions Hydrology High-Resolution Observatory For Snow And Cold Land Processes", *Proceedings of the IEEE*, vol. 98, pp. 725–765, 2010.

[ROT 14] ROTT H., NAGLER T., RIPPER E. *et al.*, "Ku- and X-Band Backscatter Analysis and SWE Retrieval for Alpine Snow", *IEEE International Geoscience and Remote Sensing Symposium (IGARSS'2014)*, pp. 2407–2410, 2014.

[SHI 94] SHI J.C., DOZIER J., ROTT H., "Snow Mapping in Alpine Regions with Synthetic-Aperture Radar", *IEEE Transactions on Geoscience and Remote Sensing*, vol. 32, pp. 152–158, 1994.

[SHI 00a] SHI J.C., DOZIER J., "Estimation of Snow Water Equivalence Using SIR-C/X-SAR, Part II: Inferring Snow Depth and Particle Size", *IEEE Transactions on Geoscience and Remote Sensing*, vol. 38, pp. 2475–2488, 2000.

[SHI 00b] SHI J.C., DOZIER J., "Estimation of Snow Water Equivalence using SIR-C/X-SAR, Part I: Inferring Snow Density and Subsurface Properties", *IEEE Transactions on Geoscience and Remote Sensing*, vol. 38, pp. 2465–2474, 2000.

[SIN 14] SINGH, G., VENKATARAMAN G., YAMAGUCHI Y. *et al.*, "Capability Assessment of Fully Polarimetric ALOS-PALSAR Data for Discriminating Wet Snow From Other Scattering Types in Mountainous Regions", *IEEE Transactions on Geoscience and Remote Sensing*, vol. 52, pp. 1177–1196, 2014.

[SMA 11] SMALL D., "Flattening Gamma: Radiometric Terrain Correction for SAR Imagery", *IEEE Transactions on Geoscience and Remote Sensing*, vol. 49, no. 8, pp. 3081–3093, 2011.

[SOL 08] SOLBERG R., HUSEBY R.B., KOREN H. *et al.*, "Time-Series Fusion of Optical and SAR Data for Snow Cover Area Mapping", *EARSeL Land Ice and Snow Special Interest Group Workshop*, Bern, Switzerland, Feb. 11–13, 2008.

[STO 84] STOGRYN A., "The bilocal approximation for the effective dielectric constant of an isotropic random medium", *IEEE Transactions on Antennas and Propagation*, vol. 32, pp. 517–520, 1984.

[STU 95] STURM M., HOLMGREN, J. LISTON G.E., "A Seasonal Snow Cover Classification System for Local to Global Applications", *Journal of Climate*, vol. 8, no. 5, pp. 1261–1283, 1995.

[STU 97] STURM M., HOLMGREN J., KÖNIG M. *et al.*, "The thermal conductivity of seasonal snow", *J. Glaciology*, vol. 43, pp. 26–41, 1997.

[SUN 15] SUN S., CHE T., WANG J. *et al.,* "Estimation and Analysis of Snow Water Equivalents Based on C-Band SAR Data and Field Measurements", *Arctic, Antarctic and Alpine Research*, vol. 47, pp. 313–326, 2015.

[TED 06] TEDESCO M., KIM E.J., "Intercomparison of Electromagnetic Models for Passive Microwave Remote Sensing of Snow", *IEEE Transactions on Geoscience and Remote Sensing*, vol. 44, pp. 2654–2666, 2006.

[TRU 09] TRUDEL M., MAGAGI R., GRANBERG H.B., "Application of Target Decomposition Theorems Over Snow-Covered Forested Areas", *IEEE Transactions on Geoscience and Remote Sensing,* vol. 47, pp. 508–512, doi:10.1109/tgrs.2008.2009122, 2009.

[TSA 80] TSANG L., KONG J.A., "Multiple scattering of electromagnetic waves by random distributions of discrete scatterers with coherent potential and quantum mechanical formalism", *Journal of Applied Physics*, vol. 51, pp. 3465–3485, 1980.

[TSA 81] TSANG L., KUBACSI M.C., KONG J.A., "Radiative transfer theory for active remote sensing of a layer of small ellipsoidal scatterers", *Radio Sci.*, vol. 16, pp. 321–329, 1981.

[TSA 07] TSANG L., PAN J., LIANG D. *et al.*, "Modeling Active Microwave Remote Sensing of Snow Using Dense Media Radiative Transfer (DMRT) Theory with Multiple-Scattering Effects", *IEEE Transactions on Geoscience and Remote Sensing*, vol. 45, pp. 990–1004, 2007.

[ULA 86] ULABY F.T., MOORE R.K., FUNG A.K., *Microwave Remote Sensing: Active and Passive*, Addison-Wesley, Reading, MA, 1986.

[ULA 96] ULANDER L.M.H., "Radiometric Slope Correction of Synthetic-Aperture Radar Images", *IEEE Transactions on Geoscience and Remote Sensing*, vol. 34, pp. 1115–1122, 1996.

[VAN 93] VAN ZYL J.J., CHAPMAN B.D., DUBOIS P. *et al.*, "The effect of topography on SAR calibration", *IEEE Transactions on Geoscience and Remote Sensing*, vol. 31, no. 5, pp. 1036–1043, 1993.

[WEN 90] WEN B., TSANG L., WINEBRENNER D.P. *et al.*, "Dense medium radiative transfer theory: Comparison with experiment and application to microwave remote sensing and polarimetry", *IEEE Transactions on Geoscience and Remote Sensing*, vol. 28, pp. 46–59, 1990.

[YAM 05] YAMAGUCHI Y., MORIYAMA T., ISHIDO M. *et al.*, "Four-component scattering model for polarimetric SAR image decomposition", *IEEE Transactions on Geoscience and Remote Sensing*, vol. 43, pp. 1699–1706, 2005.

[YUE 09] YUEH S.H., DINARDO S.J., AKGIRAY A. *et al.*, "Airborne Ku-Band Polarimetric Radar Remote Sensing of Terrestrial Snow Cover", *IEEE Transactions on Geoscience and Remote Sensing*, vol. 47, pp. 3347–3364, 2009.

6

Spatial Altimetry and Continental Waters

6.1. Introduction

Continental waters have a major impact on life on Earth as well as domestic, industrial and agricultural needs, but they also play an important role in climate variability. Nearly 73% of the worldwide demand for water is met by surface reservoirs (rivers, lakes and artificial reservoirs) and 19% is met by underground reservoirs. The rest is provided either through the treatment of wastewater or the desalination of seawater. From a societal point of view, it is highly important to estimate the changes which affect the continental water cycle. Being able to describe it in even greater detail makes it possible to make more accurate climate predictions and to refine the way in which the planet's water supply is controlled.

However, the situation regarding ground networks for monitoring the level of lakes and rivers varies greatly depending on where you are in the world. In many regions, access to these networks is limited while the political and economic situations in the countries where these lakes and rivers are located mean that they are not able to maintain existing systems or do not possess reliable networks of durable and reliable measurements.

Chapter written by Jean-François CRETAUX and Stéphane CALMANT.

For the past twenty or so years, spatial techniques have been used to study variations in water mass in large river basins over several years. The advantages of satellite systems are widely known:

– near global coverage,

– spatial and temporal coherence of data,

– archives of past data,

– guarantee of the durability and continuity of data, irrespective of the technical advances in the systems used.

A measurement carried out *in situ* in optimal conditions will intrinsically be preferable to an altimetric measurement, either in terms of precision or temporal sampling, or because a large number of lakes and rivers, if not the majority, are underequipped. This is why satellite altimetry is an attractive technique for monitoring the water levels of lakes and rivers. Essentially, this technique has benefited from a continuous service since the launch of Topex/Poseidon in 1992, and this will continue in the years to come with various new missions. Furthermore, satellite measurement is, by definition, global, given that the Earth is covered in a few days when even the spatial sampling of a mission can appear reductive. In cohabitation with several satellites, the possibilities for coverage are greatly increased. For larger lakes, precision is to within a few centimeters but for smaller lakes, reservoirs or rivers, it ranges from decametric to metric. Since the 1990s, radar altimetry (Topex/Poseidon (T/P), Jason 1&2, ERS 1&2, ENVISAT, GFO, Saral/AltiKa) or laser altimetry (ICESat) has made it possible to obtain a key parameter in continental hydrology: the water level in lakes, artificial reservoirs and rivers. Used in collaboration with other satellite techniques such as imaging (optic or radar), altimetry provides further access to volumetric data for lakes and reservoirs [GAO 12, DUA 13, ZHU 14, CRE 15], and when combined with *in situ* data and/or models, it is also possible to calculate river flow [KOU 04, ZAK 06, PAP 10, GET 12, BIR 14, PAR 15]. Measurable quantities such as level and supply of water and flow are also considered as essential climate variables by the World Meteorological Organization (WMO) and the Global Climate Observing System (GCOS) and make it possible to study the water level in drainage basins as well as links with regional climate and forcing. Data from spatial measurements such as altimetry can in turn be used to force regional or global hydrological models. In summary, using a combination of *in situ* observations of certain hydrological parameters and modeling should make it possible to address a few key questions linked to the continental water cycle, as well as water management more generally:

– small scale physical processes;

– specific problems with the hydrodynamics of river basins;

– questions relating to the impact of climate change on water supply.

Since the middle of the 1990s, there has been a noticeable rise in the use of radar and laser altimetry for continental hydrology. It is important to note that this technique was not designed for these uses, but was designed for oceanography (T/P and Jason) and then adapted for use in monitoring polar ice caps (ERS-1&2, ENVISAT, Cryosat-2). However, it very quickly became clear that it could provide considerable benefits in the study of continental waters.

Naturally, it is impossible to be exhaustive given the exponential increase in the number of applications, as evidenced in various scientific publications and international projects. We will pick out some of the leading studies which seem to highlight the irrefutable potential of this technique and which, as we will see, have led to new missions geared toward hydrology.

Indeed, various spatial, experimental or operational missions which could prove highly useful for hydrology are in preparation for the next few years. In radar altimetry, Ka band altimetry, Saral/AltiKa was launched in February 2013. 2016 will see the launch of the mission Jason-3 (T/P and Jason satellites) then in 2016/2017 the European Space Agency's Sentinel 3 A and Sentinel-3 B (SAR mode altimeters) will be launched followed by the Jason-CS/Sentinel-6A/6B mission in 2018. A laser altimetry mission (ICESat-2) is expected to be launched by NASA and the DLR (the German Aerospace Center) in 2016, following the success of the ICESat-1, which was able to carry out precise measurements of continental water levels (Table 6.1)

Finally, the Franco-American SWOT is expected to be launched in 2020. It will carry on board an interferometer in Ka band which will make it possible to create 3D maps of continental water [ROD 15]. Any river larger than 100 m will be mapped twice every 21 days (the orbital cycle of the SWOT satellite) and the international community will have free access to information on the height, slope and discharge of these rivers. For every 21-day cycle, a complete inventory of the lakes and bodies of water larger than 250 m by 250 m (6 ha) will be established along with the respective heights. The process of dynamics and the interaction between rivers and floodplains will be studied and the processes analyzed below (exchanges between the surface and the atmosphere, the role played in fluctuations by CO_2, exchanges with underground layers, the part played by floodplains, rivers and lakes in the water cycle and the contribution to the ocean). Table 6.1 summarizes previous, current and future missions. We will consider the scientific objectives of the SWOT mission in section 6.6, "Conclusions and prospects", of this chapter.

Mission	Agency	Frequency	Mode	Duration	Cycle (days)
T/P	NASA/CNES	Ku/C	LRM	1992-2005	10
Jason-1	NASA/CNES	Ku/C	LRM	2002-2011	10
Jason-2	NASA/CNES	Ku/C	LRM	2008- present	10
Jason-3	NASA/CNES/ EUMETSAT/ NOAA	Ku/C	LRM	2015	10
Jason-CS Sentinel-6A/B	NASA/CNES/ EUMETSAT/ NOAA/ESA	Ku/C	LRM & SAR simultaneously	2020 and 2026	10
GFO	NRL	Ku	LRM	2002-2008	17
ERS-2	ESA	Ku	LRM	1995-2002	35
ENVISAT	ESA	Ku/S	LRM	2002-2011	35
Saral/Altika	CNES/ISRO	Ka	LRM	2013- present	35
Sentinel-3A	ESA	Ku/C	SAR	2016	27
Sentinel-3B	ESA	Ku/C	SAR	2017	27
CryoSat-2	ESA	Ku	LRM/SAR/SARin exclusively	2010- present	369
Icesat-1	NASA	Laser	LRM	2003-2009	91
Icesat-2	NASA	Laser	LRM	2017	91
SWOT	NASA/CNES/ CSA/UKSA	Ka	LRM/SAR/SARin	2020	21

Table 6.1. *List showing past, current and future missions. Low Resolution Mode corresponds to "classic" nadir altimetry, Synthetic Aperture Radar corresponds to a burst mode using high frequency pulses to slice the ground footprint into narrow bands to increase the ground resolution, and SARin (Interferometric SAR) is a more precise interferometric mode which makes it possible to precisely localize the reflecting point in the imprint of the radar on the ground (see [CAL 16])*

6.2. Some generalities concerning the use of satellite altimetry for hydrology

The main purpose of altimetric satellites is to measure the distance between satellites and sea level (or the level of the lake or river where required) below. For that, they are equipped with a bi-frequency altimeter, an orbitography system and a microwave radiometer (for measuring the delay of propagation in the troposphere caused by water vapor). The nadir altimeter will emit pulses towards the nadir and will calculate the time taken to receive the echo bounced by the ground. The distance between the satellite and the reflective surface can thus be calculated and subtracted from the position of the satellite in the viewing direction measured by the orbitography system and we are able to deduce the height of the reflective surface with respect to a reference surface (ellipsoid). We will not go into detail here on processing and calibration but instead will refer to [CAL 16] as well as the following articles: [BIR 95, CRE 06, CAL 08, GAO 12].

There are several factors which explain why altimetry has become more frequently used for studying the water cycle – both for drainage basins and on a wider scale:

– the vast number of lakes, reservoirs and rivers on Earth makes it practically impossible to monitor them extensively on the ground;

– in 20 years, radar altimetry has gone from the experimental stage with T/P to an operational level with missions such as Jason-3 (launched in January 2016) or Sentinel-3A/3B (2016-2017) and is now considered an indispensable tool in oceanography;

– indirectly, hydrology benefits from the continuous service provided by space agencies which makes it possible to permanently maintain a mini constellation of satellites: there have never been fewer than two altimetric satellites in simultaneous operation since the launch of T/P in 1992;

– even if the precision of altimetric radar for lakes and rivers is not at the same level as that for studying oceans, it is generally sufficient to be able to use for monitoring water surfaces. This is equally true for scientific but also operational objectives [RIC 12, BIA 11, HOS 14];

– for larger lakes (lakes which cover an area of more than 100 km^2) it has been shown accuracy is possible within 4 or 5 cm [CRE 09, RIC 12], while for narrow reservoirs (a few square kilometers) or smaller lakes this accuracy is somewhere in the region of between 10 cm and 1 m [DUA 13]. However, this precision is still useful for a wide range of targets (smaller lakes or reservoirs, rivers in inaccessible areas) because these are often the only measurements available. Moreover, the different variations in water level are considerably higher than the degree of uncertainty.

For this combination of applications of altimetry on continental water, we have been able to compile a certain number of results, published or recalculated for this chapter, which try to answer a recurring question relating to the accuracy of altimetric measurements for hydrology. Is there a minimum size for calculating water level with sufficient accuracy? (Here, accuracy refers to a comparison with an external source, such as *in situ* water level). The answer to this question is complicated because we have seen in various studies and comparisons that there is no absolute rule in this respect. However, the results shown in Table 6.2 make it possible to note that the size of a lake has an impact on the quality of the results, which are also dependant on the environment around the lake (mountain, forest, the presence of snow in winter). Nevertheless, we see in this table that large lakes like Issykkul or North American lakes are very good candidates for proving that satellite altimetry is highly efficient (accurate to within a few centimeters) while for large, very narrow reservoirs with serious issues associated with drying up, such as Lake Mead or Lake Powell, accuracy is to within several decimeters.

Lake name	Continent	Surface (km^2)	Duration	RMSE (cm)	R^2
Argentino[1]	South America	1466	1992–2011	22	0.96
Athabasca[2]	North America	7900	1992–2009	28	0.91
Aydarkul[3]	Asia	3000	2002–2010	12	0.95
Baikal[4]	Asia	31500	1992–2009	11	0.94
Bratsk[4]	Asia	3100	1992–2009	41	0.98
Erie[2]	North America	25800	1992–2011	10	0.95
Guri[2]	South America	3500	2002–2010	82	0.99
General Carrera[1]	South America	1850	2002–2010	22	0.90

Huron[2]	North America	59570	1992–2011	8	0.99
Issykkul[6]	Asia	6000	2002–2014	3	0.99
Khanka[4]	Asia	4400	2000–2010	13	0.60
Ladoga[4]	Europe	18135	1992–2010	8	0.98
Mead[5]	North America	350	2001–2010	64	0.99
Michigan[2]	North America	58000	1992–2011	11	0.98
Oahe[6]	North America	700	2002–2010	45	0.99
Onega[4]	Europe	18200	1992–2010	15	0.88
Ontario[2]	North America	19000	1992–2011	6	0.98
Powell[5]	North America	400	1992–2010	85	0.99
Superior[2]	North America	82200	1992–2011	6	0.97
Tana[2]	Africa	3000	1992–2006	17	0.97
Tchad[2]	Africa	1540	1992–2008	28	0.91
Titicaca[7]	South America	7800	2000–2005	7	0.99
Volta[2]	Africa	8500	1999–2010	53	0.98
Woods[2]	North America	4350	1992–2011	27	0.81

1 Subsecretaría de Recursos Hídricos (www.hidricosargentina.gov.ar)
2 [RIC 12]
3 www.cawater-info.net
4 SHI: State Institute of Hydrology of St Petersburg, Russia
5 US Bureau of Reclamation
6 IWPB: Institute of Water Problem of Bishkek, Kyrgyzstan
7 Personnel communication of J.L. Guyot, IRD (source: Institute of Hydrology of Peru)

Table 6.2. *Comparison (RMSE and correlation) between sets of in situ data collected by a range of organizations throughout the world and altimetric data for these lakes (source: Hydroweb, Legos)*

In a recent study, Arsen *et al.* [ARS 15] have compared the performances of the Saral/AltiKa (operating in Ka band) and the ENVISAT/RA2 (operating in Ku and S bands) missions on mountain lakes. In order to do this, the results were not analyzed according to the size of the lakes but according to the length of the trace on the lake. *In situ* data were collected from around fifteen Andean lakes and were compared to data from Saral/AltiKa and data from ENVISAT. This study provided enlightening conclusions and these can be summarized in four points:

1) there is a strong connection between the length of traces cutting a lake and the accuracy observed when ENVISAT measurements are used. Using this satellite in a mountainous region makes it possible to obtain sub-metric accuracy but only for large transverse sections;

2) the opposite is true for Saral/AltiKa data where the quality does not depend on the transverse length;

3) a sizeable improvement could be observed in the Saral/AltiKa results compared to ENVISAT: on average, the accuracy with Saral/AltiKa is somewhere in the region of 10 cm while it is somewhere in the region of a meter with ENVISAT for a few very large sections;

4) for sections smaller than 3 km, no AltiKa measurements were recorded, which was not the case with ENVISAT. However, for sections smaller than 3 km the results with this satellite (ENVISAT) were hampered (several meters) and in reality are unusable from a scientific point of view.

Ka band altimetry is actually better adapted to measuring water levels in smaller sections because the pulse footprint on the ground is significantly smaller than the pulse footprint using Ku band. Furthermore, AltiKa measurements are sequenced at 40 Hz (as opposed to 20 Hz for ENVISAT) which increases the resolution of measurements along the trace and makes it possible to carry out a more precise selection of measurement points on the ground.

Before moving on to look at some case studies, it is necessary to point out that the variation of the water level of a lake is not sufficient when looking to compile a hydrological assessment for a given area. Water supply, which is a volume as opposed to a level, is the key parameter which makes it possible to understand the impact of climate.

This is why, today, scientists will try any means possible in order to obtain these variations in water supply. As a result, satellite imaging takes on a considerable importance for those looking to understand any possible links between climate and water supply for a given region.

Combined with measurements of variations in the water levels of lakes taken by satellite altimetry, these measurements enable access to information on changes in water supply, as will be seen in the case study on the Syrdarya reservoirs (Central Asia).

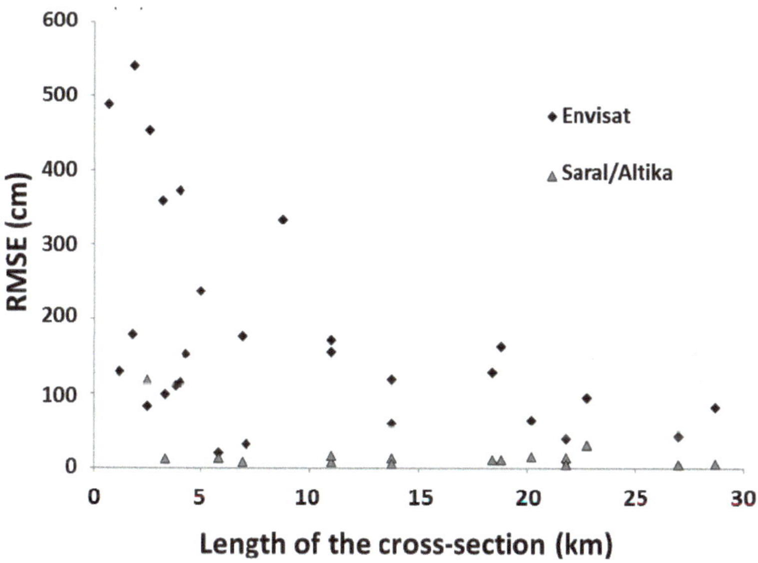

Figure 6.1. *Root Mean Squared Error (RMSE) of the water heights calculated by ENVISAT and Saral/Altika for around 15 lakes in the Andes according to the length of the track section cutting across the lake [ARS 15]*

6.3. Case studies using radar and laser altimetry

6.3.1. *Lakes in East Africa*

Obtaining an understanding of the climate variability in East Africa and the interactions between the Indian Ocean and the larger African lakes has been the subject of various publications due to its importance within the wider framework of studying regional water cycles. Continental precipitation due to evaporation over the ocean has a primary influence on the water balances of large lakes; conversely, the lakes themselves are liable to alter the regional climate. Nicholson and Y in [NIC 02] for example were able to show that precipitation over Lake Victoria is around 30% more intense than on the rest of the drainage basin. In addition, knowing the long-term impact climate has on water levels is an essential piece of information for administrators in this region [STA 05]. In fact, through the use of *in situ* data on water levels, scientists were able to demonstrate that regional climate variability has a significant impact on fish production [PLI 01, COH 06].

Given the timescale over which these effects take place, from interannual to long-term, it is essential to have access to data covering a similar period such as the water levels in large African lakes. However, the scarcity of *in situ* instruments or historical data as well as a lack of climate archives curbs the development of a more precise overview of past climate [NIC 96, VER 03]. Moreover, only a few large lakes such as Victoria or Tanganyika have been measured over several years, which significantly limits studies on the mechanisms which control the water balance of these lakes. As we can see in Figure 6.2 there are a large number of lakes of various sizes in this region which should make it possible to use them as a marker for the water cycle in East Africa and long-term interannual variability. The measurement of the main elements of the water balance of large lakes (precipitation and evaporation) remains very poor and this makes it difficult to analyze how they connect to the regional variability of the climate [NIC 02]. A sizeable proportion of these lakes have however been monitored for the past twenty years or so thanks to satellite altimetry (T/P, Jason-1, Jason-2, ENVISAT, GFO, Saral/Altika).

Several recent studies have attempted to highlight this data in order to draw conclusions on what links changes in the water levels of these lakes and the regional climate. Mercier *et al.* for example [MER 02] used seven years' worth of data from the satellite T/P as well as measurements of precipitation in East Africa and over the Indian Ocean and were able to show evidence of a significant correlation between variations in the ocean temperature and variations in the water levels of large lakes such as Victoria, Tanganyika or Nyassa-Malawi. In particular, there was evidence of synchronicity between the water levels of these lakes and precipitation over the Indian Ocean which reinforces the notion of large-scale phenomena (e.g. El Ni No: [MIS 03]) as explained in various studies in the past [OGA 88, NIC 96, ROP 96]. The role played by the Indian Ocean on water levels in large African lakes was also connected to the "Indian Ocean Dipole" (IOD) [BEL 90, HAS 93, RIC 94] which is a type of oceanic circulation linked to the atmospheric circulation around the equator [SAJ 99, BER 04, MAR 06].

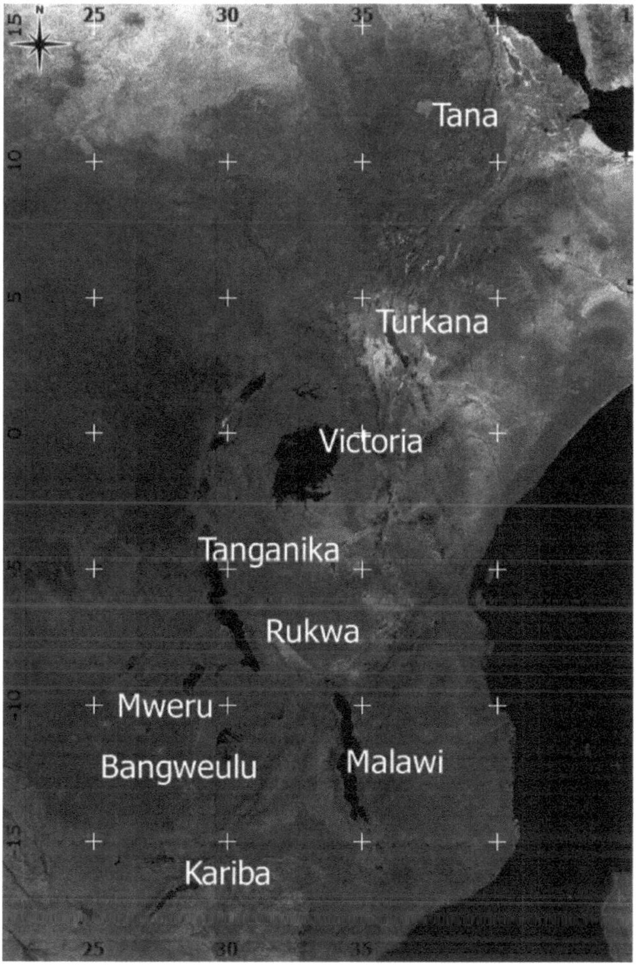

Figure 6.2. *Map showing the nine great lakes of East Africa used in Becker et al. [BEC 10]*

Recently, a study aimed at monitoring the Congo basin using satellite altimetry [BEC 14] was also able to demonstrate that the two large African lakes situated within the river's drainage basin (Tanganyika and Mweru) showed a temporal variability which correlated strongly with the anomalies in variations in water levels in the Congo's eastern section (an area called Luabala, close to Tanganyika and Mweru). An explanation for this can be found in another study published by Becker *et al.* [BEC 10] which aimed to monitor the great lakes in East Africa in relation to climate variability in the

Indian Ocean (the nine lakes indicated in Figure 6.2). In this study, variations in water supply were measured between 2002 and 2008 combining data from the spatial gravimetry mission GRACE and satellite altimetry data on African lakes. The variability of water storage measured by GRACE and for the lakes measured by altimetry showed evidence of a strong correlation with the variability of precipitations in the region. They were able to demonstrate that the major events (the minimum of 2005 for example, or the maximum in 2006–2007) were caused by the IOD on precipitations in this region as well as variations directly linked to the El-Ni no Southern Oscillation (ENSO) cycles. Figure 6.3 shows the variations in water levels for four large African lakes, and it is possible to note that the trends observed are very similar (strong increase in levels in 1998 as shown in [MER 02]). From 2003 to 2005 we notice a significant drop, followed by another increase. According to Becker *et al.* [BEC 10], this temporal variability is linked to the IOD.

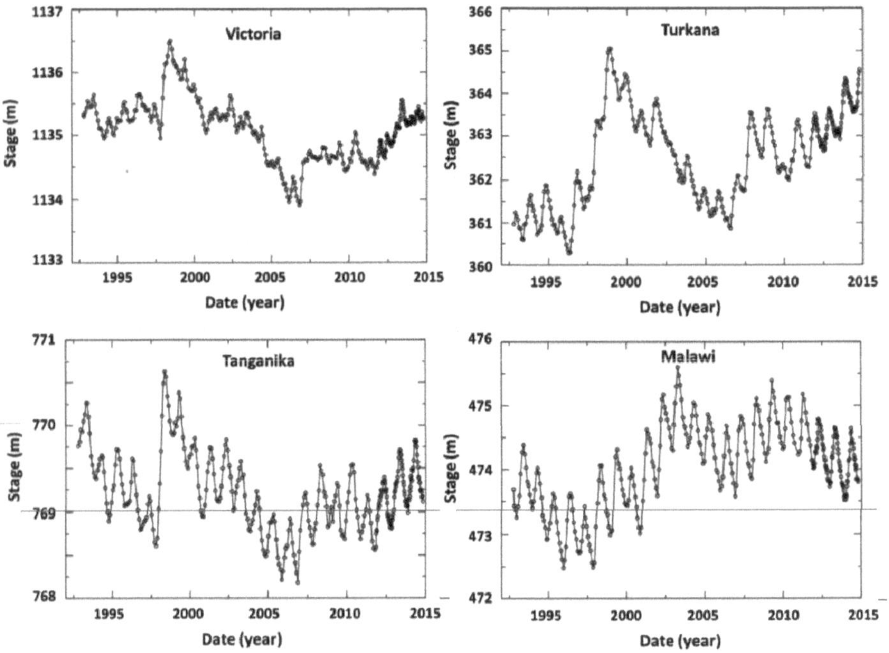

Figure 6.3. *Variations in the water levels of four large African lakes (see Figure 6.2) measured by radar altimetry using a combination of several satellites (T/P, Jason-1/2, ENVISAT and Saral/Altika)*

In another study [SWE 09] on African lakes, radar altimetry was used in conjunction with other kinds of satellite data to obtain the water balance of these lakes and to find out how this is connected to climate variability. They were able to establish that, unlike Lake Malawi or Tanganyika, where variations in water level are entirely natural and linked to regional climate variability and the Indian Ocean, Lake Victoria, whose water levels are controlled by a dyke, is only dependent on variations in levels of precipitation for around 50% of its water. Again using satellite altimetry, they were able to demonstrate that Lake Victoria saw its water level decrease roughly 40% faster than the two other lakes between 2002 and 2008. By using altimetric data for Lake Kyoga, directly downstream from and fed by Lake Victoria, they were able to conclude that only 50% of the variations in Lake Victoria's water levels could be explained by natural causes, with the other half due to waste in Lake Kyoga. Lake Malawi was also the subject of a study focusing on 80 years' worth of water level data (from 1916 to 1995) which made it possible to show evidence of various links with the spatio-temporal variability of precipitations on the drainage basin [JUR 02]. The authors of this study were able to show that the water levels in Lake Malawi, as well as flows in the Zambezi, were strongly correlated and that they displayed an energy spectrum between 2 and 10 years, consistent with the quasi biennial oscillation (QBO), ENSO and the solar cycle. Furthermore, wet years are characterized by an excess of atmospheric humidity coming from the tropical South Atlantic, especially in the Mozambique Channel.

Other phenomena play a significant role in variations in the water level of these lakes. For example, a strong correlation was found with solar activity (potentially strengthened by El Niño) for the majority of large African lakes over the past millennium [STA 05, GAR 07, STA 07]. Indian monsoons or variations in the Inter-Tropical Convergence Zone (ITCZ) can also have an impact [TIE 07].

Lastly, it is important not to overlook the effects of global warming on levels of precipitation and surface temperatures and, therefore, on the water balance of lakes [XU 05, DEW 06].

6.3.2. *Tibetan lakes*

The Tibetan Plateau is considered by climatologists as the Third Pole. It is a high plateau with an average altitude of more than 4,000 m, situated north-east of the Himalayan mountain range. It covers an area of roughly 2.5 million square kilometers and around 70% of its surface is made up of permafrost. Furthermore, it is covered by around 1,500 lakes of sizes ranging from a few hectares to several thousand square kilometers. It is one of the regions of the world most sensitive to climate change, since the average increase in temperature over the last few decades has been more than double the average for the rest of the Earth [WAN 08, LIU 98, LIU 00]. It has also been shown that the plateau itself has a significant impact, not only on the regional climate, but on the global climate, and plays an amplifying role in global warming [DUA 05, ZHO 09, KAN 10].

The climate of the Tibetan Plateau is characterized by a sizeable spatio-temporal variation in precipitation with very arid conditions in the north-west (an average of roughly 50 mm rainfall per year) and much wetter conditions in the south-east (700 mm per year). Furthermore, monsoons from India and Southeast Asia, as well as dominant winds from the west, have an impact on the region. In contrast to the almost consistent increase in the temperature of the Tibetan Plateau, the long-term variations in levels of precipitation display significant regional and temporal disparities. The north-west and the west of the plateau for example have become drier (with evidence of an acceleration over the last 30 years) while the east and the center of the plateau have become wetter [LI 07, KAN 10].

In this context, various scientific studies have highlighted that changes in climate (just like seasonal and interannual variability) have a significant impact on water supply, and in particular on water supply in lakes on the Tibetan Plateau [LIU 09, HUA 11, ZHA 11].

For several years, climatologists have been having a debate on the mechanisms which make it possible to understand the observations carried out on the variability of surface water and most notably on lakes on the Tibetan Plateau. In reality, an increase in temperature produces an increase in evaporation, which can be amplified in air which is becoming drier, but reduced in air which is becoming wetter. This same warming effect accelerates the rate at which glaciers and permafrost melt, and for lakes where the majority of the drainage basin will be covered by this type of surface, this leads to an increase in the water volume of the lakes of the Tibetan Plateau [LI 08, KAN 10]. These studies show that this warming effect is the principal factor affecting variations in the water level of these lakes, largely due to the melting of glaciers in the Himalayas and the Tibetan Plateau [WU 08, YAO 07, PHA 13]. Other studies demonstrate that these variations are more likely to be the result of changeable levels of precipitation [KAN 10, LEI 14]. Meanwhile, other authors are more cautious and maintain that, owing to a lack of *in situ* data and a lack of temporal height in the data, it is not so easy to analyze the causes of variations in the water levels of Tibetan lakes [SON 14].

For several years, scientists requiring reliable observations to substantiate their theories have turned towards spatial techniques [HUA 11, SON 14]. These are particularly essential in an environment like the Tibetan Plateau: large uninhabited areas, difficult to access with very little history of *in situ* measurements and covered by such a vast number of lakes that it is practically impossible to measure them all. Two techniques have been favored by scientists for monitoring Tibetan lakes: altimetry and satellite imaging [ZHA 10, LEE 11, LEI 14, SON 15, KLE 15]. This list is far from exhaustive, owing to the sheer range of studies on Tibetan lakes.

Given the significant spatial variability of precipitation regimes on the Tibetan Plateau, but also the distribution of glaciers, it becomes quite clear that monitoring a reduced number of lakes is insufficient to be able to understand the impact on surface waters. Indeed, because of the considerable number of lakes which cover it, the Tibetan Plateau is an excellent candidate for highlighting the potential of spatial techniques, but also for attempting to explain the interplay of mechanisms play in this region.

However, interpreting data on the variations in water supply is complicated by the fact that the response of a lake to a change in its supply (surface or through precipitation) or to the loss of water (through evaporation or runoff) is also highly dependent on its morphology [MAS 94]. In fact, any change in the water balance of a lake (in other words, when water supplied is equal to water lost) results in a new balance having to be found, which can depend on whether the lake is more or less wide, from several months or years to several decades or even centuries in extreme cases (see examples below). The wide range of situations observed on the Tibetan Plateau makes this region even more interesting, but also more complex, given that certain lakes will react very quickly to the slightest change, whereas others will react in a slower manner.

It is necessary, therefore, to observe as large a number of lakes as possible in order to cover a consistent range of situations that will make it possible to discover the source of the changes observed using ad-hoc models. For this reason, the Tibetan Plateau is an exceptional natural laboratory which makes it possible to benefit greatly from satellite observations. We are also able to understand why it is necessary to maintain spatial systems over several decades. The examples below illustrate these remarks.

Surface variations for roughly 700 lakes on the Tibetan Plateau were calculated for the decade between 2000 and 2010 using data from optical Earth observation satellites (Landsat satellites). During this period, a significant proportion of the lakes (more than 80%) saw an increase in size. It was possible to corroborate and complete these observations using satellite altimetry, which also indicated a distinct increase in water level for the majority of these lakes during the same period [PHA 12, ZHA 11].

However, when we expand the period of observation as much as possible, we notice that this increase in water level in the 2000s is neither constant nor durable. Measurements were carried out on variations in water levels over a period of around 20 years for 11 of the biggest lakes in south-west Tibet using satellite altimetry and taking into account all existing missions (from T/P to Saral/AltiKa) (Figure 6.4). Figure 6.5 shows that variations in these lakes over the last 20 years have behaved very differently.

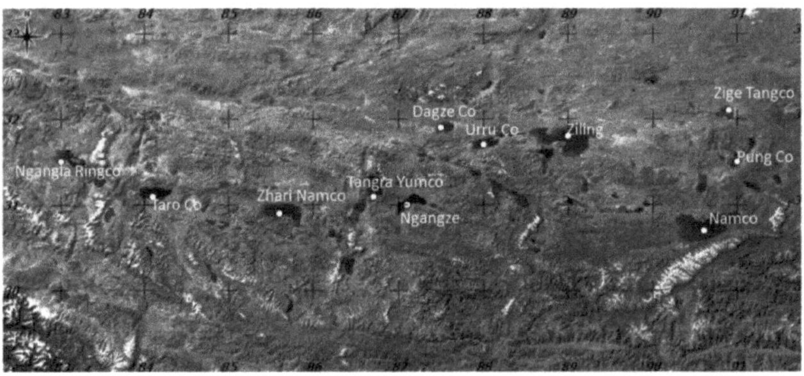

Figure 6.4. *Map of the south-west zone of the Tibetan Plateau showing the lakes for which it was possible to measure variations in levels over a period of between 12 and 20 years*

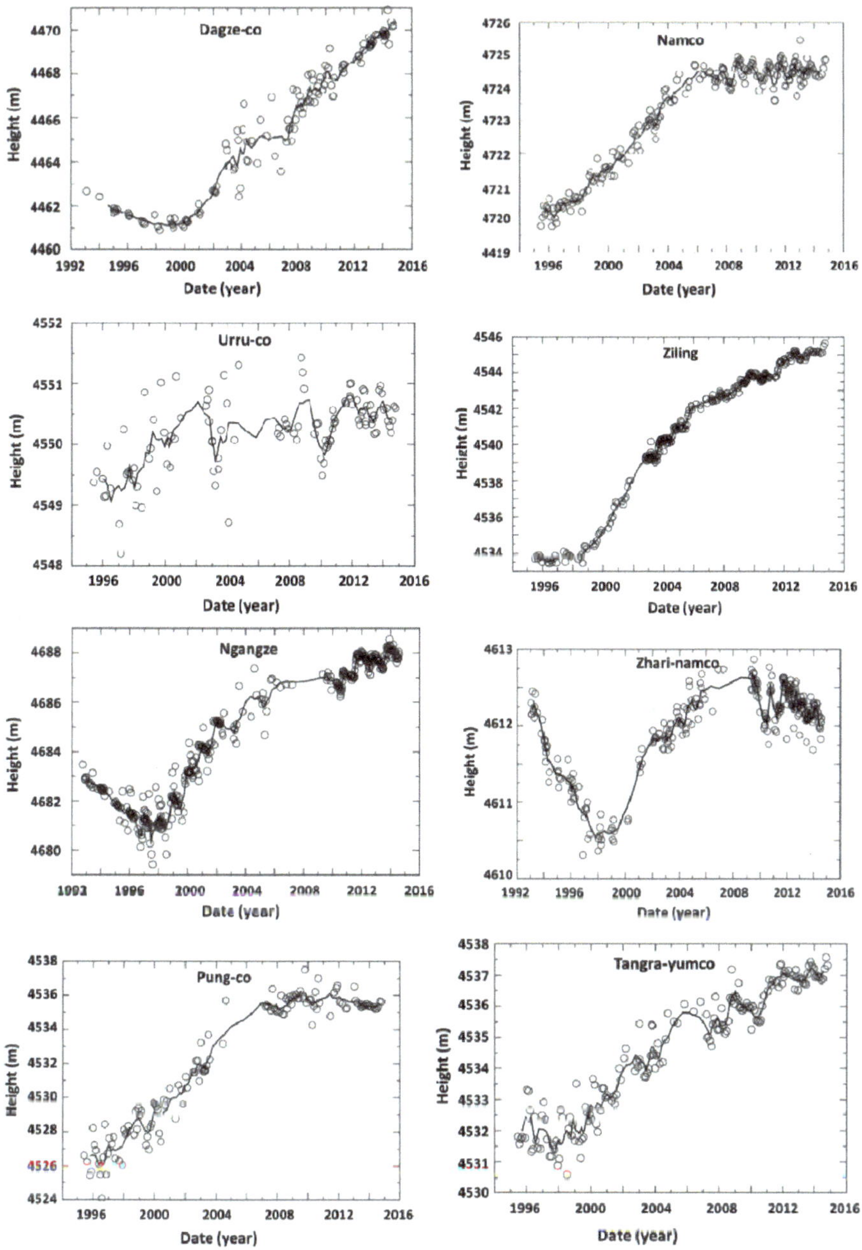
Dagze-co
Height (m)
Date (year)
Namco
Height (m)
Date (year)
Urru-co
Height (m)
Date (year)
Ziling
Height (m)
Date (year)
Ngangze
Height (m)
Date (year)
Zhari-namco
Height (m)
Date (year)
Pung-co
Height (m)
Date (year)
Tangra-yumco
Height (m)
Date (year)

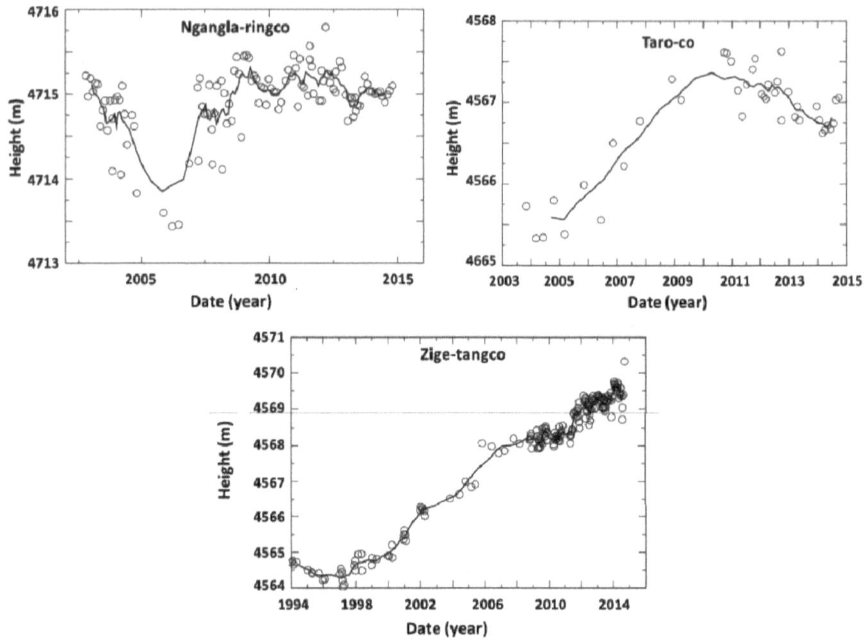

Figure 6.5. *Variations in water level measured by the satellites T/P, ERS-2, Jason-2, ICESat, ENVISAT, Cryosat-2 and Saral/AltiKa for the 11 large lakes situated in the southwestern region of the Tibetan Plateau over the period from 1992–2014 for the longer series and from 2002 to 2014 for the shorter series. Small black circles stand for actual estimates and the black line is a six-month moving average*

According to the equations which determine a lake's response to changes in one of its elements (precipitation, evaporation, entering and exiting flow), an essential parameter which makes it possible to predict the time it will take to return to an equilibrium state is the morphology of the lake – that is, the surface variation as a function of the level variation. The time it will take to return to an equilibrium state is given by the following equation [MAS 94]:

$$T_e = \frac{A_l}{\frac{dA}{dL}(E_l - P_l)} = \frac{1}{\frac{dA}{dV}(E_l - P_l)} \qquad [6.1]$$

where T_e corresponds to 66% of the time required for a lake to reach a new level of balance. In this equation *A* represents the surface of the lake, *V* represents its volume, E represents the evaporation and P represents direct precipitation into the lake. The index *l* indicates that the parameters of the balance are those with the new balance conditions that the lake is moving toward.

We can see that, aside from hydrological conditions (E_l-P_l which is the net rate of evaporation on the lake), this parameter depends on its morphology (dA/dL or dA/dV, surface variation as a function of the height or the volume). Using this equation, when one of the parameters of the balance (E_l, P_l or R[1]) undergoes the slightest change, it is possible to determine the surface and the level of the lake at any point. These equations, that we will not include here (we invite the reader to consult [MAS 94]), make it possible to predict the length and the activity of a lake which is undergoing climate change, whether this is one-off, continual or cyclical.

Using data relative to the level and the area of the large Tibetan lakes, it was stated that the range of time taken to return to an equilibrium state was very long: 520 years for Tangra Yumco, 160 years for Namco, 100 years for Ziling, 75 years for Zhari-Namco and only 24 years for Ayyakum, located to the north of the Tibetan Plateau. Because of this, two lakes which are balanced and which undergo a disturbance in their water balance will not move towards an equilibrium state in the same way. The simulations illustrated in Figure 6.6 were carried out in order to calculate the time taken to return to a equilibrium state in accordance with the morphology of the lake (represented by the term dA/dV) and for different values of the term E_l-P_l. It can be noted that, in the range 0–10 km^{-1} that is for lakes where the bathymetry is quite shallow, this time increases quickly (much higher than for a lake in a non-arid area) but that for all lakes with quite a deep bathymetry (range 10–80 km^{-1} the influence of the term E_l-P_l decreases and the time taken to return to a equilibrium state is much shorter. The vast majority of Tibetan lakes are located in a range with return times considerably higher than 50 years and because of this they react very slowly to any change in climate, irrespective of the magnitude of the change observed.

1 The equation of the balance can be given simply as dV/dt = (P–E). A + R, where R represents the sum of surface and sub-surface runoffs.

Many recent studies have put forward the hypothesis that the lakes of the Tibetan Plateau were generally in an expansion phase as a result of significant changes in the climate over several years. However, many of these studies are based on a relatively short period of observation (for the most part less than 10 years) and what is more, they were carried out in the 2000s [ZHA 11, PHA 12, WU 14 among others]. In studies mostly based on ICESat data, the authors of these publications were able to note that between 2003 and 2009, the hundred or so lakes that they had been able to measure using this satellite had seen their water levels increase by an average of 20 cm per year. We believe that it is very premature to draw such conclusions and to formulate an argument on the topic of climate change based on such a short period of time. The most striking example concerns Lake Namco (Figure 6.5) for which, for example, Wu *et al.* [WU 14] were able to develop a model seeming to demonstrate that the lake is constantly expanding and validated their model using ICESat data from 2003 to 2009. However, it is quite clear that, based on data collected from roughly 2007–2009 until the end of 2014, the lake has been stabilized to what resembles a equilibrium state around 4725 m above sea level.

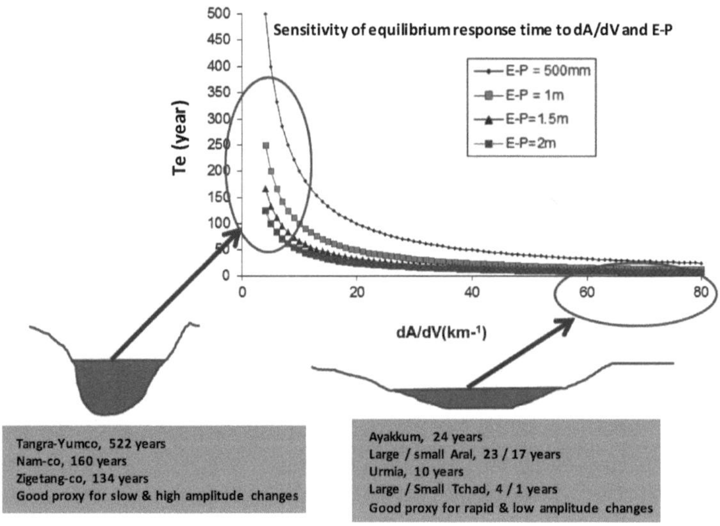

Figure 6.6. *Simulations of the time taken to return to an equilibrium state (T_e) for lakes of varying morphology, according to the surface variation versus the volume variation (dA/dV). These simulations were carried out using four different values of the term (E-P). The lakes in question are Avakkum and Urmia, in Tibet and Iran respectively*

The equations of a lake's water balance as defined in Mason *et al.* [MAS 94] show that it is often necessary to wait decades before analyzing the behavior of lakes. For all of the lakes studied with one sole exception (Ngangla-Ringco), a quite significant increase was observed between 2003 and 2009, but for at least half of them, (Namco, Urru-Co, Taro-Co, Pung-Co, Zhari-Namco) the water level stabilized around the years 2007–2009 and has even decreased since the end of the 2000s (Figure 6.5). Here we can arrive at various conclusions: either the lake has achieved an equilibrium state or it is following a multiyear cycle, as is perhaps the case with Zhari-Namco or Taro-co.

If we take the equations which characterize the evolution of a lake subject to a cyclical variability based on one or several of an environment's hydroclimatic parameters (E_l, P_l or R) we are able to predict the evolution of a lake in the form of a simple harmonic equation:

$$L(t) = \langle L \rangle + L^* \cos\left(wt + \phi(w)\right) \qquad [6.2]$$

where L(t) is the level of the lake at date *t*, ω the angular frequency of the modification of the hydroclimatic parameters (for example, a cyclical evolution of precipitations or temperature over a period of several years will lead to an oscillation in the level of the lake at the harmonic frequency ω as well as a phase difference that is to be found in the term ϕ(ω)), and L* and <L> are the maximum and average values of the level on the cycle respectively, with:

$$L^* = T_e\,(E_l - P_l)\,a(w)\frac{C^*}{\langle C \rangle} \qquad [6.3]$$

where

$$a(w) = \frac{1}{\sqrt{(1+(\omega\,T_e)^2}} \qquad [6.4]$$

$$\phi(w) = \frac{1}{\tan\,(-w\,T_e)} \qquad [6.5]$$

where C* and <C> are the index of maximum and average aridity of the lake (these indices between 0 and 1 tend towards 0 for particularly dry lakes).

These equations therefore make it possible to predict changes in a lake's levels and surface, according to a certain number of parameters when the lake is subject to cyclical changes or when a sudden change appears in the elements of the water balance.

The main difficulty of this exercise is due to the absence of *in situ* climate data that are reliable and durable over the long term for the Tibetan Plateau. Indeed, a limited number of lakes have been measured over the past few years (Namco, Kokonor). These observations only further underline the need for systems which are able to monitor lakes over a period of decades if we want to analyze the impact of climate change on water supply, and to separate cyclical patterns over different timescales from what could be long term trends.

6.4. Using altimetry to estimate river flow

The liquid flow of a river determines the evolution of the water supply in the drainage basin upstream from where the measurement is carried out. Having knowledge of this is therefore equally essential for scientific studies as it is for evaluating the impact of human activity on the water cycle or simply how to manage these activities, from the effect this has on drainage basins to the effect this has on the entire planet. However, measuring it is difficult and expensive. In [CAL 16] we outlined the drastically small amount of information gathered on river flows over the past few decades, and how it is possible to estimate flows using height and altimetric measurements via empirical relationships between height and flow. We will take a more detailed look at this in this section.

Practically speaking, it is impossible to directly measure the quantity of water passing through the cross section of a river bed. The flow, which is this quantity of water per unit of time, is directly estimated by measuring the speed of the flow, integrated over the cross section of the river:

$$Q = \int_A v\, ds \approx \sum_i v_i\, \Delta A_i, \text{ with } \sum_i \Delta A_i = A \qquad [6.6]$$

where Q is the flow through the area A of the section, ΔA_i a basic surface area in A, and v the speed of the flow. Speed varies across a section and therefore must be measured as carefully as possible (Figure 6.7). Depending on how sophisticated the equipment is, v is measured using a number of vertical profiles extending from one bank to the other and the surface at the bottom of the river (whirl method) or at any point in the section with a resolution in m^2 with an acoustic doppler current profiler (ADCP). The precision of these measurements is at best somewhere in the region of 95% of the total flow.

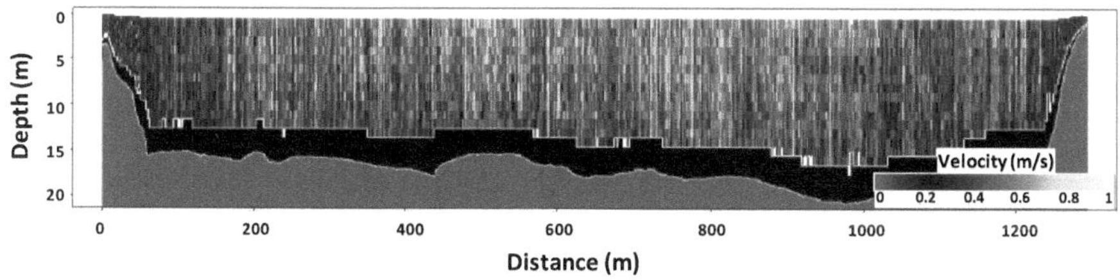

Figure 6.7. *Example showing the measurement of flow by ADCP for a section of the Rio Negro river in the Amazonian basin. Note here the absence of measurement for the bottom of the riverbed (area in black) and on the extreme left of the section (it was impossible to reach solid ground with the boat carrying the instrument). Here, the speed is interpolated, which partly explains the inaccuracy of the final measurement [PAR 15]. For a color version of this figure, see www.iste.co.uk/baghdadi/4.zip*

These measurements are costly and take a considerable amount of time (from a few hours to a whole day) and it is unfeasible to repeat them every day. In order to carry out a daily estimation of flow, operators establish empirical relationships between the flow and the height of water in the section, given that measuring water is far easier to carry out relatively inexpensively, or even to automate. These relationships between the height and flow are calculated using flow measurements, taken at as varied stages of the hydrological cycle as possible, from low water levels to high water levels. However, empirical rating curves can never be completely exact (Figure 6.8) and we can estimate that the error rate on discharge with this process is at best 15% (regularly reaching 20%).

Several methods have been developed in order to establish relationships between height and flow using altimetric spot heights. The first method involves taking flow measurements from the *in situ* station closest to the Virtual Altimetry Station being considered (see Calmant *et al.*, Volume 2, for the definition of virtual stations [CAL 08]) and to set up an empirical curve. This method has been used by Zakharova *et al.* [ZAK 06] on the Amazon, Papa *et al.* [PAP 10] on the Ganges, Birkinshaw *et al.* [BIR 14] on the Mekong and Kouraev *et al.* [KOU 04] on the Ob. Figure 6.9 shows an example of predicting flow on the Ganges and the Brahmaputra using historical altimetric data and the relationship between height and flow established using previous flow data.

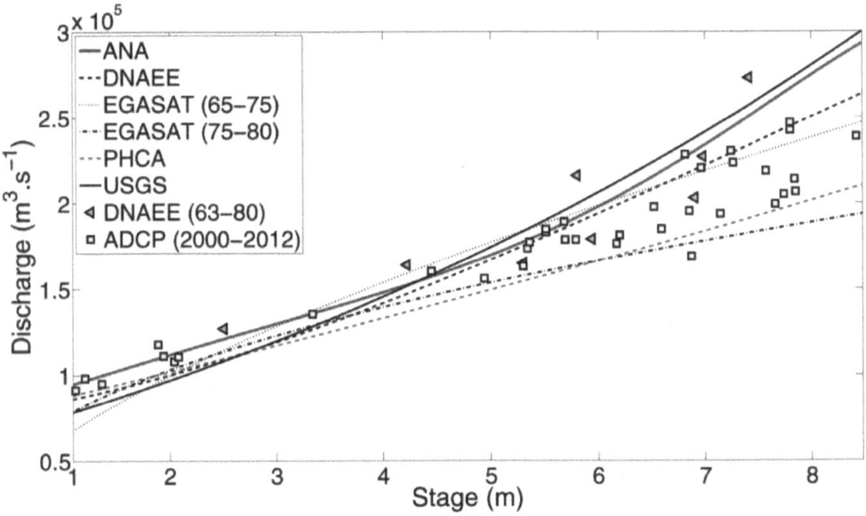

Figure 6.8. *Relationship between height and flow at Obidos on the Amazon. The red curve was obtained by adjusting the data on height and flow published by Brazil's ANA (Agencia Nacional de Aguas) on its website* hidroweb.ana.gov.br. *Height measurements are daily readings of water height carried out by an operator using a ruler positioned at the port of Obidos. The green triangles correspond to previous flow measurements, carried out using the whirl method, and made it possible to establish successive ratings curves shown on the graph. The yellow squares correspond to more recent measurements, carried out using ADCP. It can be noted that above 7 m, flows taken from the rating curve are widely overestimated. For 8 m, the error is somewhere in the region of 30 to 50,000 m³/s, or the same as the flow of the Congo, the second river in the world in terms of flow after the Amazon [JAC 87, PAR 15]. For a color version of this figure, see www.iste.co.uk/baghdadi/4.zip*

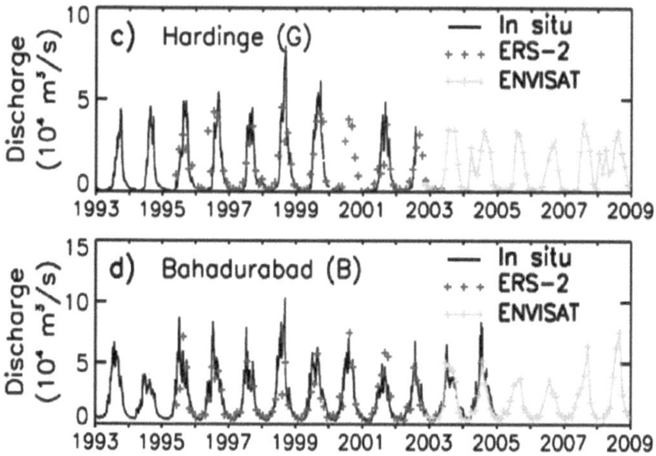

Figure 6.9. *Flow measurements carried out in situ on the Ganges (Hardinge) and the Brahmaputra (Bahadurabad) using altimetry. In this study, relationships between the height and the flow were established for a common 7-year period (1995-2002) between historical in situ flow data (in black) and the series of heights taken from the altimetric mission ERS-2 (red crosses). The flow data, measured in situ in 2005, is extended to 2009 with the height data taken from the ENVISAT mission (green crosses). The two altimetric missions followed the same orbit*

The disadvantage of this method is that it requires (1) a recovery period between the flow data and height data, and (2) that the two sets of data were collected at a sufficiently short interval in order to be able to make the hypothesis that there was no major hydrological change between the two (confluence with a tributary, phase difference between series etc.). Leon *et al.* [LEO 06] put forward a variation in order to take it out of the context of the question of proximity in a calculation of rating curves on the Rio Negro. The Rio Negro is one of the Amazon's main tributaries but its upstream section is rarely gauged. In this study, flow is routed using a hydrodynamic model (a simplified version of Mushkingum Cunge in this case [CUN 69]) calculated using all available upstream flow data and downstream of the section being studied, where possible.

A record of the flow at each virtual station can then be extracted from the output of the model and the rating curve is calculated between the altimetric series and the series of flow models (Figure 6.10).

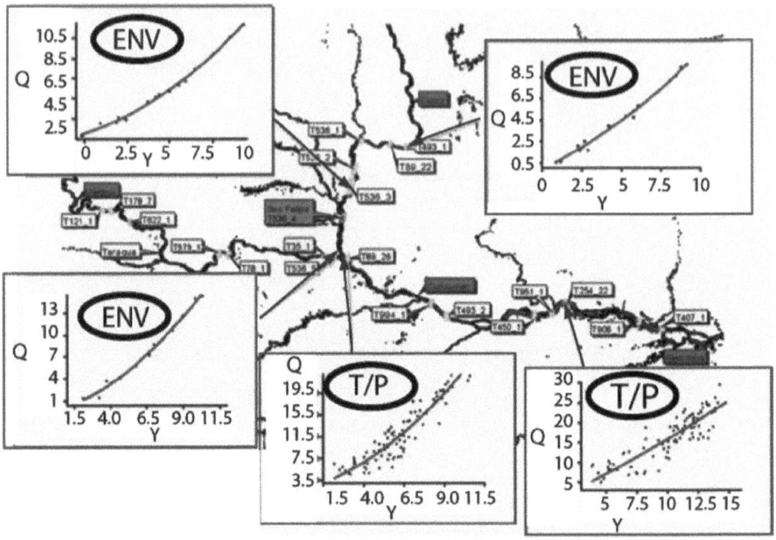

Figure 6.10. *Rating curve on the Rio Negro using Q flows (x104m^3/s) obtained through the spread of flows between measurement sites and a set of heights taken at the T/P and ENVISAT virtual stations. Heights are converted into depth Y (m) by subtracting from a reference depth Y_0 so that $Q(Y_0) = 0$. Here we can note the increase in quality between the 3 ENVISAT (ENV) series and the 2 T/P series, more populated but significantly noisier*

The second approach involves calculating flow using a rain-discharge model. These models estimate the percentage of rain captured by vegetation, the percentage which evaporates, the percentage which seeps into the ground and the percentage which reaches the river through streams or underground flow. The first study of this type was carried out by Guetirana *et al.* [GET 09] on Amazonian subbasins. The methodology was then extended to cover the whole of the Amazonian basin by Guetirana and Peters-Lidard [GET 12]. Paris's study [PAR 15] is very similar to the study carried out by Guetirana *et al.* [GET 09] in terms of the data used, the flow model and the database from virtual stations [SIL 10]. The advantage of the methodology put forward by Paris [PAR 15] is that, in addition to the flow itself, it provides an uncertainty on the value linked to uncertainties in altimetric data, in flow models and the quality of the adjustment of the relationship to the height-flow pairs. This uncertainty data is somewhat significant. We saw in Figure 6.8 that flows emanating from ratings curves can be very different from those measured. Therefore, it is important that anyone using these flow estimations has access to this information in one form or another, and the spread of uncertainties is the chief way in which information on the precision of estimation is distributed. In Figure 6.11, the rating curve from the adjustment of pairs (altimetric height, flow modeled by MGB) is superimposed on Figure 6.8. It can be seen that the uncertainty at 95% probability (equivalent to three standard deviations of a Gaussian probability function) in relation to flows predicted is relatively high, but that it is consistent with ADCP measurements (despite the fact that these are totally independent from the curve). The band of 95% probability in Figure 6.11 is not symmetrical in relation to the blue curve of the most probable value: the most probable value therefore has more chance of being overestimated than underestimated, which is not consistent with ADCP measurements. For high water level flow predicted by the ANA (Agencia Nacional de Aguas) using the red curve for heights above 7 m (the level for which the red curve leaves the grey band) there is a very strong probability of being significantly overestimated.

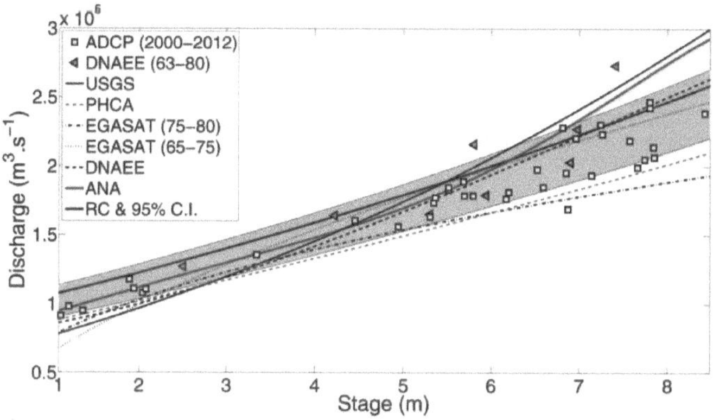

Figure 6.11. *Height–flow relationship at Obidos on the Amazon. The red curve was obtained by adjusting the measurements of height and flow published by the ANA (Agencia Nacional de Aguas) of Brazil on its website* hidroweb.ana.gov.br. *The measurements for height are daily readings carried out by an operator using a ruler positioned at the Port of Obidos. The green triangles correspond to previous flow measurements, carried out using the whirl method, and which were used to establish the ratings curves shown in this diagram. The yellow squares correspond to more recent measurements from the ADCP. The rating curve (in blue) was obtained using modeled altimetric heights and flows [PAR 15] and its realm of probability of 95% in grey. For a color version of this figure, see www.iste.co.uk/baghdadi/4.zip*

One use of satellite altimetry comes from this result: the combination of satellite altimetry with a set of independent flow measurements makes it possible to calculate ratings curves that are independent from those used by agencies in order to provide an independent element of validation. An example of this type of validation is shown in Figure 6.12.

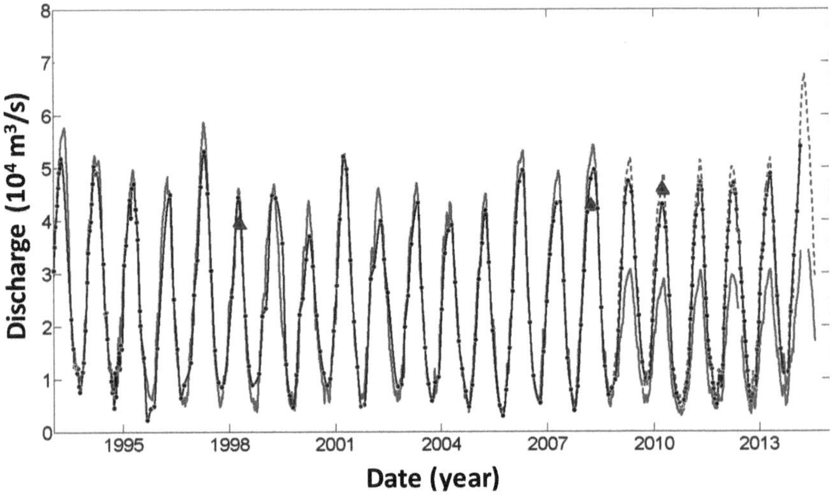

Figure 6.12. *Example showing errors in flow data from an independent satellite altimetry series for Manicoré in the Rio Madeira in the Amazonian basin. The altimetric series used in order to estimate flow is a combination of data from the ERS-2 (1995–2002), ENVISAT (2002–2010), Jason-2 (2008–2014) and Saral (2013–2014) missions. The continuous red curve corresponds to the flow data published by the ANA on its website* hidroweb ana gov. br *for the Manicoré station on the Rio Madeira. There is a clear difference between these curves from 2009 onwards. The red dotted curve was obtained by recalculating a rating curve using data on height and flow for Manicoré before 2009. We are able to state that the flow data obtained in this way conforms to flow predicted by altimetry (in black) and ADCP measurement (blue triangles* www.ore-hybam.org*) from early 2010. By making a comparison with flow data based on altimetry it is possible to show that flow data from Manicoré is incorrect from 2009 onwards*

6.5. Impact of adjustments and uses of altimetry

6.5.1. *The context and the role of spatial altimetry*

Over the course of the past 20 years a range of publications and studies have been produced relating to the use of satellite altimetry on lakes and rivers. While this method has demonstrated its ability to continuously monitor bodies of water over several hundred km^2 (lakes, reservoirs, floodplains) with decametric or centimetric precision, studies on artificial reservoirs using this technique are very rare. Certain authors have concerned themselves with the potential of satellite altimetry for forecasting high water levels within the framework of transboundary river basins. [BIA 11, PAP 12] and [HOS 14] demonstrated that it was possible to forecast high water levels downstream at the center of the delta in Bangladesh through the use of altimetric data on the Ganges basin, just upstream of the border between India and Bangladesh. To our knowledge, only the study carried out by Gao *et al.* [GAO 12] has dealt with monitoring reservoirs through a combination of altimetry and satellite imaging, applied to 34 reservoirs across the globe as well as the study undertaken by Crétaux *et al.* [CRE 15] on the Syrdarya reservoirs in Central Asia (see section 6.5.2 of this Chapter).

During the 20th Century, the number of reservoirs being built rose exponentially, reaching its peak in the 1980s [CHA 08]. The World Commission on Dams [WCD 00] states that there were around 45,000 new dams built worldwide in the 1930s. Half of all rivers on Earth have at least one dyke present in their drainage basin [WCD 00]. Because of climate change and the increase in the world's population, these reservoirs will play an increasingly important role in the years to come [VOR 05].

Owing to the transboundary nature of the vast majority of the world's great rivers, reservoirs are and will continue to be an essential political instrument for the management of water supply and will have a variety of uses (irrigation, manufacturing, hydro-electricity, flow regulation, the provision of clean drinking water, navigation, the layout of territories, etc.). Today there are around 286 transboundary river basins worldwide, draining 151 countries and corresponding to more than 40% of continental surface areas and encompassing more than 40% of the world's population (the most up-to-date information on this subject can be found at the Transboundary Water Assessment Program (TWAP), website: http://twap-rivers.org). In other words, many countries are heavily reliant on water from a country further upstream that is in charge of the regulation of water flow and the quantities of water available for neighboring countries situated downstream of the river basin. This obviously creates the potential for disputes between countries who share the same drainage basin, and has therefore led to the creation of a range of river basin commissions in order to deal with the issue of sharing water from the same river as well as the political implications for the authorities in the relevant countries. We will examine this in greater detail when we look at the example of the Syrdarya basin (Uzbekistan).

Given the necessity to control flows entering and leaving a country, and at the local level of a reservoir, one of the obvious prerequisites for any effective, unified management of the water supply available in a river basin is therefore to put in place appropriate measurement systems.

Satellite altimetry offers a precise, durable method available on a global scale in order to compensate for the absence of data in situations where it is technically difficult to maintain networks on the ground, but also for the many reservoirs located in transboundary river basins where the sharing of data between the different countries has not always been respected. As we will see in the example of Syrdarya, satellite altimetry can be used to validate databases of *in situ* data and to make it possible to compare them. Used in conjunction with *in situ* data, they also make it possible to recreate data on water supply for periods where altimetry was the only tool available. Although the majority of reservoirs located along rivers are rather narrow, making it difficult to be as precise as with large lakes, in general, the magnitude of variations in water levels in these reservoirs is at least one order of magnitude higher than the error with altimetric measurements, which are still highly useful, as will be seen with the Syrdarya. The use of measurements from future missions, such as the Sentinel-3 satellites, whose SAR is supposedly far better equipped to deal with "narrow" objects such as reservoirs, should further reinforce the role of altimetry.

6.5.2. *Case studies with radar altimetry*

A particularly interesting case study concerning the potential contribution of satellite altimetry for monitoring large, heavily anthropized river basins located in Central Asia, and here we will focus more specifically on the river basin of the Syrdarya.

The Syrdarya, located in Central Asia belongs to the drainage basin of the Aral Sea. It takes its source from the Tian Shan and the Pamir mountains (the Naryn and Karadarya rivers converge in Uzbekistan to become the

Syrdarya). The total drainage area is 485,000 km^2 and its length is 2,337 km as far as the Aral Sea that it flows into. The majority of the river basin is made up of arid steppes where annual precipitation is somewhere in the region of 100 to 150 mm per year. Since 1991 and the fall of the Soviet Union, it has been a transboundary river basin covering four countries: Kyrgyzstan and Tajikistan upstream, Uzbekistan and Kazakhstan downstream. In its upstream section, there are many artificial reservoirs along the course of the Syrdarya (the most important being, from upstream to downstream, the Toktogul, the Andijan, the Karakul, the Charvak and the Chardarya) whose purpose is to regulate the water supply downstream for the irrigation requirements of Uzbekistan and Kazakhstan (Figure 6.13). The total flow of the river is ~41 $km^3 yr^{-1}$ including 38 $km^3 yr^{-1}$ for the 1,000 km between the source and the Chardarya reservoir. The key reservoir of this system, the Naryn Syrdarya Cascade (known as the NSC) is the Toktogul because it is has the largest storage capacity (20 km^3) in addition to being the furthest upstream and therefore having total control over the flows downstream depending on how the water releases are operated. This reservoir is situated in Kyrgyzstan. The responsibility for managing the NSC is currently shared between several countries of varying needs and requirements, which leads to a number of disputes. In fact, when it was conceived, when the Syrdarya was entirely within one single country, the aim was to stockpile water in the reservoirs during winter, particularly in the Toktogul, and then to release water during the growing season that is in spring or in summer, in order to support agricultural activity, especially in Uzbekistan which suffers from a very dry climate and receives very little rain. After the fall of the Soviet Union, usage quotas for this resource were maintained by the different countries but Kyrgyzstan felt that this situation did not benefit them. They decided to modify the system, believing the Toktogul to be an indispensable tool for producing electricity, especially in winter.

As we can see in Figure 6.14, from 1992 onwards these water releases increased which created a variety of problems for the countries downstream. the other reservoirs on the Sardarya do not have the capacity to store the surplus of water caused by this decision at Toktogul and their usable quantity of water in the summer months significantly decreased, which had an impact on agricultural activity. It also caused flooding in Kazakhstan and Uzbekistan. From 1987 to 1995, water releases in winter went from roughly 1–2 to 4–5 km^3, while in summer they decreased from 6–8 to around 2–4 km^3 depending on the year (Figure 6.14). We can see that from 1993 up until the end of the 2000s, water releases during the winter were always higher than water releases during the summer. For more detailed information on the geopolitical implications surrounding the question of water in the Syrdarya river basin, a number of references can be consulted [CRE 15, WEI 06, LIB 08].

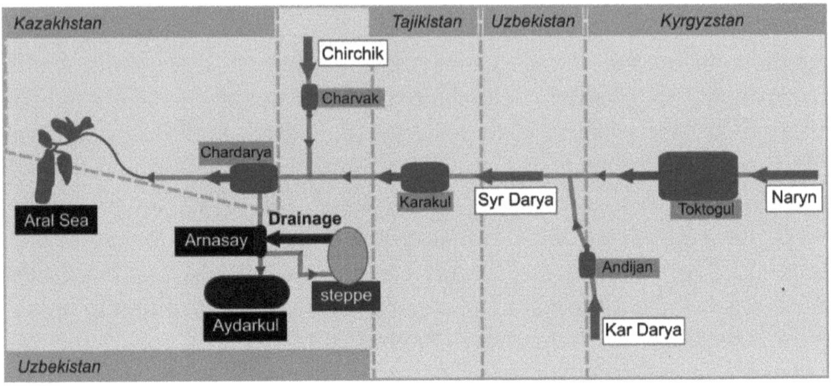

Figure 6.13. *Schematic showing the network of reservoirs on the Syrdarya*

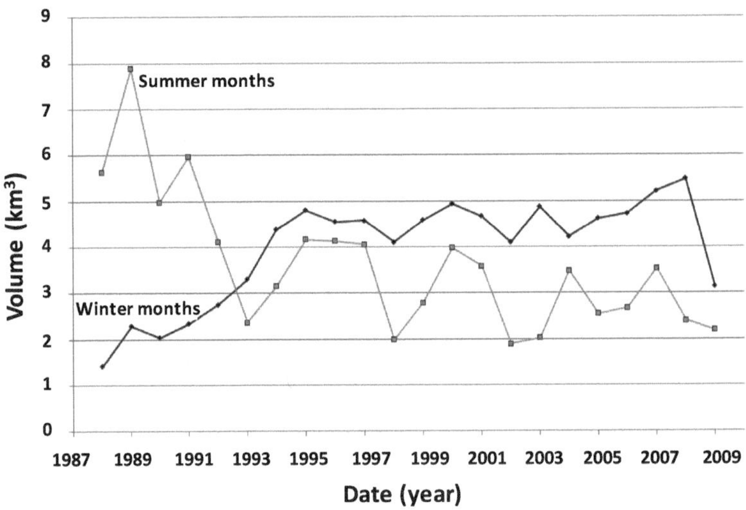

Figure 6.14. *Volume of water released from the Toktogul reservoir (source: CAWATER from 1987 to 2009, averaged out for winter (January to March) and summer (June to August)*

What is of interest to us here is to see how satellite altimetry can play a role in monitoring this river basin on a daily basis. It is clear that the different countries concerned all have an interest in reaching an agreement on the attribution of quotas as well as the introduction of integrated, coordinated management [WEI 06]. However, a source of dispute comes from a reluctance to share information, particularly information regarding the volume of water stored in each reservoir and the quantities of water released downstream over the course of a year.

At the end of the 2000s, a joint initiative between the five Central Asian countries which share the Amudarya and the Syrdarya was to put in place a database that would be freely available online and which would deliver a significant amount of quantitative information regarding water supply along the two rivers: monthly data on the volume for all the reservoirs, going back to the late 80s/early 90s, data on the amount of water entering and exiting (in km^3) all reservoirs, data on water samples for irrigation as well as metadata and maps. The data on this site, called CAWATER (www.cawater-info.net) were available for 2–3 years and access is now limited. We have shown [CRE 15] on one hand that altimetric data were precise enough in order to quantify variations in water supply in artificial reservoirs like those of the Syrdarya river basin, but also that they made it possible to act as a way of monitoring water supply in situations where *in situ* data were unavailable, particularly in the case of transboundary water basins. They also make it possible to establish rating curves, also known as hypsometry, connecting a level of water to an area and a volume, and therefore to obtain values for the surface and the volume of the reservoir using altimetric data only (Figure 6.15). On the site CAWATER, only the volumes of the reservoirs were actually available.

The same calculations were carried out for the other two large reservoirs of the NSC (the Toktogul and the Karakul) which also made it possible to recreate the variations in water supply in these reservoirs from 1992 to 2014 and on the Charvak, another reservoir that is part of the system, this time for the period from 2002 to 2010 [CRE 15].

Altimetric data were also available for the Arnasay floodplain and Lake Aydarkul (Figure 6.13) which are both supplied by water releases from the Chardarya in winter in order to relieve the surplus of water coming from the Toktogul. A complete water balance of the Chardarya/Arnasay/Aydarkul system was established, and it showed that Lake Aydarkul lost roughly

10 km^3 of water between 2011 and 2013, a period during which no *in situ* data was available. It was also possible to quantify the exchanges between the Toktogul and the Karakul between 2011 and 2013 despite the total absence of any *in situ* data (see [CRE 15] for more details on this study).

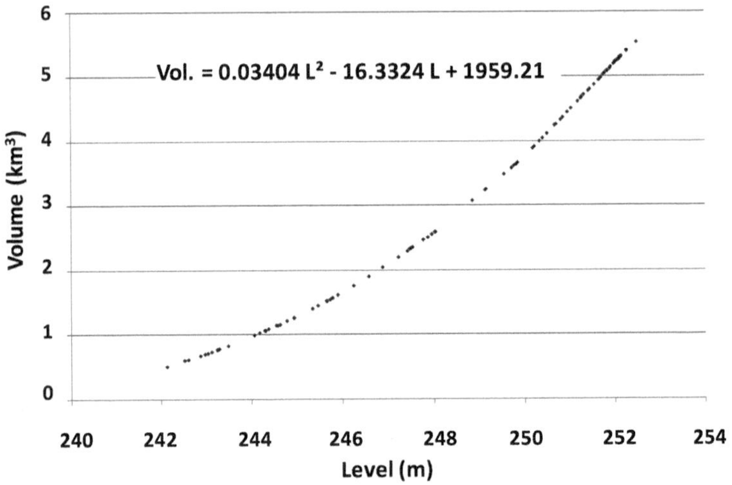

a)

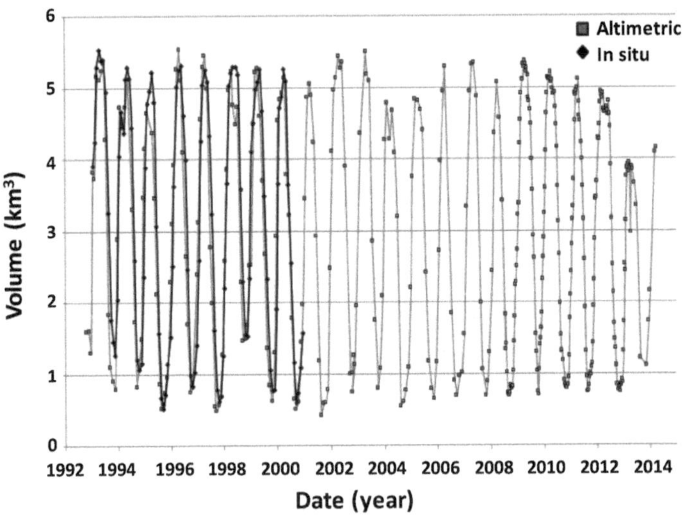

b)

Figure 6.15. *a) Hypsometric curve (height-surface relationship) obtained by combining in situ data of the volume between 1992 and 2000 (CAWATER) and altimetric data (level) over the same period for the Chardarya reservoir, and b) reconstruction of the total volume variation from 1992 to 2013 using altimetric data (T/P and Jason 2) and the level/volume relationship*

The example of the Syrdarya river basin makes it possible to show that satellite altimetry can be a tool that is highly compatible with more traditional methods in the domain of surveying large river basins. This is especially true for situations where the management of water occurs through a large network of infrastructures, as it happens, in this case, with irrigation canals controlled by a set of artificial reservoirs spread out along the river. For the Syrdarya, a transboundary river, complexity comes also from the fact that the monitoring of different samples and variations in water supplies in reservoirs is the responsibility of different countries who don't necessarily agree on the way in which they prioritize how each resource should be used, and that the responsibility for managing and maintaining networks of measurements is split up among the different countries. Of course, satellite altimetry is not a panacea, nor is it a miraculous solution that will make it possible to solve every disagreement. Nevertheless, it does have the potential to be used as an independent source of information and, moreover, the free access to data and the relative simplicity with which it can be processed will make it possible in the future to think of it as an instrument for controlling water supply for reservoirs within the reach of administrators of river basins similar to the Syrdarya. It has been shown [CRE 15] that it is possible to precisely measure the level and volume of water in NSC reservoirs. In the same study, it was also highlighted that, over the past few years, variations in the times for these variables were highly irregular and they did not comply with the quotas that the countries had decided on in the early 1990s. One key to understanding this variability is the management of the Toktogul reservoir, which is located upstream of the river basin and which possesses the largest storage capacity. Altimetric measurements for this reservoir and of the three other important reservoirs downstream have been carried out to this date using the constellation of satellites in orbit since 1992 (*in situ* data covering the same period is non-existent or very scarce).

6.6. Conclusion and prospects

As can be seen in the various examples outlined in this chapter, satellite altimetry has made it possible to raise a number of questions in the realm of hydrology. These questions can have scientific applications (monitoring the impact of climate change on water levels in lakes or on river flow), as well as more practical applications such as determining the flow of large rivers as a means of helping to manage them or monitoring the levels of water stored by artificial reservoirs. Regarding those questions which relate to the levels of water in lakes, we were able to illustrate a number of crucial points using the examples of the Tibetan Plateau and African lakes:

– lakes respond not only to changes in the climate (cyclical or longstanding) of their immediate environment but also to global changes;

– there is a wide range of behaviors associated with lakes which will be influenced by their geographical location (and the climate conditions thereof) but also their morphology;

– contrary to what has been observed with rivers, the temporal variability in the level of water in lakes is influenced over a much longer timescale.

Given these points, it is essential to have access to measurements carried out over very long periods of time to be able to analyze the causes of variations in water level for these lakes. It is necessary, therefore, to have continuous space missions are therefore necessary since they are the only means possible of providing complete global coverage.

For rivers, the essential parameter for hydrologists is not so much the water level as the flow. Satellite altimetry does not directly provide this information but by using altimetric measurements it is possible to access this variable. In order to do this, it is necessary to use rating curves (relationship between height and flow) that can be obtained either by preliminary measurements or through the use of rain–discharge models.

The different examples outlined in this chapter raise a number of questions, especially regarding the scientific community's requirement to prove how useful this method is and why it is necessary to keep these missions in orbit. To take it a step further, if the main factor influencing the decision to put into orbit radar (and laser) altimeters relates to the various studies and research carried out in oceanography, for several years the amount of publications and projects undertaken around the world concerning the use of altimetry in hydrology has had a direct impact on the development agenda of several of the world's major space agencies, leading them to develop and build this type of instrument (NASA, CNES, ESA, ISRO). Because of this, the Saral/Altika mission was the first to be launched with the specific purpose of monitoring continental areas as well as oceans or polar ice caps, with the onboard altimeter functioning in a frequency band (Ka) better suited to monitoring smaller bodies of water (reduction in the surface area of the mark on the ground). Launched in February 2013, this mission was able to demonstrate its suitability [ARS 15]. In the years to come, the outlook is ever-changing in this respect with a number of highly promising missions in the pipeline.

In 2016 the satellite Jason-3 was launched, following on from previous Jason missions. Jason-3 does not feature any technical upgrade from Jason-2; however the ESA mission Sentinel-3A, launched in 2016, is much better equipped to monitor continental surfaces. This mission will include the same type of nadir altimeter that is currently used with Cryosat-2, meaning that it will function in SAR mode (see [CAL 16] for a description of this form of altimetry). The decision to make it function in this mode was only taken after lobbying from the scientific community which preferred this option despite the potential of Cryosat-2. The orbit on which Cryosat-2 operates is not optimal for hydrology since it follows a repetitive cycle of 369 days, but the SAR measurements carried out in the few areas where it was used demonstrated its usefulness, once the targets were complex: very narrow rivers or small lakes for example. The gain in terms of footprint is somewhere in the region of 100 compared to a LRM (Low Resolution Mode) altimeter. Furthermore, in relation to the LRM, the connection between the signal and noise was improved for technical reasons. Lastly, one of the characteristics of this type of SAR is that it allows the possibility of precisely focusing the measurement on a reflective surface. This will reduce any sources of pollution that can be found when using LRM for small objects where the measurement can be affected by the surrounding environment. As a result, many targets that could have in theory been observed using altimeters in the past were ignored, due to the noisy nature of LRM data or the fact that the "tracking" system on board wasn't able to pick anything up (see [CAL 16]). The orbit of Sentinel-3A, which has a revisit time of 27 days, will cover tracks on the ground that weren't covered by previous T/P or ENVISAT altimeters. In addition, roughly two years after the launch of Sentinel-3A, a second satellite (Sentinel-3B) will be launched and set on an orbit that made it possible to overlap tracks on the ground. While this new orbit is not a huge advantage for monitoring rivers given that the virtual stations used for the past 20 years with previous missions will no longer be updated, the program of the space agency is to be spread out over a long period of time and will theoretically serve the scientific community for decades to come. Regarding lakes, this new orbit is highly interesting because it will join other orbits and will make it possible to cover a much wider range of targets. Sylvain Biancamaria of LEGOS calculated that a constellation featuring the Jason-2 and Jason-3 satellites in interleaved orbit, as well as the orbit of Saral/Altika and the two Sentinel-3 missions, will make it possible to cover 98% of the 3,720 lakes on Earth with a surface area greater than 50 km^2 and 71% of the lakes with a surface area greater than 10 km^2, which corresponds to roughly 40% of the total quantity of water stored in lakes.

However, the major advancement expected in this area comes from the SWOT mission (Surface Water and Ocean Topography) developed collectively by the CNES (French National Centre for Space Studies), NASA (United States National Aeronautics and Space Administration) and the Canadian and English space agencies, which will be launched in 2021. SWOT will be equipped with an interferometer in Ka band (called Karin) which will provide images relating to water height below swaths of 50 km on either side of the nadir. These two swaths will be separated by a band of 20 km, in the middle of which measurements using a traditional nadir altimeter will also be taken. The purpose of this mission is to be able to observe all rivers wider than 100 m as well as all lakes and reservoirs with a surface area of at least 250 m × 250 m every 21 days. Owing to its orbital configuration, the majority of areas on Earth will be observed at least twice per 21-day cycle, at least once by the ascending track and once by the descending one, and even more at higher latitudes because of swath overlaps. Most of the Earth's surface will be covered since the inclination of the satellite will be 77.6°. Karin will also make it possible to distinguish between flooded areas and land masses, making it an imaging altimeter. The expected precision for measurements is 10 cm for areas of 1 km^2, 20 cm for areas of 250m x 250 m and 45 cm for areas of a hectare.

These orbital and instrumental characteristics will make it possible, for example, to cover all 1,000 Tibetan lakes larger than 1 km^2, whereas today with LRM altimeters it is only possible to cover a hundred or so, and it is also necessary to process data from spatial imagers such as Landsat if we want to determine their surface area.

On SNC it was simulated in Crétaux *et al.* [CRE 15] that all reservoirs in this river basin would be covered by SWOT as well as the river itself.

SWOT will therefore be an asset for the complete monitoring of this type of river basin, providing consistent, reliable data for the different hydrological variables (flow, levels, supply and the area of lakes and reservoirs). It will be the same for all of the world's 286 transboundary rivers. The great diversity of geographical, climatic or geopolitical situations opens up a wide field of study and research for finding solutions for managing river basins, provided that they can be convinced of the usefulness of such a system and that they are trained in how to use and analyze this data.

One of the main purposes of SWOT will be to calculate the slope every 10 km for rivers wider than 100 m. The level of precision will be roughly 1.7 cm/km and the aim will be to use this information to produce data on the discharge. Once all of the world's rivers and the vast majority of lakes (at least those larger than 250 m × 250 m) are monitored by SWOT, it will be possible to quantify various elements of the Earth's water cycle: the global flow in rivers and the contribution they make to oceans, the water stored in reservoirs, artificial or otherwise, and lastly the water stored in large floodplains. This measurement will be carried out cycle after cycle and at the end of the three-year mission it will be possible to measure the seasonal dynamic of the global water cycle and to integrate this into models.

There are a few main questions that the scientific community involved in preparing for this mission are considering:

– what are the spatio-temporal scales of water processes that control the distribution of water on Earth, and what are their dynamics through the transport of masses of water across continents?

– what impact does human activity such as the building of large dams, samples or water releases have on continental waters?

– on different levels (regional or global) how sensitive is the distribution of bodies of water to climate, to the presence or absence of wetlands or to extreme droughts?

– how can we integrate data on the surface and level of water in large floodplains into descriptive and predictive models?

– what is the impact of introducing SWOT products in collaboration with other contemporary missions (Sentinel-1/2/3, SMOS, GRACE-FO, GPM, etc.) for the quantification of the global water cycle and the closing of the overall result?

– how can we synthesize SWOT measurements with other remote sensing measurements, particularly from altimetry, obtained since the T/P era and how will it be possible to rebuild previous banks of data in light of the new information that SWOT should bring?

These are a few of the major objectives of SWOT for hydrology and the scientific community is already preparing itself for the responses.

6.7. Key points

Two major themes have been raised in this chapter. The first concerns the use of this method for studying lakes while the second focuses on the use of altimetry for determining river flow. In this way, altimetry can be a highly effective tool for studying large African lakes or those on the Tibetan Plateau, for which a number of questions have gone unanswered over several years. Such questions relate to the characterization and the quantification of the impact of climate change on lakes in these regions, and at the same time the use of these lakes as reliable indicators of such change. Altimetry has also been used for several years in order to estimate the flow of large rivers, either alongside or as a substitute for *in situ* data. Finally, we have been able to show using the example of the Syrdarya river basin how this method can be used for monitoring artificial reservoirs. To conclude, after more than 20 years of existence, precise spatial altimetry will evolve with the arrival of new systems that are more effective, both in terms of their precision and their resolution. These new systems have been developed taking into account the demand for greater precision for continental waters. The launch of the SWOT mission in 2021 should enable the emergence of a range of new applications and areas of research in spatial hydrology, but it will be preceded by other promising missions such as Sentinel-3A (launched in February 2016), followed by, Sentinel-3B, 18 months later.

6.8. Bibliography

[ARS 15] ARSEN A., CRETAUX J.F., ABARCA-DEL-RIO R., "Use of SARAL/AltiKa over mountainous lakes, intercomparison with Envisat mission", *Marine Geodesy, Special Issue: The SARAL/AltiKa Satellite Altimetry Mission*, vol. 38, no. 1, pp. 534–548, 2015.

[BEC 10] BECKER M., LOWEL W., CAZENAVE A. *et al.*, "Recent hydrological behaviour of the East African Great Lakes region inferred from GRACE, satellite altimetry and rainfall observations", *C.R Geosciences*, vol. 342. no. 3, pp. 223–233, 2010.

[BEC 14] BECKER M., SANTOS DA SILVA J., CALMANT S. *et al.*, "Water level fluctuations in the Congo Basin derived from ENVISAT satellite altimetry", *Remote Sensing,* vol. 6, pp. 9340–9358, 2014.

[BEL 90] BELTRANDO G., CADET P., "Variabilité interannuelle de la petite saison des pluies en Afrique Orientale: relations avec la circulation atmosphérique générale", *Veille Climatique Satellitaire*, vol. 33, pp. 19–36. 1990.

[BER 04] BERGONZINI L., RICHARD Y., PETIT L. *et al.,* "Zonal circulations over the Indian and Pacific Oceans and the level of Lakes Victoria and Tanganyika", *Int. J. of Climatology*, vol. 24, no. 13, pp. 1613–1624, 2004.

[BIA 11] BIANCAMARIA S., HOSSAIN F., LETTENMAIER D.P., "Forecasting transboundary river water elevations from space", *Geophysical Research Letter*, vol. 38, no. L11401, 2011.

[BIR 95] BIRKETT C.M., "Contribution of TOPEX/POSEIDON to the global monitoring of climatically sensitive lakes", *Journal of Geophysical Research*, vol. 100, no. C12, pp. 25, 179–225, 204, 1995.

[BIR 14] BIRKINSHAW S.J., MOORE P., KILSBY C.G. *et al.*, "Daily discharge estimation at ungauged river sites using remote sensing", *Hydrol. Processes*, vol. 28, pp. 1043–1054, 2014.

[CAL 08] CALMANT S., SEYLER F., CRETAUX J.F., "Monitoring continental surface waters by satellite altimetry", *Survey in Geophysics*, *Special issue on "Hydrology from Space"*, vol. 29, no. 4-5, pp. 247–269, 2008.

[CAL 16] CALMANT S., CRÉTEAUX J-F., REMY F., "Principles of Radar Satellite Altimetry for Application on Inland Waters", in BAGHDADI N., ZRIBI M. (eds), *Microwave Remote Sensing of Land Surface*, ISTE Press Ltd, London and Elsevier Ltd, Oxford, 2016.

[CHA 08] CHAO B.F., WU Y.H., LI Y.S., "Impact of artificial reservoir water impoundment on global sea level", *Science*, vol. 320, no. 5873, pp. 212–214, 2008.

[COH 06] COHEN A.S., LEZZAR K.E., COLE J. *et al.*, "Late Holocene linkages between decade–century scale climate variability and productivity at Lake Tanganyika, Africa", *Journal of Paleolimnology*, vol. 36, pp. 189–209, 2006.

[COL 07] COLLISCHON W., ALLASIA D., SILVA B.C. *et al.*, "The MGB-IPH model for large-scale rainfall-runoff modelling", *Hydrological Sciences Journal*, vol. 52, no. 5, pp. 878–895, 2007.

[CRE 06] CRETAUX J.F., BIRKETT C., "Lake studies from satellite altimetry", *C.R. Geoscience*, vol. 338, pp. 1098–1112, 2006.

[CRE 09] CRETAUX J.F., CALMANT S., ROMANOVSKI V. *et al.*, "An absolute calibration site for radar altimeters in the continental domain: Lake Issykkul in Central Asia", *Journal of Geodesy*, vol. 83, no. 8, pp. 723–735, 2009.

[CRE 15] CRETAUX J.F., BIANCAMARIA S., ARSEN A. *et al.*, "Global surveys of reservoirs and lakes from satellites and regional application to the Syrdarya river basin", *Environmental Research Letter*, vol. 10, no. 1, 015002, 2015.

[CUN 69] CUNGE J.A., "On the subject of a flood propagation computation method (Mushkingum Method)", *Hydraulic Research*, vol. 7, no. 2, pp. 205–230, 1969.

[DUA 05] DUAN A.M., WU G.X., "Role of the Tibetan Plateau thermal forcing in the summer climate patterns over subtropical Asia", *Climate Dynamics,* vol. 24, pp. 793–807, 2005.

[DUA 13] DUAN Z., BASTIAANSSEN W.G.M., "Estimating water volume variations in lakes and reservoirs from four operational satellite altimetry databases and satellite imagery data", *Remote Sens. Environ.,* vol. 134, pp. 403–416, 2013.

[DEW 06] DE WIT M., STANKIEWICZ J., "Changes in surface water supply across Africa with predicted climate change", *Science*, vol. 311, no. 5769, pp. 1917–1921, 2006.

[GAO 12] GAO H., BIRKETT C., LETTENMEIR D.P., "Global monitoring of large reservoir storage from satellite remote sensing", *Water Resources Research,* vol. 48, p. W09504, 2012.

[GAR 07] GARCIN Y., WILLIAMSON D., BERGONZINI L. *et al.*, "Solar and anthropogenic imprints on Lake Masoko (southern Tanzania) during the last 500 years", *Journal of Paleolimnology*, vol. 37, pp. 475–490, 2007.

[GET 09] GETIRANA A.C.V., BONNET M.P., CALMANT S. *et al.*, "Hydrological monitoring of poorly gauged basins based on rainfall–runoff modeling and spatial altimetry", *Journal of Hydrology*, vol. 379, pp. 205–219. 2009.

[GET 12] GETIRANA A.C.V., PETERS-LIDARD C., "Water discharge estimates from large radar altimetry datasets in the Amazon basin", *Hydrol. Earth Syst. Sci. Discuss.*, vol. 9, pp. 7591–7611, 2012.

[HAS 93] HASTENRATH S., NICKLIS A., GREISCHAR L., "Atmospheric–hydrospheric mechanisms of climate anomalies in the western equatorial Indian Ocean", *Journal of Geophysical Research*, vol. 98, no. C11, pp. 20219–20235, 1993.

[HOS 14] HOSSAIN F., SIDDIQUE-E-AKBOR A.H., MAZUMDAR L.C. *et al.*, "Proof of concept of an altimeter-based river forecasting system for transboundary flow inside Bangladesh", *IEEE Journal of Selected Topics in Applied Earth Observations and Remote Sensing*, vol. 7, no. 2, pp. 587–601, 2014.

[HUA 11] HUANG L., LIU J., SHAO Q. *et al.*, "Changing inland lakes responding to climate warming in northern Tibetan Plateau", *Climate Change,* vol. 24, pp. 479–502, 2011.

[JAC 87] JACCON G., "Jaugeages de l'Amazone à Obidos par les méthodes du bateau mobile et des 'grands fleuves'", *Hydrol. Continent.*, vol. 2, no. 2, pp. 117–126, 1987.

[JUR 02] JURY M.R., GWAZANTINI E., "Climate variability in Malawi, Part 2: sensitivity and prediction of lake levels", *Int. J. Climatol.*, vol. 22, pp. 1303–1312, 2002.

[KAN 10] KANG S., XU Y., YOU Q. *et al.* "Review of climate and cryospheric change in the Tibetan Plateau", *Environmental Research Letters*, vol. 5, no. 015101, p. 8, 2010.

[KLE 15] KLEINHERENBRINK W., LINDENBERGH R.C., DITMAR P.G., "Monitoring of lake level changes on the Tibetan Plateau and Tian Shan by retracking Cryosat SARIn waveforms", *Journal of Hydrology*, vol. 521, pp. 119–131, 2015.

[KOU 04] KOURAEV A., ZAKHAROVA E., SAMAIN O. *et al.* "Ob river discharge from Topex/Podéidon satellite altimetry", *Remote Sens. Env.*, vol. 93, pp. 238–245, 2004.

[LEE 11] LEE H., SHUM C.K., TSENG K.H. "Present day lake level variation from Envisat altimetry over the North eastern Qinghai-Tibetan Plateau: links with precipitation and temperature", *Terr. Atmos. Ocean. Sci.,* vol. 22, no. 2, 2011.

[LEI 14] LEI Y., YANG K., WANG B. "Response of inland lake dynamics over the Tibetan Plateau to climate change", *Climate Change*, vol. 125, pp. 281–290, 2014.

[LEO 06] LEON J.G, SEYLER F., CALMANT S. "Hydrological parameters estimation for ungauged basins based on satellite altimeter data and discharge modelling, a simulation for the Caqueta River (Amazonian Basin, Colombia)", *Hydrol. and Earth System Sc Discuss.*, vol. 3, pp. 3023–3059, 2006.

[LI 07] LI X.Y., XU H.Y., SUN Y.L. "Lake-level change and water balance analysis at Lake Qinghai, west China during recent decades", *Water Resources Management*, vol. 21, pp. 1505–1516, 2007.

[LI 08] LI X., CHENG G., JIN H. "Cryosphere change in China", *Global Planet. Change*, vol. 62, pp. 210–218, 2008.

[LIB 08] LIBERT B., OROLBAEV E., STEKLOV Y., "Water and energy crisis in Central Asia", *China and Eurasia Forum Quarterly*, vol. 6, no. 3, pp. 9–20, 2008.

[LIU 00] LIU X.D., CHEN B.D., "Climatic warming in the Tibetan Plateau during recent decades", *International Journal of Climatology*, vol. 20, no. 14, pp. 1729–1742, 2000.

[LIU 09] LIU J., WANG S., YU S. "Climate warming and growth of high-elevation inland lakes on the Tibetan Plateau", *Global and Planetary Change*, vol. 67, pp. 209–217, 2009.

[LIU 98] LIU X., ZHENG M., "Contemporary climatic change over the Qinghai-Xizang plateau and its response to the green-house effect", *Chinese Geographical Science*, vol. 8, no. 4, pp. 289–298, 1998.

[MAR 06] MARCHANT R., MUMBI C., BEHERA S. "The Indian Ocean dipole – the unsung driver of climatic variability in East Africa", *African Journal of Ecology*, vol. 45, no. 1, pp. 4–16, 2006.

[MAS 94] MASON I.M, GUZKOWSKA M.A.J, RAPLEY C.G, *et al.*, "The response of lake levels and areas to climate change", *Climate Change*, vol. 27, pp. 161–197, 1994.

[MER 02] MERCIER F., CAZENAVE A., MAHEU C., "Interannual lake level fluctuations in Africa from TOPEX-Poseidon: connections with ocean-atmosphere interactions over the Indian Ocean", *Global and Planet. Change*, vol. 32, pp. 141–163, 2002.

[MIS 03] MISTRY V., CONWAY D., "Remote forcing of East African rainfall and relationships with fluctuations in levels of Lake Victoria", *Int. J. Climatol.*, vol. 23, no. 1, pp. 67–89, 2003.

[NIC 96] NICHOLSON S.E., "A review of climate dynamics and climate variability in Eastern Africa", in JOHNSON T., ODADA E. (eds) *The Limnology, Climatology and Paleoclimatology of the East African Lakes*, Gordon and Breach, pp. 25–56, 1996.

[NIC 02] NICHOLSON S.E., YIN X., "Mesoscale patterns of rainfall, cloudiness and evaporation over the Great Lakes of East Africa", in *The East African Great Lakes: Limnology, Paleolimnology and Biodiversity*, Kluwer Academic Publishers, p. 587, 2002.

[OGA 88] OGALLO L.J., "Relationships between seasonal rainfall in East Africa and the southern oscillation", *Journal of Climatology*, vol. 8, pp. 31–43, 1988.

[PAP 10] PAPA F., DURAND F., ROSSOW W. *et al.*, "Satellite altimeter-derived monthly discharge of the Ganga–Brahmaputra River and its seasonal to interannual variations from 1993 to 2008", *J. Geophys. Res.,* vol. 115, no. C12013, 2010.

[PAP 12] PAPA F., BALA S.K., PANDEY R.K. "Ganga–Brahmaputra river discharge from Jason-2 radar altimetry: An update to the long-term satellite-derived estimates of continental freshwater forcing flux into the Bay of Bengal", *Journal of Geophysical Research,* vol. 117, no. C11021, p. 13, 2012.

[PAR 15] PARIS A., Utilisation conjointe de données altimétriques et de modélisation pour le calcul de débits distribués dans le bassin Amazonien, PhD thesis, Paul Sabatier University, France, 2015.

[PHA 12] PHAN V.H., LINDENBERGH R., MENENTI M., "ICESat derived elevation changes of Tibetan lakes between 2003 and 2009", *International Journal of Applied Earth Observation and Geoinformation,* vol. 17, no. 1, pp. 12–22, 2012.

[PHA 13] PHAN V.H., LINDENBERGH R.C., MENENTI M., "Geometric dependency of Tibetan lakes on glacial runoff", *Hydrol. Earth Syst. Sci. Discuss.*, vol. 10, pp. 729–768, 2013.

[PLI 01] PLISNIER P.D., COENEN E.J., "Pulsed and dampened annual limnological fluctuations in Lake Tanganyika", in MUNAWAR M., HECKY R.E. (eds), *The Great Lakes of the World: Food-web, Health and Integrity*, Backhuys, Leiden, Netherlands, pp. 83–96, 2001.

[RIC 12] RIČKO M., BIRKETT C. M., CARTON J.A. *et al.*, "Intercomparison and validation of continental water level products derived from satellite radar altimetry", *J. of Applied Remote Sensing*, vol. 6, no. 061710, 2012.

[RIC 94] RICHARD Y., "Variabilité pluviométrique en Afrique du Sud-Est", *La Météorologie*, vol. 8, pp. 11–22, 1994.

[ROD 15] RODRIGUEZ E., "Surface Water and Ocean Topography Mission (SWOT) project", Science requirements document, JPL D-61923, February 2015.

[ROP 96] ROPELEWSKI C.F., HALPERT M.S., "Quantifying southern oscillation precipitation relationships", *Journal of Climate*, vol. 9, pp. 1043–1059, 1996.

[SAJ 99] SAJI N.H., GOSWAMI B.N., VINAYACHANDRAN P.N. *et al.*, "A dipole mode in the tropical Indian Ocean", *Nature*, vol. 401, pp. 360–363, 1999.

[SIL 10] SILVA J.S., CALMANT S., ROTUNO FILHO O.C. *et al.*, "Water levels in the Amazon basin derived from the ERS-2 and Envisat radar altimetry missions", *Remote. Sens. Env.*, vol. 114, pp. 2160–2181, 2010.

[SON 14] SONG C., HUANG B., RICHARDS K. *et al.*, "Accelerated lake expansion on the Tibetan Plateau in the 2000s: Induced by glacial melting or other processes?", *Water Resources Research*, vol. 50, pp. 3170–3186, 2014.

[SON 15] SONG C., HUANG B., KE L., "Heterogeneous change patterns of water level for inland lakes in High Mountain Asia derived from multi-mission satellite altimetry", *Hydrol. Process.*, vol. 29, pp. 2769–2781, 2015.

[STA 05] STAGER J.C., RYVES D, CUMMING B.F. *et al.*, "Solar variability and the levels of Lake Victoria, East Africa, during the last millennium", *Journal of Paleolimnology*, vol. 33, pp. 243–251, 2005.

[STA 07] STAGER J.C., RUZMAIKIN A., CONWAY D. *et al.*, "Sunspots, El Niño, and the levels of Lake Victoria, East Africa", *J. Geophys. Res.*, vol. 112, p. D15106, 2007.

[SWE 09] SWENSON S., WAHR J., "Monitoring the water balance of Lake Victoria, East Africa, from space", *J. Hydrol.*, vol. 370, pp. 163–176, 2009.

[TIE 07] TIERNEY J.E., RUSSELL J.M., "Abrupt climate change in southeast tropical Africa influenced by Indian monsoon variability and ITCZ migration", *Geophys. Res. Lett.*, vol. 34, p. L15709, 2007.

[VER 03] VERSCHUREN D., "Lake-based climate reconstruction in Africa: progress and challenges", *Hydrobiologia*, vol. 500, pp. 315–330, 2003.

[VOR 05] VÖRÖSMARTY C.J., DOUGLAS E.M., GREEN P.A. *et al.*, "Geospatial indicators of emerging water stress: An application to Africa", *Ambio*, vol. 34, no. 3, pp. 230–236, 2005.

[WAN 08] WANG B., BAO Q., HOSKINS B. "Tibetan Plateau warming and precipitation change in East Asia", *Geophys. Res. Lett.*, vol. 35, p. L14702, 2008.

[WCD 00] WORLD COMMISSION ON DAMS, "Dams and development: A new framework for decision-making", World Commission on Dams report, available at http://www.internationalrivers.org/files/attached-files/world_commission_on_dams_final_report.pdf, 2000.

[WEI 06] WEINTHAL E., "Water conflict and cooperation in Central Asia", Human Development Report, UNDP, p. 36, 2006.

[WU 08] WU Y., ZHU L., "The response of lake-glacier variations to climate change in Nam Co Catchment, central Tibetan Plateau, during 1970–2000", *Journal of Geographic Science*, vol. 18, pp. 177–189, 2008.

[WU 14] WU Y., ZHENG H., ZHANG B. *et al.*, "Long-term changes of lake level and water budget in the Nam Co Lake Basin, Central Tibetan Plateau", *Journal of Hydrometeorology*, vol. 15, pp. 1312–1322, 2014.

[XU 05] XU C.Y., WIDEN E., HALLDIN S., "Modelling hydrological consequences of climate change – progress and challenges", *Advances in Atmospheric Sciences*, vol. 22, no. 6, pp. 789–797, 2005.

[YAO 07] YAO T., PU J., LU A. "Recent glacial retreat and its impact on hydrological processes on the Tibetan Plateau, China, and surrounding regions", *Arct. Antarct. Alp. Res.*, vol. 39, no. 4, pp. 642–650, 2007.

[ZAK 06] ZAKHAROVA, E., KOURAEV A., CAZENAVE A. "Amazon River discharge estimated from T/P altimetry", *Compte Rendus Géosciences*, vol. 338, no. 3, pp. 188–196, 2006.

[ZHA 10] ZHANG G., XIE H., ZHU M., "Water level changes of two Tibetan lakes Nam Co and Selin Co from ICESat Altimetry data", *Second International Conference on Geoscience and Remote Sensing*, 2010.

[ZHA 11] ZHANG G., XIE H., KANG S. *et al.*, "Monitoring lake level changes on the Tibetan Plateau using ICESat altimetry", *Remote Sens. Environ.*, vol. 115, pp. 1733–1742, 2011.

[ZHO 09] ZHOU X., ZHAO P., CHEN J. *et al.*, "Impacts of thermodynamic processes over the Tibetan Plateau on the Northern Hemispheric climate", *Science in China Series D: Earth Sciences*, vol. 52, no. 11, pp. 1679–1693, 2009.

[ZHU 14] ZHU W., JIA S., LV A., "Monitoring the fluctuation of Lake Qinghai using multi-source remote sensing data", *Remote Sensing*, vol. 6, pp. 10457–10482, 2014.

7

Radar Altimetry for Monitoring the Antarctic Ice Sheet

7.1. Introduction

The Antarctic polar ice cap is by far the largest glacial mass in the world, the largest freshwater reservoir, and over time it could be a significant contributor to rising sea levels. The mass in Eastern Antarctica, in other words the portion spanning from 0 to 180° Longitude E, is relatively stable, but it appears that in the past couple of years, certain glaciers in Western Antarctica have become increasingly thinner. It has therefore become imperative to monitor changes in the volume of this immense cap. It is equally crucial to understand the causes of this loss in volume in order to be able to model their changes in volume, notably in terms of rising sea levels.

Surface topography is one of the most pertinent parameters for monitoring the ice cap, but also for the constraint of ice flow models. Radar altimetry on high latitude polar zones started in 1991 with the launch of the ERS-1 mission by the European Space Agency, the first mission using specialized instruments on board for the studying of poles, for which an inclined orbit of 81.6° enabled us to fly over almost all of Greenland in the Northern Hemisphere, and more than 75% of the Antarctica in the Southern

Chapter written by Frédérique REMY.

Hemisphere. In addition, radar altimetry, which is primarily used to measure distances, also provides observations on snowpack that can be linked to climatological parameters.

We will first start by familiarizing ourselves with the continent and understanding how its topography is one of the most pertinent parameters for observing and studying the ice cap through radar altimetry. We will then look at the specific features of this technique when applied outside of an oceanic context, for which it was originally designed, and will address the different subject areas for which it has a notable contribution: climatology, ice dynamics and lastly the monitoring of its variations in volume.

7.2. Antarctica

7.2.1. *The continent of superlatives*

The polar ice caps, and especially Antarctica's, play a very important role for the climate because they are both the world's glacial archives and a sensitive indicator of current fluctuations in climate. Antarctica (Figure 7.1) has a surface of 14 million km^2 and an average elevation of 2,000 m, that can sometimes exceed 4,000 m. It represents 90% of terrestrial ice and contains the equivalent of 66 m of ocean levels. The Transantarctic Mountain Range separates the continent into two parts, east and west. Western Antarctica is smaller and characterized by having the majority of its bedrock, which the ice rests upon, below sea level. Between these two sections, the Ross and the Ronne-Fichner ice shelves, each measuring around 500,000 km^2 (the surface area of France), serve as reinforcement for the ice contained inside.

Antarctica is the coldest, the highest on average, the driest and the windiest continent of planet Earth. The temperature decreases from the coast going inwards from -15°C to -60°C on average, with record freezing temperatures approaching -90°C. The air is very dry at this temperature and it snows less here than it rains in the Sahara! The cold and dense air from the center of the continent descends the slopes creating katabatic winds (from the Greek word *katabatikos* which means going down the slope), strong and persistent winds.

At Dumont d'Urville, their average speed nears 40 km/h, the maximum attaining easily three or four times this value. These winds sculpt the surface of the snow from a centimeter scale (microroughness) to a metric scale (sastrugi). They move enormous amounts of snow, making the possibility of having direct measurements for snow accumulation rates unrealistic over a small period of time, and are also the cause of one of the difficulties in obtaining indirect measurements by remote sensing. In fact, the majority of sensors used operate in the microwave range, with a wavelength typically ranging from around 10 cm to a couple of millimeters, commensurable with the surface's microroughness that therefore plays a major role in the behavior of this radar signal in this frequency range. The size of the continent and its climatic conditions explain the major contribution of remote sensing observations. However, the interaction between the radar wavelength and the rough surface, low density and cold, is complex. *In situ* measurement allowing us to describe the environment, at least in statistical terms, remain therefore essential for understanding and modeling this interaction.

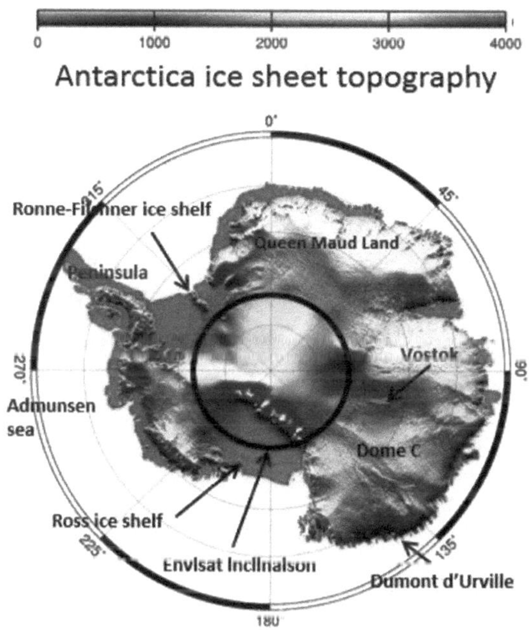

Figure 7.1. *The topography of Antarctica obtained from the geodetic orbit ERS-1. The central circle indicates the limit of the altimeter's visibility and the names of the cited places in this chapter are indicated. In situ data were used to extrapolate the topography that is out of ERS-1's reach*

7.2.2. *The Antarctic machine*

Snowfalls are very light; expressed in the water equivalent, they vary by a couple of centimeters per year in the center to a few tens near the coast. Accross all of Antarctica, this represents close to 2,200 km^3/year, a volume that, distributed on the ocean surface, represents a sea level variation of 6 mm. In other words, a slight imbalance of snowfall can contribute to the variations in the sea level. Over time, new snowfalls accumulate in layer formations that sink, and gradually transform into ice. This ice, as a result of its own weight, flows from upstream to downstream in the downslope direction of the surface. The velocity of flow is quite weak: a couple of meters per year in the center of the continent and several tens near the coast, or even more for certain outlet glaciers on the coast. Thus, snow takes several thousand years to return the ocean, a large majority in the form of icebergs. This long period of residence turns these polar ice caps into "Planet Earth's glacial archives." Thus, core drilling in Vostok allowed us to retrieve climatic terrestrial history going back several hundreds of thousands of years, or several glacial cycles each about one hundred thousand years old. It showed without ambiguity the link between greenhouse gases and the temperature. The European EPICA C Dome core drilling allowed us to go back around 800,000 years, or eight climatic cycles.

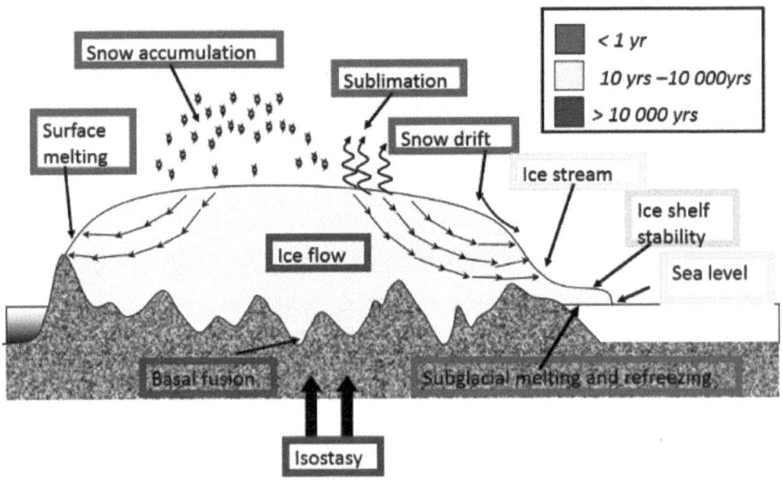

Figure 7.2. *The functioning of a polar ice cap and the reaction time of the different components*

The volume of Antarctica is controlled by the balance between the gains, the precipitation and losses of snow, essential due to the discharge of ice, melting on the surface being negligible. The behavior of the ice cap is thus determined by a range of different meteorological, dynamic or mechanical processes that all depend on the climate at different time scales (Figure 7.2). On a large scale, the profile of the ice cap's surface is quasi-parabolic and shows the balance of these processes. In the center, the slope is very weak, less than a few meters per kilometer, and increases towards the coast. The atmospheric processes (snowfalls, sublimation, melting on the surface) react instantaneously to climate variations. For example, for warmer air storing more humidity, warming will create an excess of precipitation. Antarctica is also sensitive to boundary conditions at the sea level, to the stability of the ice shelves, etc., that react according to longer time scales, a thousand or even tens of thousands of years. Nevertheless, the flow of glaciers in the Amundsen Sea currently seems to be accelerating on a ten year time scale. Lastly, snow is an excellent insulator. A "wave" of temperature will take a significant amount of time before substantially changing the temperature of the ice and in turn the flow velocity. A variation in the volume of the ice will change the way in which it presses on the bedrock, which will in turn, adjust by isostasy and will modify the boundary conditions. The latter processes react very slowly with regards to the climate. The ice cap thus lives and reacts to the rhythm of all climatic fluctuations, from a seasonal scale to the scale of major glacial cycles [RIT 01]. Mathematically, the ice cap cannot be balanced and the imbalances need to be measured.

7.2.3. *What and how to observe?*

In order to date core drilling, constrain the discharge of ice and evolution models, establish the mass balance or to anticipate the effects of current warming, numerous characteristics are necessary. The laws of mechanics, the physical processes that allow the ice to deform or to slide, the role of boundary conditions on the coast or at the base of the cap, and the effect of longitudinal constraints are the subject of many debates. The spatio-temporal distribution of the snow's accumulation rates is not highly constrained. The snowpack holds much information, the size of snow grains, the density of the snow, the stratification; enough meteorological parameters necessary to find the accumulation rates and to measure changes. The ultimate question is obviously the health status of the ice sheet. Does it grow due to a surplus in the accumulation rate because warmer air contains more humidity and precipitation? Or does it lose mass due to the acceleration of outlet glaciers as we are starting to observe in certain locations?

On board satellites, numerous sensors are used to understand the different unknowns. These measuring instruments operate in the microwave, infrared or visible ranges. The snow is cold, dry and not very dense, to such an extent that microwaves penetrate and emit or send back information about the snowpacks' surface area and depth. For example, the scatterometer allows us to measure the direction of katabatic winds through the surface anisotropy that they create. The radar interferometry is very important in measuring the speed of discharge from outlet glaciers. The optical or near infrared sensors are sensitive to albedo, the size of the snow grains or the roughness of the surface. The strongest limitation of data interpretation originates from the sensors on board satellites, for which the altimeter is the indirect correlation between all of the parameters that control measurements. The surface altitude, the velocity of wind, the surface slope, the surface roughness, the snow precipitation, the temperature, and the surface undulations increase from the center towards the coast. All of these parameters are strongly geographically correlated.

The altimeter is one of the most pertinent sensors because it measures the surface topography which contains all of the processes that play a role, as much meteorological as dynamic. Moreover, successive observations allow us to estimate the mass balance with precision. However, the altimeter measures the surface and therefore the volume of the ice cap and its variations. We cannot distinguish a precipitation deficit from a loss of mass through discharge. For this it is necessary to have coupled measurements with gravimetric measurements (see [RAM 16] and Chapter 5 in this book). For example, the coupled data from the satellite GRACE and from the altimeter ENVISAT, that took flight around the same time, allow us to separate the losses or the gains due to variations in precipitation from those due to variations in discharge [MEM 14]. On the other hand, the radar altimeter is not very precise near the coast, where changes in volume are much more important. Optical stereography techniques therefore replace the altimeter, this is why a project dedicated to the construction of a polar glacier topographic map using images from an HRS on SPOT-5 high-resolution spatial sensor was made possible during the International Polar Year [KOR 09].

7.3. Polar altimetry

7.3.1. *Some specifics of altimetry on polar ice caps*

The altimetry technique was presented in the chapter from [CAL 16], here, we shall only be discussing about specifics linked to measurements on the polar ice caps. The first amongst these is obviously the satellite's orbit. The altimeters classically dedicated to oceanography, the French-American families Topex-Poseïdon followed by the Jason series are on an orbit with an inclination of 66° that only allows for flight over southern Greenland and only barely touches upon Antarctica. Therefore altimeters aboard the European satellites ERS-1 (1991–2000), ERS-2 (1995-2003) and ENVISAT (2003-2010) must be used, all of them inclining at 81.6° and on the same orbit for 35 days, theoretically assuring the repeated monitoring of the surface on the same track for close to 20 years. ERS-1 theoretically had a geodesic orbit from April 1994 until March 1995 (two cycles of 168 days), a drifting orbit that had provided a very dense spatial network. The second specific is the slope error. Altimetry does not measure the distance between the satellite and the sub-satellite point (the nadir), but the smallest distance between the satellite and the ground. If the surface is sloppy, the point of impact where the measurement is taken is shifted upwards. This error depends on the square of the slope, therefore it very quickly becomes prohibitive. Poor repetitivity and the presence of a perpendicular slope also cause errors in height measurement.

A third specific comes from strong topography, in relation to topography for oceans for which the altimeter is optimized. The tracking loop is not able to monitor topographic fluctuations, to such an extent that altimetric waveform is no longer centered in the receiving window. The data must therefore be revised to calculate the height of the surface. This process, named retracking, benefits from several different algorithms dedicated to polar ice caps. Ice-1 calculates the center of gravity of the waveform and searches for the rising edge's position for which the energy is proportional to the center of gravity. Ice-2 adjusts the rising edge by an error function and uses its center as a corresponding point to the surface. The advantage of this algorithm is that, in addition to estimating the height, it estimates the width of the leading edge and the slope of the trailing edge, as well as the total backscatter.

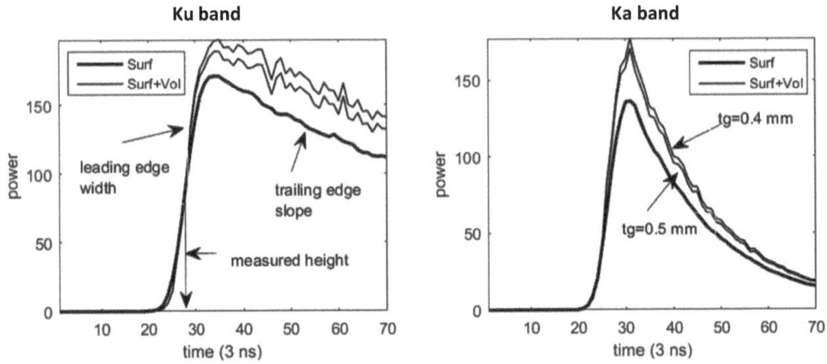

Figure 7.3. *Simulated altimetric waveform on a Ku-band (13.6 GHz to the left) and on the Ka-band (36 GHz to the right) and the parameters coming from the retracking algorithm "Ice-2." The surface slope is 1 m/km. The volume simulations were created using two grain sizes: tg = 0.4 and 0.5 mm*

To finish, the fourth specific, by far the most critical, is the error of penetration, because the radar wavelength penetrates the dry and cold snow. The penetration depth depends both on the sensor's frequency and the dielectric properties of the environment, that are generally little or poorly known. Depending on the environment at hand, the "volume" echo is more or less commensurable with the surface echo and the surface is seen as being farther away than it is. This additional echo is represented on Figure 7.3. The error of the distance measurement depends on the relationship between the intensity of the surface echo and that of the volume, the two echoes being eminently variable. The first is very sensitive to the surface roughness, which varies with the wind, and the second depends on snowpack on a seasonal scale. This variability creates fluctuations in the height measured concerning the average depth. Moreover the properties of snowpack can evolve and therefore create a bias on the height measurements that would indicate artificial variation in volume [REM 09].

Since March 2013, the French National Center of Spatial Studies (CNES) and the Indian agency ISRO (Indian Space Research Organization) have put on this same orbit, inclined at 81.6°, the SARAL satellite (Satellite with ARgos and ALtika, or "simple" in Hindi) that took on board the AltiKa altimeter. This altimeter operates on the Ka band (35.75 GHz) instead of the Ku band (13.6 GHz) from previous altimeters. At this frequency, the ionospheric effect causing propagation delays is lower, and we thus do not need the presence of another instrument aboard, functioning at a low-frequency, to correct the induced error (C-Band "5.6 GHz" for the Jason series and the S-band "3.6 GHz" for the European satellite ENVISAT). Moreover, there may be a smaller antenna present for the same resolution as altimeters on the Ku-band. This instrument, lighter and less expensive, was originally designed to move towards an altimetric constellation. The theory of electromagnetic scattering suggests that the volume echo should be lower on the Ka-band than the Ku-band, due to weaker wave penetration (Figure 7.3). The first results showed that the volume echo on the Ka-band is stronger than the Ku-band, but only comes from the first tens of centimeters underneath the surface, and thus has a lesser effect on the height restitution [REM 15].

7.3.2. *Methodology of evaluating a time series for measuring height*

Two major sources of error complicate the process of evaluating a time series. The first is a slope perpendicular to the track that allows for a different understanding of the ground altitude from one passage to another. In fact, a track from a cycle goes back to ± 1 km of the nominal track, to such an extent that for a slope with an average of a couple m/km, the impact on the height measurement is prohibitive. Classically, to avoid this error, the variation of the height is calculated by using the cross points from the satellite's ascending and descending tracks. At the crossover point of the tracks, the slope is identical and cancels out by subtraction. In order not to limit ourselves to these points, a team from LEGOS in Toulouse currently use data from all along the track, and take advantage of poor repetition from one cycle to another to estimate the surface slope. This dataset is 20 times denser and above all allows us to visualize the continuous profiles. This density will allow us to describe in detail all of the causes for change on the surface [REM 14] through abstaining from this first source of error.

The second source of error for evaluation of a time series is the time variation of the radar wavelength's penetration within the snow. When backscatter increases through added echo volume, we can note that the height of the surface is observed to be lower, the leading edge stretches out and the trailing edge slope rises (Figure 7.3). We therefore apply an empirical correction while adjusting the height fluctuations and those from retracking parameters.

When using altimetric series, the error due to the poor repetition from the satellite ground track and from the presence of a slope must be corrected. Flament and Rémy [FLA 12] have shown that we must adjust a biquadratic form of unknowns x and y, or six parameters (a constant, the two terms in x and y, as well as those from the second degree, x^2, y^2, xy). In order to reduce the height variations due to wavelengths penetrating into snowpack, we therefore use three classical parameters from the waveform and we calculate the sensitivity of the measured height for each of the three. For temporal variations, we extract the tendency, but also the seasonal variations, in other words three additional parameters (the time as well as a cosine and a sine for a year-long period). We therefore have an equation with 12 unknowns to solve, more constrained than the temporal series is long. Remy *et al.* [REM 14] show that after around 50 cycles, the error on the locally measured height trend is a few percentage points below the rate of snow accumulation. The satellite ENVISAT flew for eight years, offering 85 repetitive cycles, which represents a large quantity of data. The inversion of all these parameters is thus very robust.

As the outcome of these inversions, the temporal trends observed re-establish height measurements and for backscatter, their seasonal variations, and an excellent estimation of the surface slope on a kilometric scale are re-established.

7.4. Contribution to climatology

Altimetry does not play a major role in a climatological context. However, it offers additional observations from other high-frequency sensors, such as the radiometer or the scatterometer. Moreover, to correctly interpret the height variations and to detect all possible artifices that could affect the measured trends, it is also as important to thoroughly understand the "physics of measurement." The waveform is controlled by the intensity of the volume and surface echo. The wavelength's penetration in an environment leads to the leading edge stretching out and the lifting of the trailing edge (Figure 7.3). The comparison between altimetric radar data on the Ku band and the data from the ICESat laser altimeter, that only "sees" the surface of the snow, as well as the inversion of waveforms, indicate an average penetration depth of 2 to 5 m [MIC 14]. This is principally controlled by volume scattering from snow grains. Larger grains increase the loss by scattering, which diminishes the penetration depth. On a seasonal scale, the variations in the waveform's parameters essentially come from the densification of snowpack and from the growth in snow grains. The signal to up to several meters deep can still contribute to the volume echo, but after reprocessing waveforms the error induced on the height is only about a meter.

The successive layers of snow evolve throughout the seasons, accumulateing so the snowpack is thus stratified. This stratification induces dielectric discontinuities at each stratum. The importance of the dielectric constant's variability on the echo depends on the altimeter frequency. We can show on the S-band (3.2 GHz) the variability of the dielectric constant strongly influences the shape of the echo, is insignificant on the Ka-band (35.75 GHz) and comparable to snow grains scattering on the Ku-band (13.6 GHz), the altimeters's nominal frequency [REM 15].

Just as on the ocean, wind creates several scales of roughness on the snow. The microroughness essentially plays a role in the backscatter which is also affected by the surface's slope via the antenna gain. The sastrugi also stretch out the leading edge of the altimeter. Moreover, the scale of roughness to which the radar is sensitive depends on the length of the wavelength used.

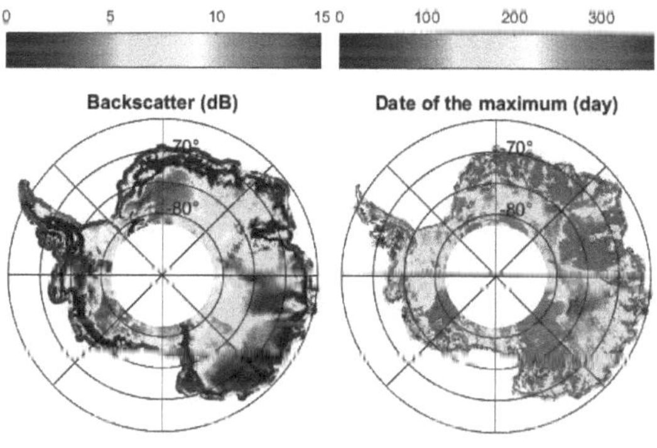

Figure 7.4. *The backscatter coefficient (to the left) varies greatly, close to 15 dB (or close to a factor of 30 in power). To the right, the spatial distribution of the maximum backscatter date shows a bimodal distribution with an extremely marked geography transition. For a color version of this figure, see www.iste.co.uk/baghdadi/4.zip*

To finish, we can observe on a seasonal scale, one of the mysterious phenomena of the altimeter in Antarctica. The spatial distribution of the maximum backscatter date is clearly a bimodal distribution and above all geographically well marked (Figure 7.4). The maximum sometimes takes place in the beginning of April at the end of the austral summer, sometimes in the beginning of October, at the end of winter. It is difficult to imagine a climatologic forcing with the same signature. It seems more reasonable to imagine that from one environment to another, the altimeter reacts in opposite phases with the same forcing… what's left to determine is which one and why! We have started to gather a few clues. The altimeter on the S-band ("lower" frequency, 3.2 GHz) from ENVISAT has operated for close to five years. The maximum backscatter seems to systematically take place in the beginning of April. However, the evaluation of the three first years of AltiKa backscatter on the Ka-band ("higher" frequency), suggests that the maximum backscatter takes place in October. We can see a connection with the altimeter's frequency. A different understanding of volume or of the roughness depends on the wavelength? The problem remains unresolved.

Even with the help of other sensors, the problem largely remains undetermined. There is a clear need to extract the geophysical parameters and to better constrain the measurement, for in situ measurements in order to support the electromagnetic models.

Let's cite another mystery of the altimeter for ice that strongly illustrates the complexity of the interaction between the electromagnetic wave, the random rough surface, and the more or less dense volume. Structures related to the slope's orientation, or to the macroroughness orientation, clearly appear when the difference between the measurements of an ascending track and that of a descending track is mapped. The altimeters's antenna has an unidirectional linear polarization. The wavelength's capacity to penetrate, more or less, in the snow depends on the angle between the direction of the antenna's polarization and that of the surface's macroroughness, so from the geometric configuration between the track and that of the roughness! The altimeter's behavior of the surface between the ascending and descending track is therefore different without us really understanding why.

7.5. Antarctica in a stationary state

We suppose here that that Antarctica is in a stationary state, meaning that it is in perfect equilibrium and as a result the surface does not vary with time. Here, we are looking to collect main characteristics of ice dynamics.

7.5.1. *Surface topography*

From a kilometer to hundreds of kilometers or even on a continental scale, the ice sheet shape is very much marked by all of the current or past, climatological or dynamic processes that control change (Figure 7.1), to such an extent that the deduced topography from the geodesic orbit, ERS-1, provides vast amounts of information. Firstly and on a global scale, the profile of the polar ice cap is quasi-parabolic, like a plastic body on an uneven base. We have seen this previously, it reflects the balance between the snow precipitations and the evacuation of ice and the boundary conditions. On a scale of a few hundreds of kilometers, structures appear very stretched, barely visible to the naked eye and only a mathematic transformation allow us to bring it to light. For this, we calculate the curvature either in the direction of the slope x, or in the perpendicular direction y, in other words $\partial h^2/\partial x\partial x$ and $\partial h^2/\partial y\partial y$ respectively (h being the elevation). Two families of very coherent structures are therefore clearly distinguishable. The perpendicular curvature at the slope (Figure 7.5) clearly shows the discharge of ice and separates the different basins. This curvature is mathematically related to the expression showing the convergence or the local divergence of the discharge, either $1/l(x)$ $\partial l/dx$ or $l(x)$ is the width of the channel.

Intuitively, two parallel channels receive the same amount of snow and have the same outflow. If one discharges more slowly, it will be slightly thicker and vice versa. This irregularity of height is controlled by valleys in the bedrock near the coast, suggesting that it is caused here and propagates upstream to the summit, thus allowing a visualization of the drainage and the main basins. The most surprising observation is the alternation between the rapid channels and slow channels at the coast with a wavelength of 250 km. This unexplained phenomenon could be due to a complex feedback effect between the erosion of the base and the ice velocity [REM 97]. The parallel curvature at the surface slope also shows very coherent stretched structures that could possibly be caused by the bedrock's topography, but nothing has really been explained concerning this topic yet.

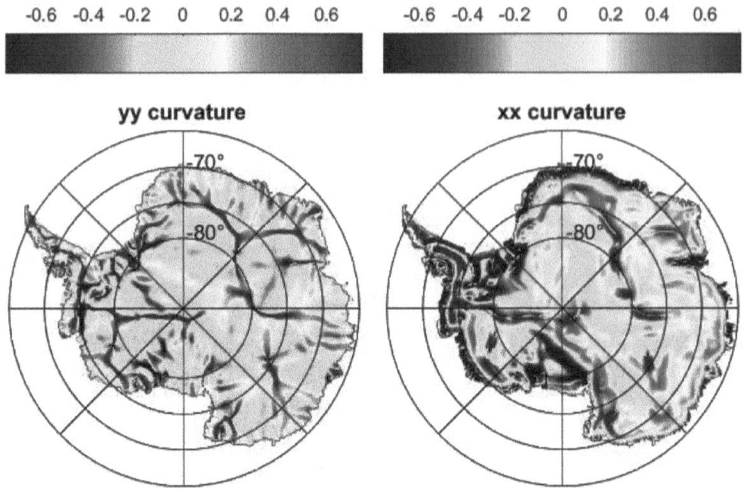

Figure 7.5. *Curvature at a scale of 100 km, expressed in cm/km², perpendicular to the slope (to the left) and parallel to the slope (to the right). In both cases, extremely coherent stretched structures appear: to the left they show the discharge and allow us to visualize the different basins, to the right, they are possibly caused by the bedrock and remain unexplained. For a color version of this figure, see www.iste.co.uk/baghdadi/4.zip*

On a scale of a few tens of kilometers, we see some features directly linked with ice dynamics. We can see a smooth and very flat zone that is a couple of hundreds of kilometers long and a few dozen kilometers wide near Vostok (Figure 7.1). This is the sign of a sub-glacial lake. The Vostok Lake is the largest sub-glacial lake in Antarctica, but there are numerous others. In fact, the bottom of the ice cap near the bedrock is isolated from atmospheric cold by thousands of meters of ice and heated by the geothermal flow to such an extent that, according to the shape of the bedrock, the melted water is discharged and stored to the point of forming lakes. The best-known amongst them is the Vostok Lake, whose existence was suspected in the 1960's because echoes obtained through low-frequency radars that were used to estimate the thickness of the bedrock had unusual intensity. Its existence was confirmed after launching ERS-1, given the signatures of the lake on the surface. The sliding of the ice under water creates a flat and smooth zone, contrary to what is produced on the bedrock. Since, several hundreds of lakes have been indexed either by echoes from low-frequency radars or by their altimetric signature on the surface. Some of these lakes seem to be aligned, and on a scale of a few kilometers the curvature of the surface has made very thin channels visible, some of which are several hundreds of kilometers long and link the lakes to each other. We will see that the "emptying" of these lakes can be detected and monitored at a low altitude.

Lastly, the signature that is probably the most characteristic of the surface is the presence of undulations with wavelengths of 10–20 km long, with an amplitude of a few meters that cover the entirety of the continent (Figure 7.1). These undulations, weak in the center and stronger at the edge, are due to the effect of the bedrock on ice flow, softened to a degree depending on the thickness of the ice.

7.5.2. *Constraints of the models*

The topography of the surface is also a large constraint for ice flow models. The direction of the largest slope dictates the direction of the flow, the angle between the two directions of the largest slope makes way for the flow convergence or divergence, and the intensity of the slope controls the basal constraint, an important parameter of the flow. Thus, a very simple model with precise topography allows us to deduce the velocity of the ice's flow, supposing that the ice cap is in equilibrium. The model assumes that ice is incompressible and that, between two flows of ice discharge, the outgoing flow of ice is equal to the entering flow added with precipitation. The topography provides the direction and the convergence of the flow, the knowledge of snow accumulation rates and the density of the ice allow us to reconstruct the velocity field. Until this point, Antarctica seemed to be a continent relatively inert and with uniform flow. In early 2000, we noticed that this was not true [TES 03]. The ice on average deforms extremely slowly. Over more than 80% of the ice cap, the velocities are less than 10 m/year! However, on several occurrences outlet glaciers have attained velocities of 100 m/year (Figure 7.6). In fact, the polar ice cap is drained by several major glaciers that only represent a small percentage of the Antarctic coast. This illustrates the need to monitor certain outlet glaciers; because they control Antarctica's volume. Each year, the density of the ice, the flow of ice, and the internal ice stratifications are measured in different locations on the terrain. This data, coupled with spatial observations, allows us to thoroughly understand how these glaciers operate, to follow their evolution given climatic change, and to constrain these models [LEM 14].

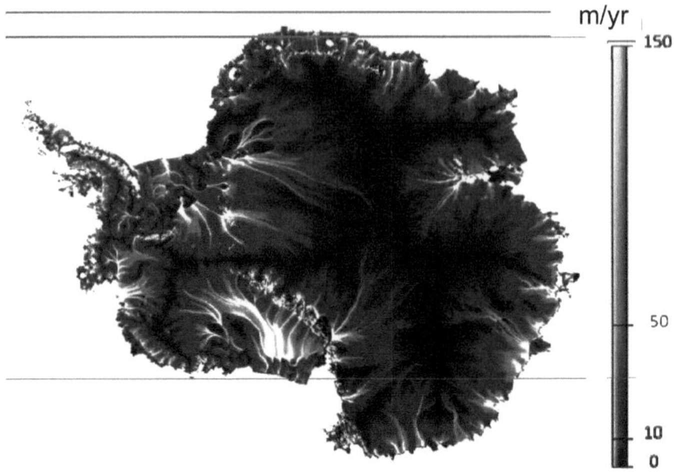

Figure 7.6. *Velocity of ice flow, on a logarithmic scale. The majority of the continent's velocities are less than 10 m/year. However, they can be several orders of magnitude higher for some outlet glaciers.They represent a small percentage of the coast and drain the majority of the ice cap. For a color version of this figure, see www.iste.co.uk/baghdadi/4.zip*

The models used by glaciologists are extremely complex [RIT 01] and take into account physics and boundary conditions that are increasingly representative of reality. Numerous satellite observations and *in situ* observations are necessary to parameterize the different processes and to force the current initial and past conditions. The topography is used for boundary conditions, to test these models or for the initial conditions of forecasting models. For example, it allows us to better constrain the models for dating core drilling [PAR 04], especially if they are not situated on a dome. For this, we must know where the ice comes from upstream and the path it has taken. The precise topography, as well as density measurements of the ice, has allowed us to "age" certain layers of the core drilling from Vostok and to better agree upon the chronology of other core drillings.

7.6. Temporal variations

7.6.1. *Very different times and signals*

Figure 7.2 suggests that, mathematically, Antarctica cannot be in equilibrium, as long as the dynamic and climatic processes that react to its volume are numerous and react on very different time scales. We have seen this, as Antarctica responds very quickly to all fluctuations in accumulation rates. We expect that in the future these accumulation rates will increase because warmer air can contain more humidity. The warming of coastal sea waters has already brought about a noticeable acceleration in velocity from outlet glaciers in the Amundsen Sea. However, the heat wave from the beginning of the Holocene, 15,000 years ago, has still not reached the base of the ice cap where there was a deformation of ice. The central part of the Antarctic still discharges in a glacial rhythm, and should continue to increase in volume. Finally, after a minuscule perturbation, Antarctica has taken time to come back to its state of equilibrium: this is the relaxation time. This perturbation can potentially be a local fluctuation of the accumulation rates of snow. This time can be calculated and only depends on the ice velocity, the surface slope and the snow density [REM 02]. It varies by a couple of hundred years at the coast at the same levels of outlet glaciers, up to a couple of dozen or hundreds of thousands of years in the center. In other words, the surface of the ice sheet cannot be balanced between two fluctuating accumulation rates for which the variability is estimated at 10% of the average value over approximately 30 years. The surface of the polar ice cap varies randomly during snow falls that are of varying importance, fluctuating around the average height but with the potential to vary greatly. We therefore show that the probability of an artificial decrease in Antarctica's volume, provoking a 0.5 mm/year increase of sea levels, is 12%!

We are therefore waiting for strongly different spatio-temporal signatures, but this would be without counting on the variable behavior of numerous phenomena independent of climatic change. Before establishing the global mass balance, we will look at an example of local variability, independent of climate: the draining of sub-glacial lakes.

7.6.2. *Lake draining*

When we evaluate a time series obtained through remote sensing, the problem concerning data editing arises. In fact, depending on the sensor used, we sometimes observe isolated data that clearly go beyond the error bar. Classically, we remove data that differs from the average greater than 3 σ. Nevertheless, this data somtimes corresponds to a specific geophysical signal. Thus, if we look at ENVISAT's height maps every 35 days, the variation should be weak and continuous. However, Flament *et al.* [FLA 14] have discovered and isolated very strong variations in height on these temporal maps. These "abnormal" variations seem to be aligned and spread along the surface in a coherent pattern. They have detected the draining of a sub-glacial lake and a created waterfall almost 400 km long.

The impact of the lake draining on the surface is so substantial and created a hole so large that one cannot see the bottom. For this, we have used images from the HRS sensor on SPOT-5 and have shown that the surface has caved in by 70 m over a possible 200 km². The lake has therefore lost 6.4 km^3 of water. Following this, the water has flowed towards other glacier cavities, raising their surface along more than 400 km (Figure 7.7). The drainage took two years and its discharge under the ice also took almost close to two years. For the moment, it seems that the freed water has remained trapped under the ice cap and has still not reached the ocean.

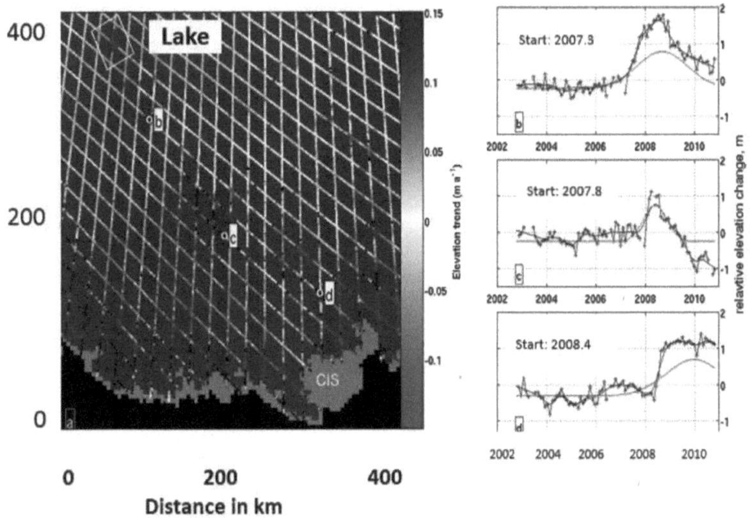

Figure 7.7. *To the left, varying height abnormalities appear well aligned on the ice's discharge lines. To the right, we see that the wave at point b) arrives in the beginning of 2007, at c) the middle of 2008 and at d) the end of 2008. For a color version of this figure, see www.iste.co.uk/baghdadi/4.zip*

Such lake drainages can only have a large effect on the discharge of ice, through lubrication of the base. It is therefore imperative to try and detect these phenomena and above all quantify their impact in terms of acceleration of ice flow and in loss of mass. Beyond the anecdotal evidence, this example also brings about a methodological problem. How can we filter data without removing such signals? It is clear that we must proceed on a case by case basis to avoid missing interesting events. This example also illustrates the importance in processing the data along the track and not simply just those that are at the crossing points.

7.6.3. *Global mass balance*

Since the 1990s, numerous satellites have been used to estimate changes in volume or mass. In altimetry, the series ERS-1, ERS-2 and ENVISAT have longevity, offering almost 20 years of continuous and homogeneous data.

Unfortunately, they only measure the change in volume, they are affected by the penetration error from radar wavelengths that can potentially vary in time, and are not adapted to coastal regions. Data from laser altimetry do not have the same limitations as those coming from a radar, but are only available on shorter periods. Finally, the gravimetric satellite GRACE that directly measures the mass and its changes, has reasonable longevity as it was launched in 2002 and is still active in 2016. However, it has very bad spatial resolution and is affected by post-glacial rebound that is difficult to correct. The scientific community currently agrees that, from 1992 to 2011, Antarctica lost 71 Gt/year ± 53, contributing to an increase in sea level of 0.2 mm/year. The eastern portion, is still relatively stable and will have likely gained mass of 14 Gt ± 43, whilst the maximum loss is concentrated in the western portion, thought to be 85 Gt ± 40 [SHE 12].

To finish, let's look more closely at spatio-temporal height variations. These are represented by Figure 7.8, for the period ERS-2 (1995-2003) and that of ENVISAT (2003-2010). On average, the variations are between ± 0.15 m/year. On the Eastern Antarctic coast, we can see signals with opposing signs. For example, in Queen Maud Land (Figure 7.1), we can clearly see the trend changes sign. Help from gravimetric measurements shows that these changes in height correspond to surpluses and then to deficits in snow for a period of 4.7 years caused by an atmospheric perturbation rotating around Antarctica. It is in this way that Mémin *et al.* [MEM 15] showed the impact of the circumpolar wave on Antarctica's records.

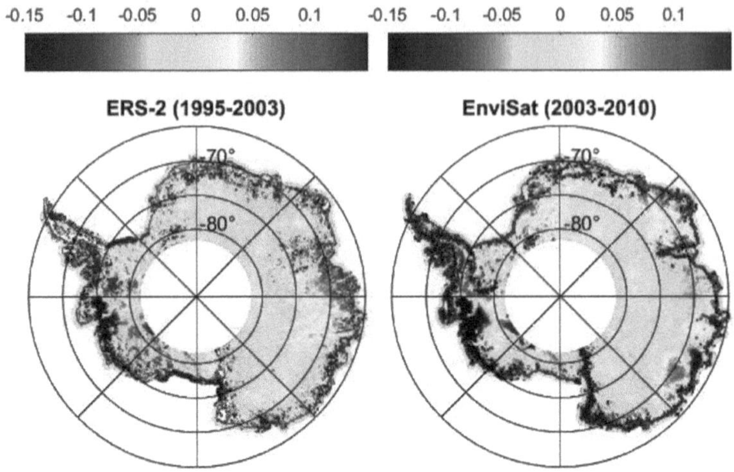

Figure 7.8. *Changes in Antarctic's height, measured during the period ERS-2 (left) and ENVISAT (right), expressed in m/year. We notice the systematic loss in mass from glaciers in the west of the Amundsen Sea and inverse signals from the eastern side. For a color version of this figure, see www.iste.co.uk/baghdadi/4.zip*

7.6.4. *Acceleration of outlet glaciers: can they bolt?*

On the other hand, we can clearly see the systematic loss from glaciers from the western side. This will continue to accelerate at an increasing rate [FLA 12]. The loss in volume from the glacier's basin in Pine Island has gone from 7 km^3/year during the period from 2002 to 2006, to 48 km^3/an in 2006 to 2010. The loss in mass is transferred upstream and is currently visible 300 km away from the coast. This sector of Antarctica attracts the attention of numerous researchers because the measured acceleration could eventually become catastrophic.

These rapid changes in the dynamics of certain glaciers, essentially those flowing into the Amundsen Sea, are only partially understood and will diametrically stand in opposition to the idea that an ice cap with uniform velocities, from a dynamic point of view, only reacts over long time scales. In Greenland, the acceleration of certain glaciers is possibly due to the melting of snow on the surface that infiltrates the moulins and lubricates the base of the ice cap. Melting on the surface is rare in Antarctica and it's more of a question of an intrusion of warmer oceanic waters underneath the glacial tongue. Up to now, models that have been designed to examine the changes for large periods of time are moving towards models that are more geared towards what is happening on the floating line, where the glacial tongue is able to float (Figure 7.2). The position of this line controls the volume of the ice cap going upstream. What happens if the positioning of this line is changed? We now know that this essentially depends on the bedrock's slope [DUR 09]. New questions therefore arise concerning the laws to be used with these models, the boundary conditions at the floating line, and so on. Altimetry contributes to improving the knowledge of these outlet systems, not by finely describing the topography of the surface, but by describing its temporal evolution.

7.7. Summary and perspective

Altimetry for polar ice caps started in 1991 with the launch of the European satellite ERS-1. In almost 25 years, this instrument has allowed us to make progress on numerous subjects. However, relatively cumbersome processes of different complexity were necessary, from reprocessing of the waveform to the construction of a topographic map. In particular, evaluating a time series requires special attention due to the radar wavelength penetrating into the unknown snow, and the weak repetition of the satellite ground track from one cycle to the next, on the surface of the slope. We can show with around fifty repetitive cycles that all of the errors can be corrected, or at least reduced. ENVISAT's 85 cycles have allowed us to carry out a robust inversion providing the seasonal signal and trends over time. The instrument detected sub-glacial lakes, coherent structures with low-frequencies linked to the ice flow, thin and stretched structures connecting the lakes to each other, etc., and offers good constraints for ice flow models. We were able to detect drainage of sub-glacial lakes, observe height variations related to climatic phenomena such as the circumpolar wave, observe acceleration of outlet glaciers from the Amundsen Sea, and with the help of gravimetric measurements, measure the mass balance of the ice sheet. Lastly, the extraction of parameters from snowpack, such as the size of snow grains and the internal stratification or the roughness is still in development. Since March 2013, the French-Indian satellite SARAL with its AltiKa altimeter on board has flown Antarctica on the same orbit as the previous European satellites, but with different technical features. Not only do we hope to extend the time series started in 1991 but also, thanks to the Ka-band, we hope to be able to develop and inverse electromagnetic models to better simulate the surface and volume scattering, and to better constrain the physics of the measurements. The European satellite Sentinel-3, operating on the Ku-band, was put on a polar orbit in February of 2016 that is different from previous satellites, with an altimeter that has a synthetic aperture radar mode that will provide better spatial resolution. New reprocessing algorithms for waveforms must be developed. This will thus extend the altimetric observations in Antarctica.

7.8. Key points

The altimeter, even though it was designed for observing the ocean, allows us to reproduce the height of Antarctica on almost 75% of the surface. Penetration by the radar wavelength in snowpack and the slope error are two main problems. However, disregarding the coast, where the slope of the surface is too steep, these observations also inform us about the characteristics of the snowpack: the size of the grains, the roughness of the surface, or stratification. The topography deduced from the geodesic orbit ERS-1 has offered numerous amounts of information about the dynamic processes that occur in Antarctica and allows to constrain the models. Above all, the 20 years of continuous and coherent measurements allow us, with help from other sensors, to measure the mass balance, to visualize the acceleration of certain glaciers, and to better understand the causes of temporal variability of an ice cap.

7.9. Bibliography

[CAL 16] CALMANT S., CRETAUX J-F., RÉMY F., "Principles of radar satellite altimetry for application on inland waters", in BAGHDADI N., ZRIBI M. (eds), *Microwave Remote Sensing of Land Surfaces*, ISTE Press, London and Elsevier, Oxford, 2016

[DUR 09] DURAND G., GAGLIARDINI O., DE FLEURIAN B. *et al.*, "Marine ice sheet dynamics: Hysteresis and neutral equilibrium", *Journal of Geophysical Research*, vol. 114, pp. 1–10, 2009.

[FLA 12] FLAMENT T., REMY F., "Dynamic thinning of Antarctic glaciers from along-track repeat radar altimetry", *Journal of Glaciology*, vol. 58, no. 211, pp. 830–840, 2012.

[FLA 14] FLAMENT T., BERTHIER E., REMY F., "Cascading water underneath the Antarctic ice sheet (Wilkes Land) observed using altimetry and digital elevation models", *The Cryosphere*, vol. 8, pp. 1–15, 2014.

[KOR 09] KORONA J., BERTHIER E., BERNARD M. *et al.*, "SPIRIT: SPOT 5 Stereoscopic Survey of Polar Ice: Reference Images and Topographies during the International Polar Year", *ISPRS Journal of Photogrametry and Remote Sensing*, vol. 64, pp. 204–212, 2009.

[LEM 14] LE MEUR E., SACCHETTINI M., GARAMBOIS S. *et al.*, "Two independent methods for mapping the grounding line of an outlet glacier – an example from the Astrolabe Glacier, Terre Adélie, Antarctica", *The Cryosphere*, vol. 8, no. 4, pp. 1331–1346, 2014.

[MIC 14] MICHEL A., FLAMENT T., REMY F., "Study of the penetration bias of Envisat Altimeter observations over Antarctica in comparison to ICESat observations", *Remote Sensing*, vol. 6, pp. 9412–9434, 2014.

[MÉM 14] MÉMIN A., FLAMENT T., REMY F. *et al.*, "Snow and height changes in Antarctica from satellite gravimetry and altimetry data", *Earth and Planetary Science Letters*, vol. 404, pp. 344–353, 2014.

[MÉM 15] MÉMIN A., FLAMENT T., ALIZIER B. *et al.*, "Detection of the Antarctic Circumpolar Wave in a combined analysis of satellite gravimetry and altimetry data", *Earth and Planetary Science Letters*, vol. 422, pp. 150–156, 2015.

[PAR 04] PARRENIN F., RÉMY F., RITZ C. *et al.*, "New modelling of the Vostok ice flow line and the implication fort the glaciological chronology of the Vostok ice core", *Journal of Geophys. Res*, vol. 109, p. D20102, 2004.

[REM 97] REMY F., MINSTER J.F., "Antarctica ice sheet curvature and its relation with ice flow and boundary conditions", *Geophysical Research Letters*, vol. 24, no. 9, pp. 1039–1042, 1997.

[REM 02] REMY F., TESTUT L., LEGRÉSY B., "Random fluctuations of snow accumulation over Antarctica and its relation with sea level change", *Climate Dynamics*, vol. 19, pp. 267–276, 2002.

[REM 09] REMY F., PAROUTY S., "Antarctic ice sheet and radar altimetry: A review", *Remote Sensing*, vol. 1, pp. 1212–1239, 2009.

[REM 14] REMY F., FLAMENT T., MICHEL A. *et al.*, "Ice sheet survey over Antarctica with satellite altimetry: ERS-2, EnviSat, SARAL/AltiKa, the key importance of continuous observations along the same repeat orbit", *International Journal of Remote Sensing*, vol. 35, no. 14, pp. 5497–5512, 2014.

[RAM 16] RAMILLIEN G., FRAPPART F., SEOANE L., "Space gravimetry using grace satellite mission: basic concepts", in BAGHDADI N., ZRIBI M., *Microwave Remote Sensing of Land Surfaces*, ISTE Press, London and Elsevier, Oxford, 2016.

[REM 15] REMY F., FLAMENT T., MICHEL A. *et al.*, "Envisat and SARAL/AltiKa observations of the Antarctic ice sheet: a comparison between the Ku-band and the Ka-band", *Marine Geodesy,* vol. 38, no. 1, pp. 510–521, 2015.

[RIT 01] RITZ C., ROMMELAERE V., DUMAS C., "Modeling the evolution of Antarctic ice sheet over the last 420 000 years: Implications for altitude changes in the Vostok region", *Journal of Geophysical Research*, vol. 106, no. D23, pp. 31943–31964, 2001.

[SHE 12] SHEPERD A., IVINS E.R., GERUO A. *et. al.*, "A reconcilited estimate of ice-sheet mass balance", *Science*, vol. 338, pp. 1183–1189, 2012.

[TES 03] TESTUT L., COLEMAN R., REMY F. *et al.*, "Precise drainage pattern of Antarctica derived from high resolution topography", *Annals of Glaciology*, vol. 37, pp. 337–343, 2003.

8

Monitoring Water Mass Redistributions on Land and Polar Ice Sheets Using the GRACE Gravimetry from Space Mission

8.1. Introduction

Water on continents is a fundamental reservoir of the global hydrological cycle. It plays a major role for Earth's climate via its exchanges with the atmosphere (precipitation and evapotranspiration) and the oceans (outflow of rivers), the regulation of energy flow and biogeochemical flow. Despite its importance, estimations of water stored on land remains uncertain, as much on a regional scale as a global one. This is due to a lack of *in situ* measurement networks which prevents continuous monitoring of the different hydrological reservoirs. Even though such monitoring is possible thanks to a few local measurement networks, our current knowledge of the water cycle mainly originates from global hydrological models, for which the capabilities of simulating the exchanges between the different reservoirs from regional to global scales suffer from significant limitations. This is mainly due to the absence of certain hydrological reservoirs in modeling, such as floodplains and groundwater tables, as well as forcing errors. Notably, forcing from rain, that has inhomogeneous quality depending on the region of the world considered, is a major issue. In addition this is the shortage of information, on a small scale, for certain

Chapter written by Frédéric FRAPPART, Guillaume RAMILLIEN and Lucía SEOANE.

parameters, such as the nature of the soil and vegetation cover and their evolution.

The classical techniques used in remote sensing from space, such as satellite imagery and altimetry, provide information on the extent of wetlands and the spatial distribution of soil moisture, or on water levels for rivers, lakes and floodplains. None of these techniques is able to give us information about the water content in sub-surface reservoirs. Since its launch in March 2002, the Gravity Recovery and Climate Experiment (GRACE) gravimetry mission from space measures the spatio-temporal variations of the Earth gravity field that can be related to variations of terrestrial water storage. This corresponds to the sum of surface waters, to water contained in the root zone, the groundwater tables, or is present in the form of snow or ice [TAP 04], offering a new source of unseen information for hydrologists and water cycle modelers

The principle of measurement of the GRACE mission and the link between the observables and the Earth gravity field were described in [RAM 16]. In this chapter, we firstly present the main post-processing techniques of data from the GRACE mission, then the primary applications of the derived products for hydrology and glaciology.

8.2. Post-processing techniques for global solutions

The most commonly used GRACE data in hydrology and glaciology are the gravity anomalies from the Level-2 GRACE solutions, that are provided as Stokes coefficients (see [RAM 16] for more details on the measurement principle of the GRACE mission and the nature of the data made available). The Level-2 GRACE solutions are corrected from the known effects of gravitational forces (atmosphere, oceans, polar motion, periodic tidal forces) through considering *a priori* models [RAM 16] Amongst these, some describe tidal phenomena (oceanic, atmospheric, terrestrial) and others model the contributions to the gravity potential of atmospheric and oceanic circulations not related to tides, at periods lower than the temporal resolution of Stokes coefficients. These models are imperfect, especially those describing high frequency variability and called "de-aliasing." The improper correction of high frequency variations in gravitational potential leads to significant errors due to aliasing effects, manifested as north-south stripes, amplified by the orbital configuration of the GRACE mission which is also oriented north-south. As we can see in Figure 8.1(a), if the Stokes coefficient is not filtered, the de-aliasing errors dominate the anomalies of water mass from the GRACE missions' measurements. Since the first Level-2 GRACE solutions were made available, different filtering techniques were proposed to eliminate the unrealistic undulations due to *striping*. Examples of filtering results are presented on the Figures 8.1(b), 8.2(b), 8.3(a) and 8.3(b).

8.2.1. *Empirical methods*

8.2.1.1. *Isotropic and non-isotropic filters*

In earlier studies made before the launch of the GRACE mission, simple low-pass isotropic Gaussian filters were proposed to reduce the effects of striping. They were then applied to real Level-2 solutions to demonstrate the capability of GRACE's very precise inter-satellite measurement – *K-Band microwave ranging*, or KBR – operating K (f=26 GHz) and Ka (f=32 GHz) bands to restore the temporal variations of land water mass [TAP 04]. The disadvantages of the Gaussian type filtering are:

– the arbitrary choice of the filtering radius (few hundreds of kilometers);

– the inability of such a filter to discriminate the geophysical signals of noise.

Thereafter, more elaborate filters appeared to reduce the errors in GRACE solutions. The spatial averages obtained in large basins were based on a trade-off between the mitigation of the hydrological signal and the leakage errors [RAM 16], the optimal filter having been obtained thanks to the Lagrange multipliers and to the relation between an exact solution and a smoothed solution without constraints [SWE 02] and with constraints (form of the symmetrical covariance function depending on the azimuth) [SEO 05]. To take into account meridian structures of noise, several non-isotropic smoothings were proposed, such as the one using a series of Legendre functions depending on the degree and the order [HAN 05], or using a Tikhonov type regularization [KUS 07].

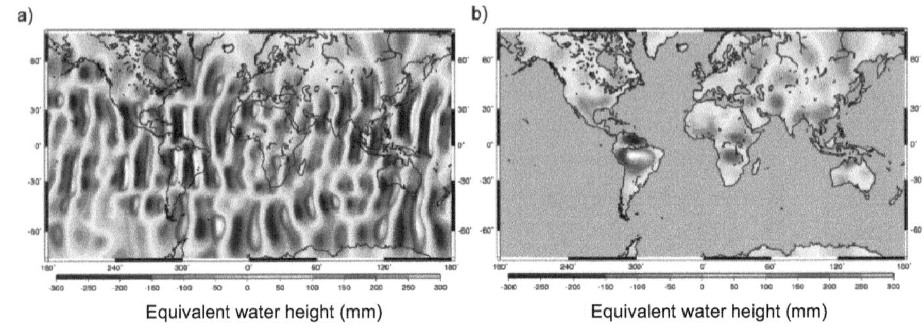

Figure 8.1. *Map of mass anomalies measured by the GRACE mission in April 2003: a) spatial representation of Stokes coefficients: and b) results of filtering using an least-squares based inverse method. Adapted from Ramillien et al. [RAM 05]. For a color version of this figure, see www.iste.co.uk/baghdadi/4.zip*

8.2.1.2. *The method of destriping*

The techniques of destriping remove the errors correlated between the sectoral and *quasi*-sectoral Stokes coefficients [RAM 16] responsible for the north–south undulation in GRACE solutions. For any given harmonic order *m*, the Stokes coefficients are smoothed using a quadratic polynomial in a sliding window with a fixed width, and centered on the degree *n*. The adjustment is performed with coefficients of the same parity as *n* [SWE 06]. This method provides better results for high latitudes than close to the equator, because of residual errors from the *quasi*-sectorial coefficients ($n \sim m$) that are not completely eliminated [SWE 06, KLE 08]. Monthly data, expressed in equivalent water height, obtained by using this method and covering the period of 2002–2010 are available for download on the website: http://grace.jpl.nasa.gov.

8.2.1.3. *The stabilization method*

The laser data from the geodesic missions LAGEOS 1 and 2 are used to determine the temporal variations for low degree harmonics ($l \leq 2$) from the gravity field, that complete the GRACE solutions provided from the Space Geodesy Research Group (*GRGS* in French) [BRU 10]. The solutions from GRGS are obtained after empirical stabilization that consists of adding constraints, mostly on high degree Stokes coefficients, when the amplitude of these coefficients is greater than that of a product of power laws on a static field of gravity's variance depending on the degree and the order

[BRU 10]. Currently, a new version of GRGS solutions (RL03) is available. For the purpose of improving the gravity field's estimations with respects to previous versions, several improvements were provided in the processing of observations from GRACE. The aliasing effects [RAM 16] are corrected using more recent versions of atmospheric and oceanic models. The inversion of Stokes coefficients is now achieved by using a method of singular value decomposition, allowing the performance of regularization by truncation of the explained variance. The monthly data from GRGS (releases 01 and 03) and at 10 day intervals (release 02), obtained by applying this method of stabilization, are available for download on the website: http://grgs.obs-mip.fr.

8.2.2. *Statistical methods*

8.2.2.1. *Wiener filtering*

Isotropic Wiener filtering is based on the least-square minimization of the difference between modeled and filtered signals, and only depends on the power of degree of the signal and the noise. Using hydrology and ocean models for simulating the degree power spectrum of the GRACE signal, the formal error of the GRACE coefficients, and the GRACE coefficients themselves, Sasgen *et al.* [SAS 06] have determined the optimal coefficients of the Wiener filter through minimization in the least-squares sense of gravity field's power spectrum depending on the harmonic degree using the second Kaula empirical law (decrease of the gravity field with the harmonic degree following a power law [KAU 66]). These authors have shown that this method was reliable for low-pass filtering of GRACE solutions, and without the specification of a smoothing radius.

8.2.2.2. *Principal component analysis*

Principal component analysis (PCA) was applied to a multi-year series on grids of equivalent water height and Stokes coefficients. The PCA allowed to separate the different time and spatial modes of geophysical contributions from those corresponding to north-south undulations. Applied directly to the Stokes coefficients, the PCA allowed suppression of the meridian undulations present in the GRACE solutions. After Gaussian pre-filtering using a 6–7° radius, the first three modes of the PCA explained more than of 70% of the observed variance, and are related to continental hydrology. The residual modes contain S2 aliasing errors and a semiannual continental hydrology signal contained in the Global Land Data Assimilation System (GLDAS) model [SCH 07].

8.2.2.3. *Independent component analysis*

The independent component analysis (ICA) approach is used to extract hydrological signals from the noise in the Level-2 GRACE solutions by considering completely objective constraints, so that the gravity component of the observed signals is forced to be uncorrelated numerically through the combination of GRACE solutions from three computing centers forming the Science Data System (SDS): Centre for Space Research (CSR in Austin, Texas), the Geo Forschungs Zentrum (GFZ in Potsdam, Germany) and the Jet Propulsion Laboratory (JPL in Pasadena, California). The blind separation is based on the statistical independence of elementary signals that compose the observations, that is geophysical signals and spurious noise, and does not require supplementary *a priori* information. At release it provides modes containing the geophysical signal and others containing striping (Figure 8.2). Comparisons on a global scale show that ICA solutions contain less striping on continents than the filtered "destriped" solutions, and more realistic hydrological structures than the destriped solutions in the tropics. In addition, these solutions do not contain Sumatra 2004 tsunami-type signals and present a negative trend in Scandinavia and the Hudson Bay in Canada. The rapid variation of mass caused by the tsunami can be found in the two modes containing north–south bands. The slow signals observed in northern latitudes, seem to be corrected from the post-glacial rebound effect (or the isostatic adjustment, in other words the lifting of terrestrial masses due to the melting of glacial ice caps), but this does not appear very clearly when using an independent mode, especially because it is surely masked by north-sound stripes. At basin-scale, the ICA-based solutions allowed us to filter out the unrealistic peaks present in the time series of TWS obtained using classical filtering for basins with areas lower than 1 million km^2 [FRA 2011a]. The major drawback of this approach is that it cannot directly be applied to the GRACE Level-2 raw data, as a first step of (Gaussian) prefiltering is required. Monthly CSR, GFZ and JPL ICA-filtered TWS grids of 1-degree spatial resolution are available on the Toulouse GRGS website: http://grgs.obs-mip.fr.

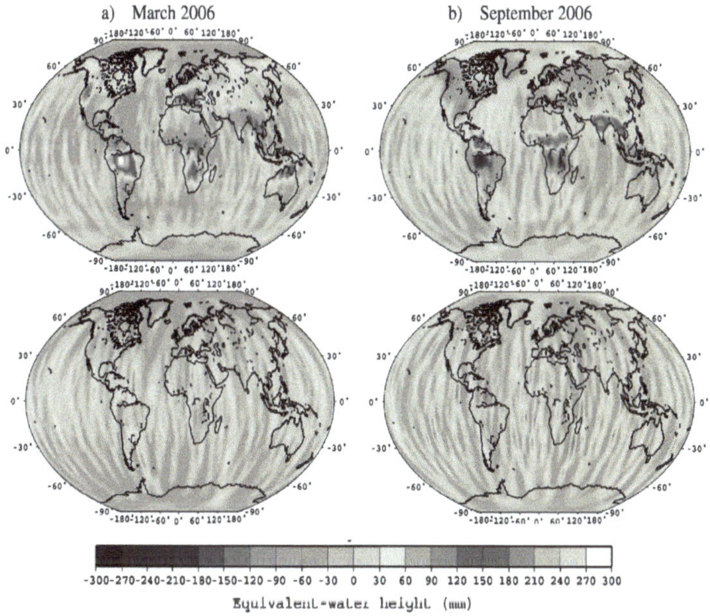

Figure 8.2. *GRACE water storage from GFZ filtered with a Gaussian filter of 400 km radius. Top – first ICA component corresponding to land hydrology and ocean mass. Bottom – sum of the second and third components corresponding to the north–south stripes: a) March 2006; b) September 2006. Units are millimeters of EWH. Adapted from Frappart et al. [FRA 11a]. For a color version of this figure, see www.iste.co.uk/baghdadi/4.zip*

8.2.2.4. *Smoothing by the Kalman filter*

Daily solutions have been derived from GRACE observations using a Kalman filter approach. They are estimated using a Gauss Markov model, the solution at day t+1 slightly differing from the solution at day t from the noise prediction (first-order Markov process) estimated using a Kalman filter. *A priori* information on the hydrological patterns from the WGHM model was introduced in the covariance matrix to compensate for the limited number of available satellite observations each day [KUR 12]. This approach allows improved temporal resolution at the expense of spatial resolution (harmonic degree of 40 or 500 km) of the GRACE solutions. It must be noted that the time series, even though daily, are very smooth and do not allow detection of sudden hydrological variations. The Stokes coefficients obtained in applying this method and on a period from 2002–2010 are available on the website: http://www.igg.uni-bonn.de/apmg.

8.2.3. *The inverse methods*

8.2.3.1. *Iterative adjustment using least-squares*

This method is derived from the matrix formalization of the generalized least-squares and consists in estimating separately the spherical harmonics coefficients relative to the different hydrological contribution of four water mass reservoirs (atmosphere, ocean, snow and liquid water) from monthly Level-2 GRACE solutions, using external information on the dynamic of each reservoir provided by models. For each reservoir, the anomaly of mass is determined via a linear combination of Stokes coefficients, as well as covariance matrices that describe the spatio-temporal dynamics of each reservoir from hydrological models [RAM 05]. An example of an anomaly of total water storage obtained with the inverse method is presented in Figure 8.3. If we compare the results obtained with the inverse method and with a classical Gaussian filtering, we notice a better localization of hydrological structures with the inverse method, with extrema located along the Amazon central corridor, and secondary minima along the Orinoco mainstream (in blue) and the confluence of the Parana and Paraguay Rivers and the Parana and Uruguay Rivers (in orange, in the south-eastern of the map obtained with the inverse method).

8.2.3.2. *Optimal filtering*

This method takes into account the statistical information present in each GRACE monthly solution. The noise and full signal variance–covariance matrix is used to tailor the filter to the error characteristics of a particular monthly solution. The resulting filter was found to be both anisotropic and non-symmetric to accommodate noise of an arbitrary shape as the north–south stripes present in the GRACE monthly solutions. It is optimal as it minimizes the difference between the signal and the filtered GRACE estimate in the least-squares sense. This filter was found to perform better than any isotropic, or anisotropic and symmetric filters, preserving better the hydrological signal amplitudes [KLE 08]. Monthly GRACE-derived TWS grids of 1-degree spatial resolution computed using the optimal filter are available at: http://www.citg.tudelft.nl/en/about-faculty/departments/geoscience-and-remote-sensing/research-themes/models-data/champgracegoce-gravity-models-data/dmt-1b/.

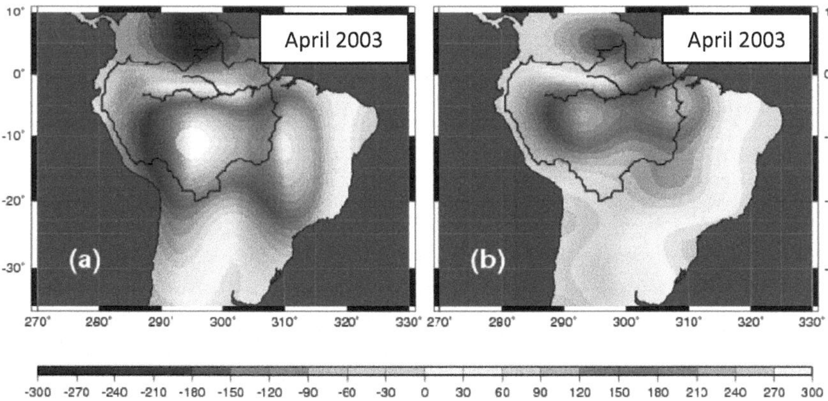

Figure 8.3. *Continental water storage variations over the Amazon basin for April 2003 derived using two approaches: a) Gaussian filtering; and b) the inverse method. Adapted from Ramillien et al. [RAM 05]. For a color version of this figure, see www.iste.co.uk/baghdadi/4.zip*

8.3. Regional approaches

Since the launch of the new generation of low altitude geodesic satellites (CHAMP, GRACE, GOCE), dedicated to the precise determination of the Earth's gravity field, the estimation of local variations of water mass on a network of elementary surfaces, often rectangular, representing the surface of Earth is a promising alternative over study areas where hydrological changes are extreme, such as the regions subject to recurrent floods and droughts, or for the monitoring of continued melting from continental glaciers. The so-called "regional" approaches represent an alternative to problems related to the classical global representation in spherical harmonics by choosing a suitable geometry of surface tiles [RAM 16].

8.3.1. *Mascons*

In the *Mascons* approach (derived from *mass concentrations*) developed by NASA, variations of water mass are determined in rectangular surface elements of 4° by 4° and 2° by 2°, based on radar telemetry KBR, along the satellite tracks of GRACE, above the study area [RAM 16]. The size of these elements correspond to the maximal spatial resolution *a priori* of data from

GRACE, in other words 300-400 km. The adjustment method of "regional" Stokes coefficients uses spatial and temporal constraints that induce a smoothing effect. *Mascons* type solutions have provided more information on the inter-annual variability in these particular regions than the classical global approach.

However, the global and *mascons* approaches tend to provide equivalent information on the variations of equivalent water heights when no constraint is taken into account [ROW 10]. They also suffer from inherent spectral truncation errors in the representation of underlying spherical harmonics (omission errors, see [RAM 16]). Ten-day *mascons* time series solutions by continental zone are available on the website: http://grace.sgt-inc.com/V2/Global.html.

8.3.2. *The regional method*

Another type of regional approach that was recently proposed is to adjust the surface densities of mass based on Level-1 data from the GRACE mission [RAM 16]. More specifically it is based on precise measurements on the K-band and the variations of speed between the two satellites. Contrary to the *mascons* approach, the mode of representation in spherical harmonics is not used here. The regional solutions therefore do not suffer from spectral truncation effects, and offer a better geographical location of hydrological structures [RAM 12]. After correction of known accelerations (static fields, atmosphere, oceans, tidal periods), the residues of the KBR speed are converted into variations of kinetic energy difference (or potential as a result of the conservation of total energy) along the orbits, that mainly represent the contribution of continental hydrology (see the principle in Figure 8.4). Newton's First Law allows construction of the matrix for the passage between the water heights and the potential differences, using the positions of potential residue and those of the surface elements, meaning the linear equation system that is then resolved using regularization techniques. The regional method was improved, taking into account geographical radius correlation type spatial constraints between surface elements [RAM 12]. Pluri-annual series from successive regional solutions were calculated on 10-day time intervals over South America, Australia and Africa, with a spatial resolution of 200 km, where slow changes where detected in stored water such as drought, significant seasonal variations in major drainage basins and faster hydrological processes. The time series of regional solutions over Africa [RAM 14] from June 2007 to May 2008 is presented in Figure 8.5. The series of maps allow us to monitor the variations in mass relative to the water cycle in Africa during a hydrological year. We notice mainly that seasonal variations on the Sahelian and the Sahelo–Sudanian band between 0 and 15°N, presenting a maximum at the end of summer and the beginning of autumn, in basins of the Congo, Zambezi and the Orange River, and the Eastern-African Great Lakes in the beginning of spring. Secondary extrema are also observable in the Nile Valley and in the Great East-Saharan Aquifer located in the south of Libya. Stable zones are also present and located in the Sahara Desert, in the north and in Kalahari, in the south, between the Zambezi Basin and the Orange River. Finally, Ramillien *et al.* [RAM 15] have proposed a new method using a Kalman filter providing daily regional solutions by the progressive accumulation of information provided by differences in residual potential along the GRACE orbit.

8.4. Applications

The time variations of the gravity field measured by the GRACE mission come from redistributions of mass in the Earth. They provide unprecedented information on the variations of ocean mass and allow a better understanding of the oceanic currents and their variability as well as the heat transportations in oceans, the variations of pressure at the ocean bottom, on the mass balance of polar ice caps and glaciers, the variations of stocked water and snow on continents, as well as on the transfers of water between the different hydrological reservoirs. In the following, we focus on the data from GRACE for application to continental hydrology and glaciology.

8.4.1. *Applications in land hydrology*

8.4.1.1. *Variations of mass in large river basins*

The analysis of Level-2 GRACE solutions on land brought to light important seasonal variations of total water storage on the continents. The GRACE solutions allowed detection of the evolution of stored water on continents, and highlighted the annual cycle and the inter-annual variability of resources in large river basins such as the Amazon or the

Ganges (Figure 8.6). The largest variations of water stored on land were observed in South America, Africa, Southeast Asia and in northern latitudes (Alaska, Northern Canada, Siberia, etc.). These structures are comparable to those obtained by global hydrological models but with larger amplitude [TAP 04, RAM 05]. They allowed characterization of the impact of extreme climatic events on the total water storage. For example, Andersen *et al.* [AND 05] showed that Central Europe had lost, on average, the equivalent of 7.8 cm of water on its whole surface during the heat wave that occurred in summer 2003. Frappart *et al.* [FRA 13] spatialized the deficits of total water content in the Amazon Basin, during major droughts of 2005 and 2010, as well as surpluses, during the flood of 2009 (Figure 8.7).

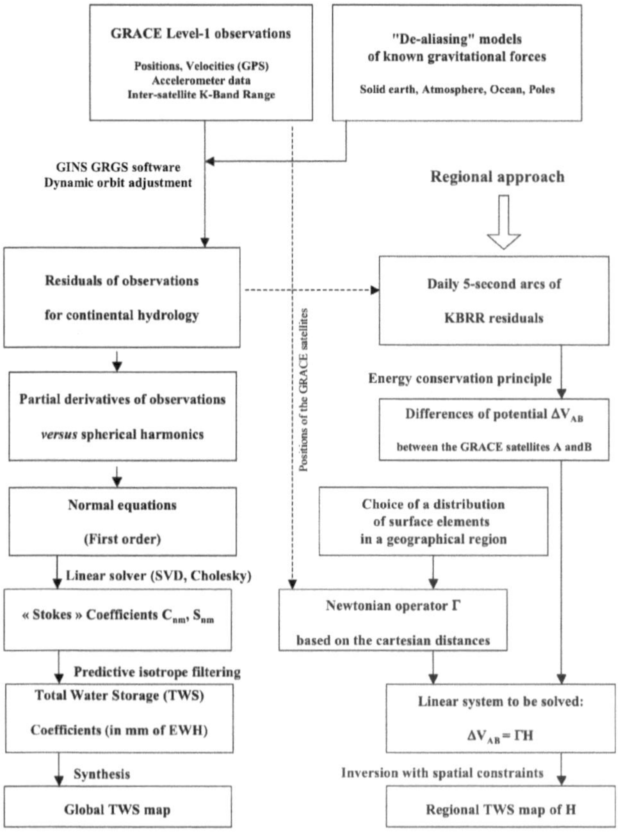

Figure 8.4. *Diagram summarizing the principles of the global approach through stabilization [BRU 10] and the regional approach [RAM 12] developed by the Space Geodesy Research Group (GRGS). H represents the variation of mass due to hydrology. Adapted from [RAM 14]*

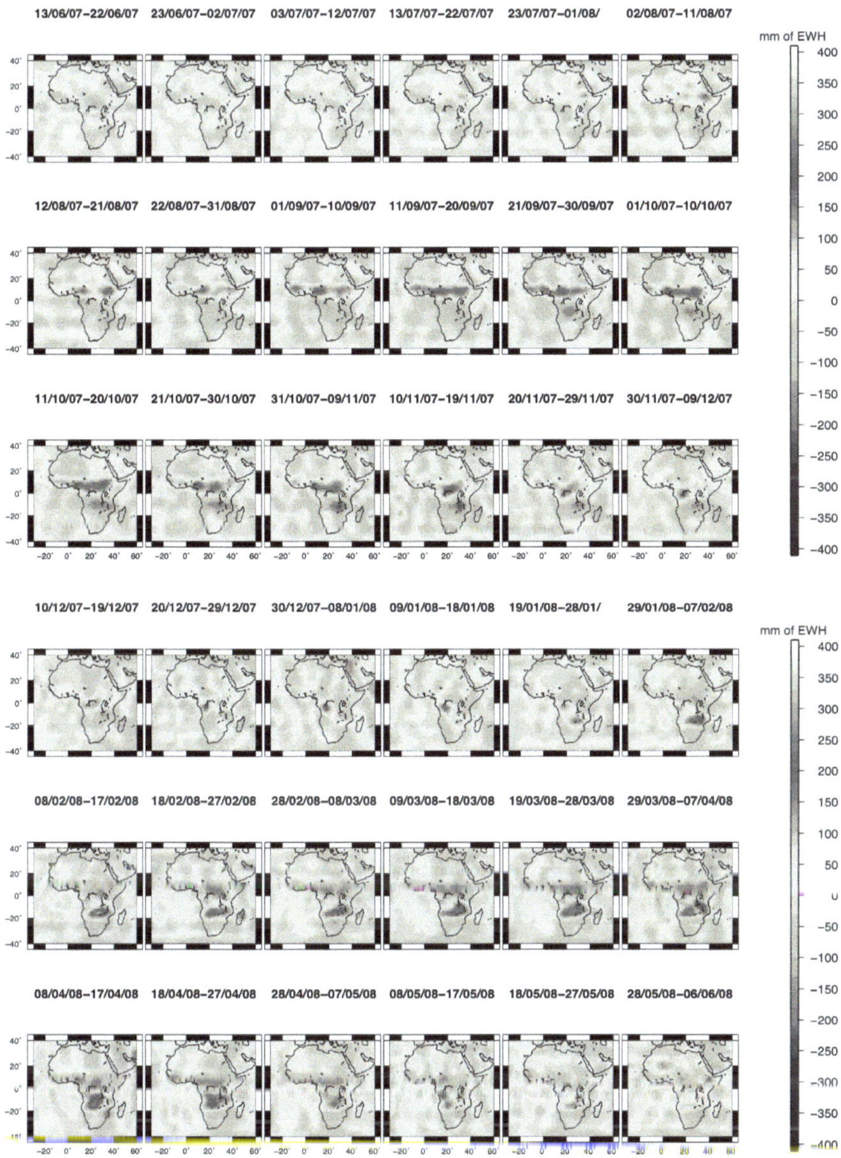

Figure 8.5. *Example of a regional solution time series at a spatial resolution of 2 degrees by 2 degrees over Africa, and determined using data from the GRACE mission at ten day time intervals, for the period of June 2007–May 2008 (equivalent water height). We see the seasonal alternation of water masses caused by the African monsoon. Adapted from [RAM 14]. For a color version of this figure, see www.iste.co.uk/baghdadi/4.zip*

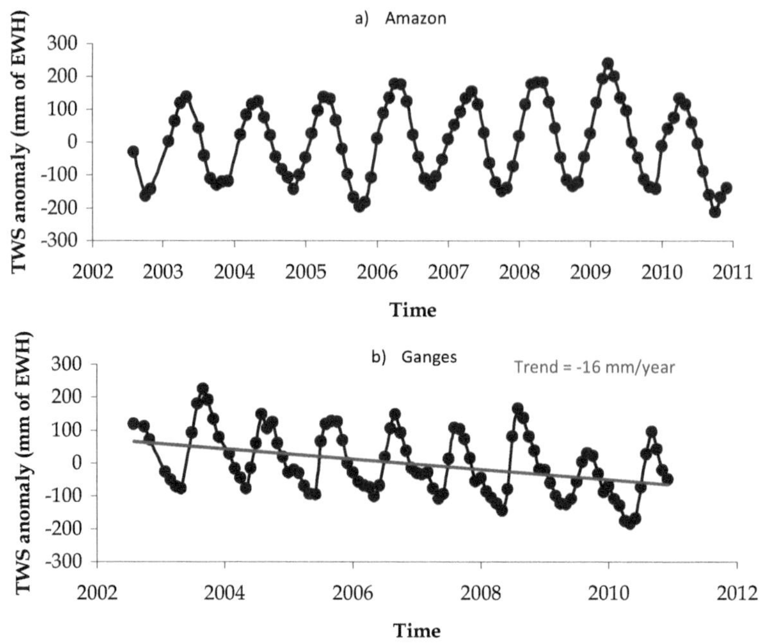

Figure 8.6. *Time series of the average equivalent water height (mm) for basins in a) the Amazon: and b) the Ganges; obtained from Level-2, release 04 solutions from GFZ and filtered by ICA [FRA 11a]*

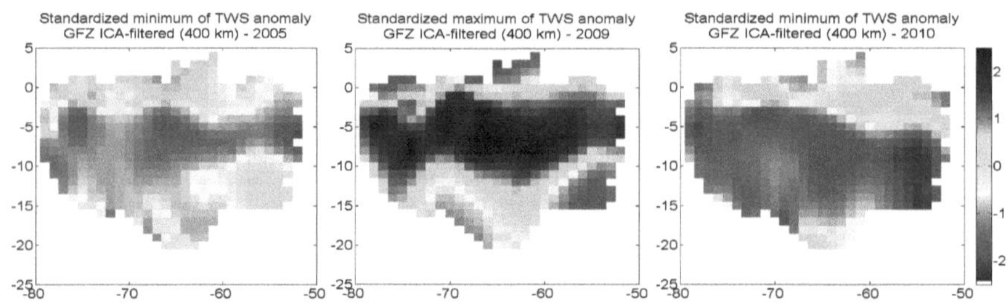

Figure 8.7. *Signatures of extreme climatic events that occurred in the Amazon Basin during the last ten years in terms of total water storage, derived from GRACE observations (expressed in anomalies of annual minima during dry years and maxima during wet years over 2003–2010, normalized by the standard deviation of considered extrema during the same period): drought of 2005 (left), flood of 2009 (center), drought of 2010 (right). Adapted from [FRA 13]. For a color version of this figure, see www.iste.co.uk/baghdadi/4.zip*

Figure 8.7 represents the difference between the extremum of each year with the presence of an extreme climatic event (minima in 2005 and 2010 and maximum in 2009) and the average between 2003 and 2010 of the GRACE data for the month when the extremum occurred divided by the standard deviation of the anomalies of water storage from the same month between 2003 and 2010. If we compare the droughts from 2005 and 2010, we notice that 2010 was both more intense and affected a larger zone than that of 2005, in conformity with what was observed using other environmental variables [FRA 13].

8.4.1.2. *Detection of water mass variations in aquifers*

The variations of water mass in groundwater can be directly estimated, using anomalies of water storage measured by GRACE, combined with external data of water storage in other hydrological reservoirs. The time variations of total water storage are equal to the sum of contributions from the different hydrological reservoirs in a drainage basin:

$$\Delta TWS = \Delta SW + \Delta SN + \Delta TSS \text{ where } \Delta TSS = \Delta SM + \Delta GW + \Delta PE \qquad [8.1]$$

where ΔTWS, ΔSW, ΔSN, ΔTSS represents the time variations of total water storage, surface water (rivers, lakes, reservoirs, floodplains), of snow and of sub-surface water storage. The latter terms are the sum of the soil moisture component (ΔSM), water content in aquifers (ΔGW) and permafrost (ΔPE). This method was applied with success to monitor the effects of climatic variations and of anthropic origin for water resources at regional to global scales. It allowed the study of the recent drought that affected the Murray-Darling Basin in Southern Australia [LEB 09], the depletions of aquifers from Northern India [ROD 09] removing the soil moisture from the total water storage in global hydrological models. On a seasonal scale, strong correlations (R ~ 0.8 – 0.9) were found between the estimates of an aquifer's water content based on GRACE solutions and *in situ* measurements in Illinois [YEH 06] or the Great Plains [STR 07]. For large rivers covered with extensive floodplains, such as the Ob River Basin in Siberia, the variations of an aquifer's water content are estimated as the difference between:

– the total water storage measured by GRACE;

– the surface component, estimated using inundated areas estimated using satellite imagery and surface water levels provided by radar altimetry;

– the soil moisture simulated by global hydrological models [FRA 10].

8.4.1.3. *Estimation of hydrological variables on a regional scale*

The measurements from the GRACE mission offered a unique opportunity for estimating the hydrological fluxes between the continental surfaces and the ocean or the atmosphere, from a regional scale to a global scale. They are mainly used to estimate river runoff and evaporation of large river basins.

8.4.1.3.1. River discharge

River discharge estimates are based on the resolution of the water balance equation at drainage basin scale deriving against time the measurements of the total water storage from GRACE:

$$\frac{\partial TWS}{\partial t} = P - ET - R \qquad [8.2]$$

where P and ET are the vertical fluxes of precipitation and of evapotranspiration of the whole basin, and R the discharge at the exit of the drainage basin that includes the surface and sub-surface runoff components.

The quantity ET is poorly known, so it is possible to eliminate it as well as P using the water balance equation of the atmospheric part of the water cycle:

$$\frac{\partial W}{\partial t} = ET - P - div\vec{Q} \qquad [8.3]$$

Where W is the vertically-integrated precipitable water content, $div\ \boldsymbol{Q}$ is the divergence of the vertically-integrated average atmospheric moisture flux vector. These two terms are computed as follows:

$$W = \int_{p_T}^{p_S} \frac{q}{g} dp \qquad [8.4]$$

$$\vec{Q} = \int_{p_T}^{p_S} \frac{q}{g} \vec{V} dp \qquad [8.5]$$

where p_S and p_T are the pressure on the surface and at the top of the atmosphere, q the specific humidity, g the acceleration of gravity and $\boldsymbol{V}$ the horizontal component of the wind vector.

This approach allows estimation of the total volume of water transferred from continents to oceans via continental outflows, that were estimated at 30,354 ± 1,212 km^3/year [SYE 09].

8.4.1.3.2. Evapotranspiration

The estimations of total water storage from GRACE can also be used to determine the variations in the evapotranspiration flux by solving the water balance [8.2]. These variations were estimated in the Mississippi Basin combining Level-2 GRACE solutions with *in situ* measurements of rainfall and water runoff [ROD 04]. Using the same approach of solving the water balance equation, Ramillien *et al.* [RAM 06] provided evapotranspiration estimations on the 16 largest drainage basins using rainfall from the Global Precipitation Climatology Centre (GPCC) and runoff from the WGHM hydrological model. Their estimations were of the same order of magnitude as those of the four global hydrological models.

8.4.1.4. *The transfer of water masses from continents to oceans*

Continents and oceans continually exchange water masses via the outflow of rivers and the evaporation from oceans transforming into rainfall on land. The residency time of water in the atmosphere being less than 10 days, it is possible to characterize the mass transfers between continents and the ocean by calculating the opposite of the relationship between the anomalies of water mass in the study area and that of all oceans according to the principle of mass conservation. The anomalies of mass are calculated as the product of the water (fresh or salt) density by the considered surface and the anomaly of the equivalent water height:

$$\Delta h_{ESL} = -\frac{\rho_{fresh-water} S_{drainage}}{\rho_{salt-water} S_{ocean}} \Delta h_{TWS} \qquad [8.6]$$

where $\rho_{fresh-water}$ = 1,000 kg.m^{-3} and $\rho_{salt-water}$ = 1,030 kg.m^{-3} are the densities of fresh and salty water, $S_{drainage}$ and S_{ocean} = 360 million km² are the surfaces of the drainage region and the oceans. The mass variations expressed in sea level and equivalent water height are Δh_{ESL} et Δh_{TWS} respectively.

Using GRACE regional solutions over Africa, presented in Figure 8.3, it is possible to estimate the contribution from continental waters present on the African continent (Figure 8.8(a)) to the water mass variations in oceans using [8.6]. The results are presented in Figures 8.8(b) and 8.8(c) for large river basins and drainage regions. The positive sea level contributions come from the Nile Valley, from Congo and from the Orange River and represent 0.044 mm/year, whereas the negative contribution has a total value (-0.123 mm/year), essentially due to gains in mass in the Zambezi Basin (-1.1 mm/year). The contribution from drainage regions toward the Atlantic and Indian Oceans and the Mediterranean Sea present annual cycles with amplitudes of 2, 1 and 0.05 mm respectively. The Indian Ocean's variations are in opposite phase in comparison to the two others. The contributions of these three zones are -0.026, 0.031 and -0.094 mm/year between 2004 and 2012 respectively, indicating the continent's gain in mass with respect to the oceans. The loss of mass of the Indian Ocean was strongly accelerated from -0.034 mm/year to -0.123 mm/year [RAM 14].

8.4.2. *Applications in glaciology*

8.4.2.1. *Mass balance of polar ice sheets*

Level-2 GRACE solutions have been widely used to estimate the mass balances of Arctic and Antarctic polar ice sheets. Using the Release 05 of GRACE solutions from CSR from January 2003 to November 2012, Velicogna and Wahr [VEL 13] showed that Greenland has lost -258 ± 41 Gt/year of ice mass, whereas the mass in Antarctica was reduced by -83 ± 49 Gt/year or by -147 ± 41 Gt/year, either the derived variations of mass from GRACE are corrected from the post-glacial reborn model IJ05_R2 [IVI 13] or ICE5G [PEL 04]. These models represent the uplift of continental masses in response to the withdrawal of polar ice sheets during the last deglaciation. Analysis of the GRACE solutions series showed that the melting of polar ice caps is not stationary. In Greenland, an increase in melting occurred in 2004, that accelerated between 2006 and 2008, before a first deceleration phase between 2008 and 2009, and a second after 2010 [BER 12] (Figure 8.9). The most important melting rate occurs on coastal glaciers in the south-west and in the Western Greenland. The strong decreases in mass in Antarctica are located on the Antarctic Peninsula, to the west, and Wilkes Land and Victoria Land, to the east, whereas an increase of mass can be observed in Queen Maud Land [VEL 13]. However the results obtained are affected by significant errors due to poorly modeled phenomena such as post-glacial rebound (especially in Antarctica where the measurements related to this phenomenon are in insufficient number), the aliasing of atmospheric phenomena, the measurement errors, the leakage effects and the difference in data processing from each center. The latter point is important to keep in mind, because Baur *et al.* [BAU 09] showed that according to the chosen data source, Greenland's mass balance can vary by a factor of three.

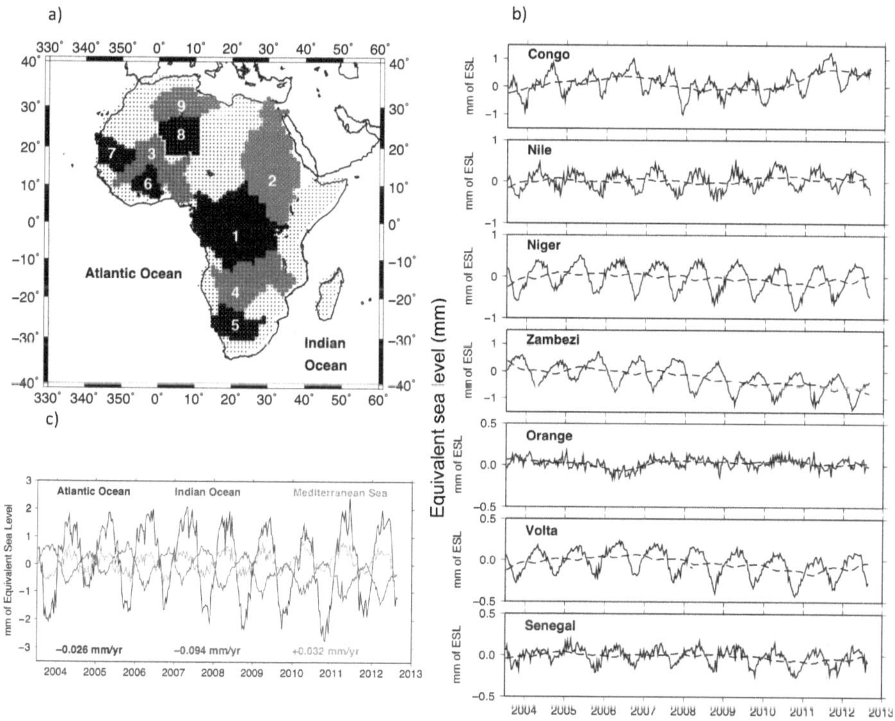

Figure 8.8. *a) Time variations of water mass anomalies transferred from the African continent; b) to oceans for the main drainage basins and c) by drainage region. On the map of Africa (a), the main drainage basins are numbered as follows and their surface areas in millions of km² are given between parenthesis: 1) Congo (~ 4), 2) Nile (~ 3,4), 3) Niger (~ 2,1), 4) Zambezi (~ 1,4), 5) Orange (~ 0,97), 6) Volta (~ 0,41), 7) Senegal (~ 0,27). The desert zones are numbered as follows and their surface areas in millions of km² are given between parenthesis: 8) Sahara Desert in Southern Algeria (~ 1.1) and 9) Aquifer in North-Western Sahara (1.6). The drainage zones towards the Atlantic Ocean, the Indian Ocean and the Mediterranean Sea are represented with the blue, red, and green dots, and the endorheic regions with the pink dots. Adapted from [RAM 14]. For a color version of this figure, see www.iste.co.uk/baghdadi/4.zip*

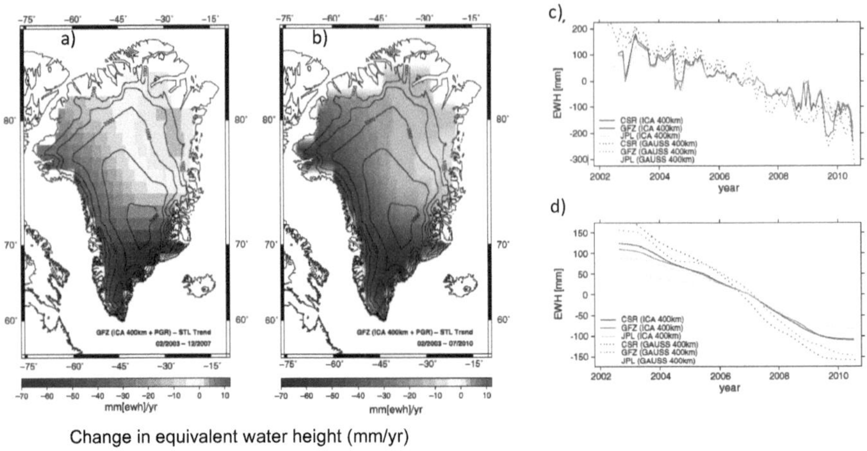

Figure 8.9. *a) Maps of trends for the periods 02/2003-12/2007; and b) 02/2003-07/2010; c) Mass variation time series using level 2 solutions from CSR, GFZ, JPL post-processed by ICA and the Gaussian filter; d) and associated trends. Adapted from [BER 12]. For a color version of this figure, see www.iste.co.uk/baghdadi/4.zip*

8.4.2.2. *Inter-annual variations of glaciers*

GRACE solutions allowed quantification of the recent decrease in mass from glaciers in Western Antarctica, Patagonia [CHE 07] and in the Gulf of Alaska [ARE 13], in response to climate variability. Chen *et al.* [CHE 07] observed a decrease in ice mass of -27.9 ± 11 Gt/year for glaciers in Patagonia between April 2002 and December 2007. The mass balance of glaciers in Alaska presents a decrease of -61 ± 11 Gt/year from October 2003 until October 2009, that agrees well with a hypsometric extrapolation of -65 ± 12 Gt/year provided by variations in topography from glaciers measured by the ICESat laser altimetry mission [ARE 13].

8.4.2.3. *Monitoring of mass variations of snow in periarctic basins*

Using a least-squares-based separation method, Ramillien *et al.* [RAM 05] demonstrated the possibility to realistically separate the variations of mass in different hydrological reservoirs, and more specifically of snow in high latitudes (section 8.2.3.1). The snow mass estimates derived from monthly GRACE solutions are in good agreement with those provided by SSM/I microwave sensor, global hydrological model outputs and climatologies (Figure 8.10). Root-mean-square difference of 10 to 20 mm were obtained on a large periarctic drainage basins such as the Yenissey, the

Ob, the Mckenzie or the Yukon. These comparisons at drainage basin scale also demonstrated the strong correlation between snow mass river discharges at inter-annual time-scale [FRA 11b].

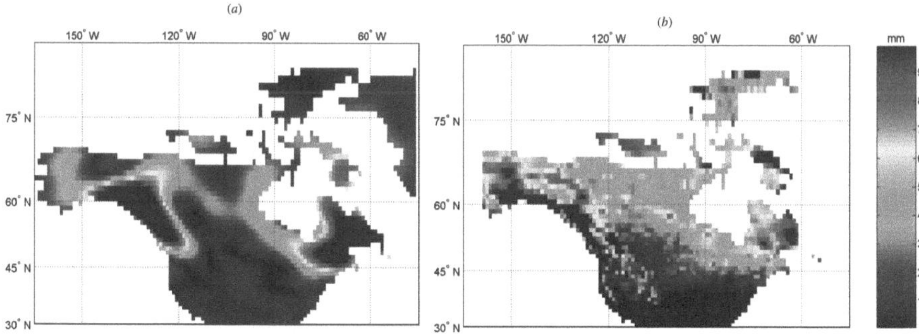

a) b)

Figure 8.10. *Maps of the annual amplitude of snow mass variations in Northern America expressed in height equivalent: provided by the inversion of data a) from GRACE between 2003 and 2006 using the approach from [RAM 05], and b) from USAF/ETAC climatology between 1950 and 1980. Adapted from [FRA 11b]. For a color version of this figure, see www.iste.co.uk/baghdadi/4.zip*

8.5. Perspectives

The GRACE gravimetry from space mission is about to end. To ensure the continuity of the monitoring of the mass balance of Earth's hydrological reservoirs, NASA and GFZ have decided to put the GRACE Follow-On mission into orbit, in 2017. Before the launch of the Follow-On mission, regional approaches should allow access to spatial (better localization of structures by construction) and temporal (through daily updates using approaches such as the Kalman filter) scales lower than those from global solutions. Through combining these new estimates of total water storage with data from other spatial missions (especially the maps of water level from the future interferometric altimetry mission Surface Water and Ocean Topography, or SWOT, that should be launched by CNES and NASA around 2020), and hydrological model outputs, it will be possible to estimate the variations of water mass in sub-surface reservoirs. These estimations will be very useful to calibrate hydrological models and constrain their operations through assimilation techniques.

8.6. Key points

Since its launch in March 2002, the GRACE gravimetry from space mission, composed of two satellites following each other on the same orbit at a distance of around 220 km, give access to the large-scale spatio-temporal variations on the Earth's gravity field. On time scales from a few months to a few years, these variations are mostly caused by the redistributions of water masses in fluid reservoirs (atmosphere, ocean, continental hydrology and polar ice caps). On land and the polar ice sheets, this new and unique source of information has allowed estimation of the mass balance of large river basins and polar ice sheets, with a precision of a few tenths of millimeters of equivalent water height to spatial resolutions larger than 300 km. Over land, the total water storage derived from GRACE measurements represents the integrated content of water of the soil column, in other words the sums of the masses of water contained in the difference hydrological reservoirs (surface water, sub-surface, snow, etc.).

8.7. Bibliography

[AND 05] ANDERSEN O.B., SENEVIRATNE S.I., HINDERER J. *et al.*, "GRACE-derived terrestrial water storage depletion associated with the European heat wave", *Geophysical Research Letters*, vol. 32, no. 18, p. L18405, 2005.

[ARE 13] ARENDT A., LUTHCKE S., GARDNER A. *et al.*, "Analysis of a GRACE global mascon solution for Gulf of Alaska glaciers", *Journal of Glaciology*, vol. 59, no. 217, pp. 913–924, 2013.

[BAU 09] BAUR O., KUHN M., FEATHERSTONE W.E., "GRACE-derived ice mass variations over Greenland by accounting for leakage effects", *Journal of Geophysical Research*, vol. 114, no. B13, p. B06407, 2009.

[BER 12] BERGMANN I., RAMILLIEN G., FRAPPART F., "Climate-driven interrannual variations of the mass balance of Greenland", *Global and Planetary Change*, vol. 82-83, pp. 1–11, 2012.

[BRU 10] BRUINSMA S., LEMOINE J.M., BIANCALE R. *et al.*, "CNES/GRGS 10-day gravity field models (release 2) and their evaluation", *Advances in Space Research*, vol. 45, no. 4, pp. 587–601, 2010.

[CHE 07] CHEN J.L., WILSON C.R., TAPLEY B.D. *et al.*, "Patagonia Icefield melting observed by Gravity Recovery and Climate Experiment (GRACE)", *Geophysical Research Letters*, vol. 34, no. 22, p. L22501, 2007.

[FRA 10] FRAPPART F., PAPA F., GÜNTNER A. *et al.*, "Interrannual variations of the terrestrial water storage in the Lower Ob basin from a multisatellite approach", *Hydrology and Earth System Sciences*, vol. 14, no. 12, pp. 2443–2453, 2010.

[FRA 11a] FRAPPART F., RAMILLIEN G., LEBLANC M., "An independent component analysis filtering approach for estimating continental hydrology in the GRACE gravity data", *Remote Sensing of Environment,* vol. 115, no. 1, pp. 187–204, 2011.

[FRA 11b] FRAPPART F., RAMILLIEN G., FAMIGLIETTI J.S., "Water balance of the Arctic drainage system using GRACE gravimetry products", *International Journal of Remote Sensing*, vol. 32, no. 2, pp. 431–453, 2011.

[FRA 13] FRAPPART F., RAMILLIEN G., RONCHAIL J., "Changes in terrestrial water storage vs. rainfall and discharges in the Amazon basin", *International Journal of Climatology*, vol. 33, no. 14, pp. 3029–3046, 2013.

[HAN 05] HAN S.C., SHUM C., JEKELI C. *et al.*, "Non-isotropic filtering of grace temporal gravity for geophysical signal enhancement", *Geophysical Journal International*, vol. 163, no. 1, pp. 18–25, 2005.

[IVI 13] IVINS E.R., JAMES T.S., WAHR J. *et al.,* "Antarctic contribution to sea level rise observed by GRACE with improved GIA correction", *Journal of Geophysical Research*, vol. B118, pp. 3126–3141, 2013.

[KAU 66] KAULA W., *Theory of Satellite Geodesy: Applications of Satellites to Geodesy*, Blaisdell Publishing, Waltham, MA, 1966.

[KLE 08] KLEES R., REVTOVA E.A., GUNTER B.C. *et al.*, "The design of an optimal filter for monthly GRACE gravity models", *Geophysical Journal International*, vol. 175, no. 2, pp. 417–432, 2008.

[KUR 12] KURTENBACH E., EICKER A., MAYER GÜRR T. "Improved daily GRACE gravity field solutions using a Kalman smoother", *Journal of Geodynamics*, vol. 59-60, pp. 39–48, 2012.

[KUS 07] KUSCHE J., "Approximate decorrelation and non-isotropic smoothing of time variable GRACE-type gravity field models", *Journal of Geodesy*, vol. 81, no. 11, pp. 733–749, 2007.

[LEB 09] LEBLANC M.J., TREGONING P., RAMILLIEN G. "Basin-scale integrated observations of the early 21st century multiyear drought in southeast Australia", *Water Resources Research*, vol. 45, no. 4, p. W04408, 2009.

[PEL 04] PELTIER W.R., "Global glacial isostasy and the surface of the ice-age Earth: the ICE-5G (VM2) model and GRACE", *Annual Review of Earth and Planetary Sciences*, vol. 32, pp. 111–149, 2004.

[RAM 05] RAMILLIEN G., FRAPPART F., CAZENAVE A. *et al.*, "Time variations of land water storage from the inversion of 2 years of GRACE geoids", *Earth and Planetary Science Letters*, vol. 235, nos. 1–2, pp. 283–301, 2005.

[RAM 06] RAMILLIEN G., FRAPPART F., GÜNTNER A. *et al.*, "Time-variations of the regional evapotranspiration rate from Gravimetry Recovery and Climate Experiment (GRACE) satellite gravimetry", *Water Resources Research*, vol. 42, no. 10, p. W10403, 2006.

[RAM 12] RAMILLIEN G., SEOANE L., FRAPPART F. *et al.*, "Constrained regional recovery of continental water mass time-variations from GRACE-based geopotential anomalies", *Surveys in Geophysics*, vol. 33, no. 5, pp. 87–905, 2012.

[RAM 14] RAMILLIEN G., FRAPPART F., SEOANE L., "Application of the regional water mass variations from GRACE satellite gravimetry to large-scale water management in Africa", *Remote Sensing*, vol. 6, no. 8, pp. 7379–7405, 2014.

[RAM 15] RAMILLIEN G., FRAPPART F., GRATTON S. "Sequential estimation of surface water mass changes from daily satellite gravimetry data", *Journal of Geodesy*, vol. 89, no. 3, pp. 259–282, 2015.

[RAM 16] RAMILLIEN G., FRAPPART F., SEOANE L., "Space gravimetry using GRACE satellite mission data: basic concepts", in BAGHDADI N., ZRIBI M. (eds), *Microwave Remote Sensing of Land Surface*, ISTE Ltd, London and Elsevier Ltd, Oxford, 2016.

[ROD 04] RODELL M., FAMIGLIETTI J.S., CHEN J. *et al.*, "Basin scale estimates of evapotranspiration using GRACE and other observations", *Geophysical Research Letters*, vol. 31, no. 20, p. L20504, 2004.

[ROD 09] RODELL M., VELICOGNA I., FAMIGLIETTI J., "Satellite-based estimates of groundwater depletion in India", *Nature*, vol. 460, no. 7257, pp. 999–1003, 2009.

[ROW 10] ROWLANDS D.D., LUTHCKE S.B., MCCARTHY J.J. *et al.*, "Global mass flux solutions from GRACE: a comparison of parameter estimation strategies – mass concentrations versus Stokes coefficients", *Journal of Geophysical Research*, vol. 115, no. B1, p. B01403, 2010.

[SAS 06] SASGEN I., MARTINEC Z., FLEMING K., "Wiener optimal filtering of GRACE data", *Studia Geophysica et Geodaetica*, vol. 50, no. 4, pp. 499–508, 2006.

[SCH 07] SCHRAMA E.J.O., WOUTERS B., LAVALLÉE D.A., "Signal and noise in Gravity Recovery and Climate Experiment (GRACE) observed surface mass variations", *Journal of Geophysical Research*, vol. 112, no. B8, p. B08407, 2007.

[SEO 05] SEO K.W., WILSON C.R., "Simulated estimation of hydrological loads from GRACE", *Journal of Geodesy*, vol. 78, nos. 7–8, pp. 442–456, 2005.

[STR 07] STRASSBERG G., SCANLON B.R., RODELL M., "Comparison of seasonal terrestrial water storage variations from GRACE with groundwater-level measurements from the High Plains Aquifer (USA)", *Geophysical Research Letters*, vol. 34, no. 14, p. L14402, 2007.

[SWE 02] SWENSON S., WAHR J., "Methods for inferring regional surface-mass anomalies from Gravity Recovery and Climate Experiment (GRACE) measurements of time variable gravity", *Journal of Geophysical Research*, vol. 107, no. B9, pp. ETG 3-1–ETG 3-13, 2002.

[SWE 06] SWENSON S., WAHR J., "Post-processing removal of correlated errors in GRACE data", *Geophysical Research Letters*, vol. 33, no. 8, p. L08402, 2006.

[SYE 09] SYED T.H., FAMIGLIETTI J.S., CHAMBERS D.P., "GRACE-based estimates of terrestrial freshwater discharge from basin to continental scales", *Journal of Hydrometeorology*, vol. 10, no. 1, pp. 22–40, 2009.

[TAP 04] TAPLEY B.D., BETTADPUR S., RIES J.C. *et al.*, "GRACE measurements of mass variability in the Earth system", *Science*, vol. 305, no. 5683, pp. 503–505, 2004.

[VEL 13] VELICOGNA I., WAHR J., "Time-variable gravity observations of ice sheet mass balance: Precision and limitations of the GRACE satellite data", *Geophysical Research Letters*, vol. 40, no. 12, pp. 3055–3063, 2013.

[YEH 06] YEH P.J.F., SWENSON S.C., FAMIGLIETTI J.S. *et al.*, "Remote sensing of groundwater storage changes in Illinois using the Gravity Recovery and Climate Experiment (GRACE)", *Water Resources Research*, vol. 42, no. 12, p. W12203, 2006.

9

Applications of GNSS-R in Continental Hydrology

9.1. Introduction

9.1.1. *History*

GNSS (Global Navigation Satellite System) Reflectometry (GNSS-R) is a bistatic radar remote sensing technology (transmitters and receivers are not in the same place) that uses microwave signals of opportunity from radio navigation such as the Global Positioning System (GPS). It extends and complements existing techniques for observation of the Earth by multiplying the number of transmitters (satellites in different positioning constellations) and receivers through the use of ground measurement networks, such as the permanent GNSS network in France (Réseau permanent GNSS, RGP, (http://rgp.ign.fr)) or the Plate Boundary Observatory (PBO, http://pboweb.unavco.org) in the western United States, and allowing new measuring geometries (see [DAR 16]).

In 1988, Hall and Cordey [HAL 88] proposed the concept of a "multistatic scatterometer" to estimate surface winds using a space radiometer measuring the signals from the GPS constellation reflected by oceans. In 1992, when tested on a French military plane, detection of GPS interference caused by reflections on the ocean surface led to the first analysis of GNSS-R multistatic scattering [AUB 94]. In 1993, the concept of

Chapter written by Erwan MOTTE, Alejandro EGIDO, Nicolas ROUSSEL, Karen BONIFACE and Frédéric FRAPPART.

a spatial or airborne altimeter using GPS satellites reflections on the ocean surface came into existence [MAR 93].

In 1994, the first reflected GNSS signals were measured from space by the SIR-C device [LOW 02]. From then on, numerous studies were carried out to evaluate the potential of the technique in characterizing the Earth. Firstly, they considered oceanographic applications, which were more obvious given the high level of reflectivity of the sea (altimetry, sea state, surface winds), then they opened themselves up to studying the cryosphere (ice topography, thickness of the sea ice) and land surfaces (altimetry of inland waters, soil moisture, biomass estimation). In 2003, the UK-DMC satellite carried out a mission to evaluate the potential of GNSS-R from space using a dedicated device. The results were convincing and the first constellation of satellites dedicated to GNSS-R reflectometry should be launched in October 2016 [RUF 12].

9.1.2. *Chapter outline*

In this chapter, we will explain how reflectometry through signals of opportunity can be used to measure geophysical properties over land surfaces. We will begin with a brief review of dual antenna waveform analysis techniques, whether these be either "conventional" (cGNSS-R, section 9.2.1.1) or "interferometric" (iGNSS-R, section 9.2.1.2) using dedicated GNSS-R receivers, as well as the mono-antenna technique that is based on the study of interference between direct and reflected contributions (Interference Pattern Technique (IPT)) via a standard GNSS receiver.

We will then discuss the applications of reflectometry over continental areas through the water cycle, which is a key parameter for monitoring recent and future climate change (Figure 9.1). We will start at the source, in mountainous areas, where we will consider height variations of the snowpack in order to estimate the stock present in the monitored mountainous area. Then, we will move downstream to water reservoirs like lakes and ponds, for which drastic variations in height are reflected in periods of drought or heavy rainfall. We will end our journey by seeking to address where soil moisture changes in the alluvial plains play a vital role in soil–vegetation–atmosphere processes.

These areas that show water cycle evolution from upstream to downstream will also be addressed based on reflectometry methods. Thus, applications linked to continental hydrology (measuring snow and water reservoir heights) will be based on altimetry, while moisture and vegetation cover measurements in plains will be based on scatterometry (studying the perturbation of the signal reflected by the reflective medium). Finally, we will discuss the prospects of the GNSS-R technique, including ongoing and future space missions.

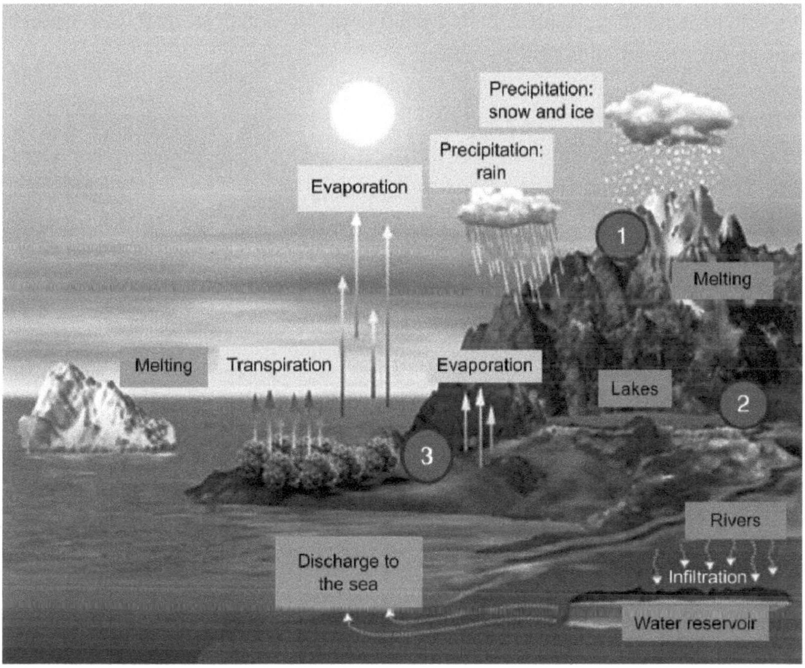

Figure 9.1. *Monitoring continental water cycles via GNSS-R measurements: 1) snow depth, 2) heights of reservoirs, 3) soil moisture variations and monitoring of plant growth*

9.2. Background on measurement and GNSS-R observable techniques

GNSS-R measurement techniques are discussed in detail in [DAR 16]. However, we will present a brief reminder to highlight the important tools and metrics for understanding its applications.

9.2.1. *GNSS-R by waveform analysis*

GNSS-R by waveform analysis techniques are based on in situ, airborne or spaceborne receivers, equipped with one or several (if the receiver works by multi-polarization or multi-frequency) antenna pairs, followed by one or several pairs of analog-digital converters, preferably synchronized to the same clock signal. Each pair corresponds to a channel that measures the direct signal with its up-looking antenna and another channel measuring the signal reflected by the surface with its down-looking antenna. The reflection point minimizing the distance between the transmitter and the receiver is called the specular point (Figure 9.2).

Two groups of methods exist for processing and/or combining the signals: cGNSS-R is based on correlation of received signal with a specific code replica generated by the tool and iGNSS-R, where the reflected signal is correlated with the direct signal, considered to be the "clean" code version. The aim of these techniques, as well as the advantages and disadvantages, are summarized below. Other hybrid processing methods for GNSS-R by waveforms analysis such as the reconstructed GNSS-R (rGNSS-R) and partial interferometric GNSS-R (piGNSS-R) have been developed to try to get the best of both worlds, however, these complex and experimental techniques are not detailed in this section.

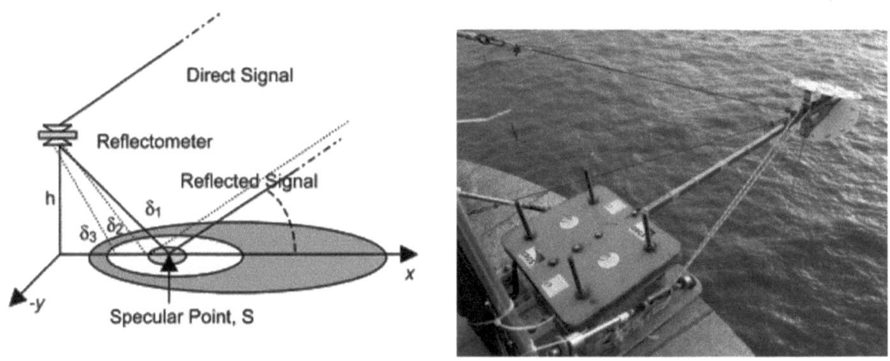

Figure 9.2. *Left: geometry of a GNSS-R dual-antenna receiver. Right: receiver located on the Cordouan Lighthouse, France. From [LES 16]*

9.2.1.1. *Measurements by conventional waveform or cGNSS-R*

In the same manner as commercial receivers, cGNSS-R type receivers use the properties of GNSS signals (see [DAR 16]) to estimate the initial value (acquisition) and variations (tracking) of parameters in order to decode the signal:

– the position of the beginning of the code (called delay t_0);

– the Doppler frequency f_d.

These operations are performed by seeking the maximum correlation, also called *correlation peak*, between the received signal (direct or reflected) *x(t)* and the replica of the locally generated code *a(t–τ)* that is shifted in the temporal domain and in the frequency domain.

Coherent integration improves the Signal-to-Noise Ratio (SNR) by performing a correlation over a period T_C, during which time it is estimated that t_0 and f_d are stable.

The correlation amplitude Y^c between the received signal and the replica of the carrier is finally given by the following expression:

$$Y^c(t_0,\tau, f_d)= \frac{1}{T_c}\int_{t_0}^{t_0+T_c} x(t)a^*(t-\tau)\, exp^{-j2\pi(f_c+f_d)t}\, dt \qquad [9.1]$$

where f_c is the carrier frequency and τ is the temporal code shift.

The amplitude $Y^c(\tau)$ for the Doppler frequency f_d that maximizes correlation is also called *waveform*. The maximum of the waveform is called the *correlation peak.*

For the direct signal received by the up-looking antenna, the signal-to-noise ratio is usually high enough for algorithms to detect the correlation peak and decode the message with a relatively low integration time (~1 ms). Many tools can carry out this standard GPS process (for example [FER 11]) with an estimation for Y^c on a limited number of correlators.

With regard to the reflected signal, which is received by the down-looking antenna, the surface scattering process can cause acquisition and tracking mechanisms to be ineffective, even with extended coherent integration time. In order to overcome this limitation, a master–slave processing scheme is usually used in the case of ground-based and low altitude aircraft receivers. In this case, correlation parameters f_d and t_0 calculated for the direct path are used and are blindly applied to the reflected signal, assuming them to be identical (satellite at the same relative speed). In the case of high-altitude and space-borne receivers, the correlator delay τ and the Doppler frequency shift must be estimated to accommodate for the additional path delay and different relative velocity of the receiver with respect to the specular point position.

9.2.1.2. *Measurement by interferometric waveform or iGNSS-R*

In iGNSS-R type receivers, the reflected signal is not correlated to a replica, but rather to the direct signal from the GNSS satellite. In this case, it is assumed that the direct signal SNR is sufficient to demodulate the reflected signal. The undeniable advantage of iGNSS-R compared to cGNSS-R is its ability to work with all existing signals without prior knowledge of their modulation or coding. This enables the use of higher bandwidths than GNSS public code signals (GPS L1 C/A) such as the GPS L1 P(Y) and also satellite television signals. The major constraint is that without the possibility of selecting the satellite through a local code replica, the direct source of the signal must be selected using a high gain antenna that is able to isolate and track the transmitting satellite. This is generally possible using beam forming antennas [MAR 11].

9.2.1.3. *Polarization*

GNSS signals are generated in circular polarization, which gives them a better immunity to atmospheric conditions. Since the first systems were designed, Right Hand Circular Polarization (RHCP) was conventionally chosen for signal transmission. The up-looking antennas are therefore RHCP antennas in order to capture the direct signal.

When the GNSS signal is reflected on the Earth's surface, its polarization is reversed. For a perfectly reflecting surface (specular reflection) with a low angle of incidence, polarization of the signal is inverted at the reflection, thus becoming a Left Hand Circular Polarization (LHCP). For GNSS-R measurements by waveform analysis, the down-looking antennas are therefore generally LHCP antennas.

However, for partially reflecting surfaces (rough, covered with vegetation or with high angles of incidence), the relative contribution of the two polarizations in the reflected signal may vary (see sections 9.4 and 9.5 in this chapter).

For this reason, some systems integrate two reflected channels, LHCP and RHCP, in order to measure the relative contribution and extract information about the surface.

9.2.1.4. *Ground footprints*

For specular reflections, which occur in the case of low surface roughness, it is generally considered that the majority of the reflected signal power comes from the *first Fresnel zone*. This zone is defined by the set of points for which the trajectory difference (relative to the point of specular reflection) is less than $\lambda/2$.

The first Fresnel zone corresponds to a semi-major axis ellipse (r_a) and semi-minor axis ellipse (r_b) that can be calculated by [LAR 13]:

$$r_b = \sqrt{\frac{\lambda h}{\sin(\theta')} + \left(\frac{\lambda}{2\sin(\theta')}\right)^2} \quad [9.2]$$

$$r_a = \frac{r_b}{\sin(\theta')} \quad [9.3]$$

where λ is the wavelength of the signal, h is the height of the receiver and θ' is the elevation angle of the satellite relative to the surface plane passing through the specular point.

Roussel *et al.* [ROU 14] developed a geometric model to take the effects of sphericity, heterogeneity of the surface of the Earth and the troposphere into account in order to accurately calculate the ground footprints' positions in different measurement configurations.

Figure 9.3(a) shows a simulation of the coverage of the signal reflection of footprints for 24 consecutive hours of measurements (one measurement every 15 minutes) from the Cordouan Lighthouse in France (h = 60 m). Figure 9.3(b) shows a simulation of the coverage for a geodetic network type of snow measuring station (antenna 2 m above the ground) with six different azimuths and five elevation angles. In this case, the total spatial coverage is ~1000 m², which decreases when snow accumulates in the vicinity of the antenna.

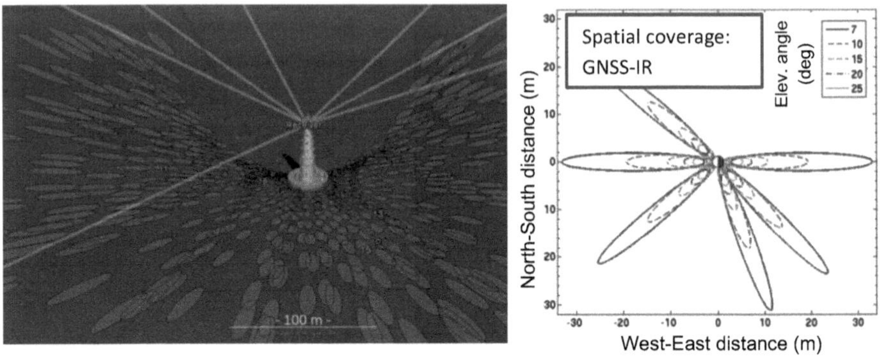

Figure 9.3. *a) Simulation of the first Fresnel zone for the Cordouan Lighthouse, France (24 hours, one measurement every 15 minutes, $\theta' > 5°$), adapted from [ROU 14]; b) smulation of the first Fresnel zone for six different azimuths and five elevations for a receiver located 2 m high. For a color version of this figure, see www.iste.co.uk/baghdadi/4.zip*

9.2.1.5. *Basic observables in GNSS-R by waveform analysis*

9.2.1.5.1. Waveform

For a reflected signal, waveform analysis can be used to accurately point out the reflected correlation peak delay relative to the direct signal in order to extract altimetry information. The scatterometric characteristics can be assessed from the measurement of the reflected correlation peak broadening in the temporal domain and the waveform amplitude variations (Figure 9.4) linked to soil properties (moisture, roughness) or vegetation cover. The processing of GNSS-R data for these applications will be discussed in sections 9.4.1 and 9.5.1.

In order to reduce "speckle" type noise, we generally carry out a so-called "incoherent" integration of N_i waveforms. The averaged waveform expression looks like:

$$< \left|Y^i(\tau, f_d)\right|^2 >= \frac{1}{N_i}\sum_{n=1}^{N_i}\left|Y^i(t_n, \tau, f_d)\right|^2 \qquad [9.4]$$

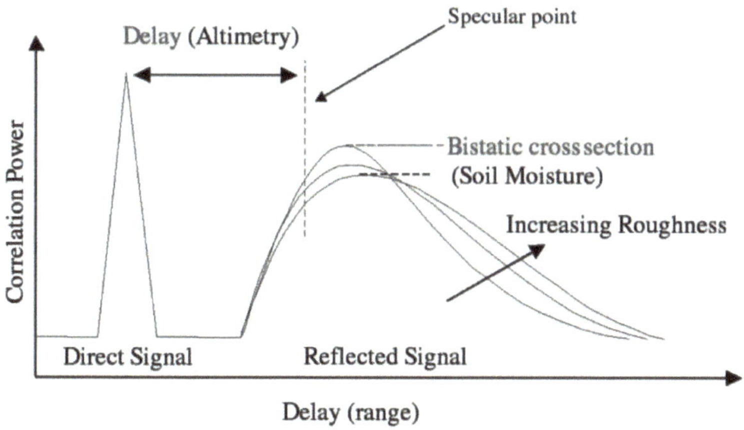

Figure 9.4. *Superimposed direct and reflected waveforms, based on [MAS 04]*

9.2.1.5.2. Interferometric complex field (ICF)

The interferometric complex field can be defined as the ratio of the maximum of the reflected waveform $Y_r^c(\tau)$ to the maximum of the direct waveform $Y_d^c(\tau)$:

$$ICF = \frac{\max(Y_r^c(\tau))}{\max(Y_d^c(\tau))} \quad [9.5]$$

For low altitude observations (< 1 km), this ratio will undo effects related to propagation, such as those induced by the troposphere and the ionosphere, and will have calibrated strength information if we consider planar antennas, which are identical and symmetrical to the horizontal plane.

9.2.2. *GNSS-R by analysis of changes in the signal-to-noise ratio*

9.2.2.1. *Interference pattern technique (IPT)*

In addition to classic positioning information, commercial GNSS receivers are able to give an indication of the received signal quality, through the so-called Signal to Noise Ratio (SNR) linked to the correlation amplitude between the received signal and a replica of the code generated by the device.

In the IPT technique, direct and reflected GNSS signals are received by a single omnidirectional antenna located between 1 m and a few hundred meters above the ground (Figure 9.5). While most of the signal comes directly to the antenna in RHCP polarization, the surrounding planar surface (water, vegetation, soil) reflects a portion of the incident GNSS signal by inverting its polarization to LHCP. The antenna combines direct and reflected signals, which are then generally processed by a standard GNSS receiver.

Depending on the geometry and nature of the terrain, the power of direct and reflected contributions will be more or less significant and the signals will be combined constructively or destructively, creating interferometric patterns (hence the name IPT).

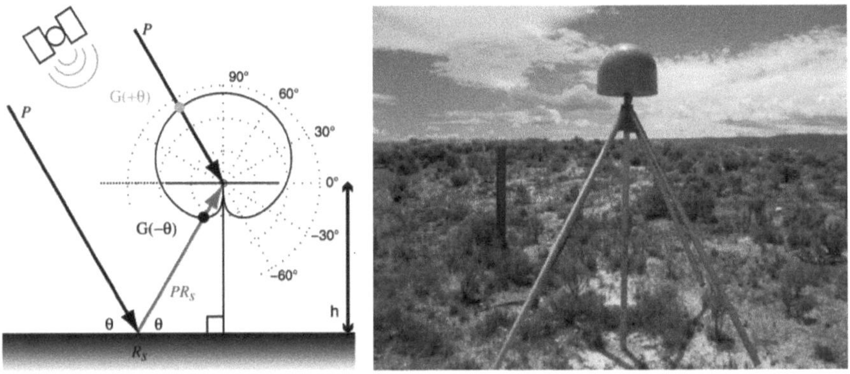

a) b)

Figure 9.5. *a) Geometry of a direct signal P from a satellite at elevation θ and its specular reflection PRs captured by a mono-antenna (modeled by a solid line arbitrary radiation diagram). We see that the reflected signal component will have gain $G(-\theta)$ that is generally smaller than the direct signal $G(\theta)$ (adapted from [LAR 08]). b) Illustration of a GNSS measurement station that can be used for reflectometry measurements*

9.2.2.2. *Basic observables in IPT*

The basic IPT technique observable is the time series of the reflected signal to noise ratio $SNR_r(t) = SNR_t - SNR_d$ where SNR_t represents the IPT–*SNR* that combines direct and reflected signals and the SNR_d represents the *SNR* of the direct signal (generally approximated by a second order polynomial, [LAR 08]). For a given height h, $SNR_r(t)$ is a periodic function of the carrier phase ϕ_m, which can be written [LAR 08] as:

$$SNR_r(t) = A_m(t)cos\left(\frac{4\pi h}{\lambda}sin(\theta) + \phi_m(t)\right) \quad [9.6]$$

where $A_m(t)$ depends on the intensity of the reflective surface and combines the antenna gain diagram and the intensity of the response linked to the reflective surface. Both are dependent on elevation θ of the satellite (Figure 9.5) and wavelength λ of the signal.

$A_m(t)$, $\phi_m(t)$ and h are the three metrics that can be inverted in the IPT technique; it has been shown that variations were strongly correlated with the effective height of the receiver above the reflective surface, soil moisture levels and vegetation cover.

To invert these metrics, SNR_r is measured for a range of elevations θ. Analysis of the amplitude, frequency and phase of these variations is directly related to $A_m(t)$, $\phi_m(t)$ and h (sections 9.3.1.2, 9.4.3 and 9.5.3).

The height of the antenna and the surrounding obstacles define the coverage and spatial resolution of the measurement.

9.3. Altimetry

GNSS signals are a suitable resource for altimetry, especially over surfaces with high reflectivity such as water bodies. Moreover, for surfaces with low roughness (standard deviation of heights σ_{surf} less than $1/10^{th}$ of the wavelength λ of the signal), there is a significant dominance of coherent scattering that results in specular reflection of the signal, which results in a phase shift between the direct and the reflected signal that can be used to improve altimetric measurements. As part of this work dedicated to continental surfaces, we will focus on snow levels for which roughness of the surface is less than $1/10^{th}$ of the wavelength of the signal, which is about 2 cm in L-band. We will then look at the heights of water reservoirs (lakes, ponds) for which surface reflection can be seen as a mirror plane with zero roughness when there is no wind. For questions related to the retrieval of altimetric parameters in the event of severe roughness (oceanographic issues), we encourage the reader to read Jin *et al.* [JIN 14].

In GNSS reflectometry, altimetry information is contained within two types of observables that quantify the delay of the reflected signal relative to the direct signal:

– the relative delay between the direct path and the reflected path of the digital message or the "code" transmitted by the one or more positioning satellites;

– the difference in carrier phase in the case of coherent scattering.

The following sections illustrate three GNSS-R altimetry applications based on these observables, generated using the different techniques described above: IPT processing with commercial mono-antenna receivers [DAR 16], and cGNSS-R with dedicated 2-antenna receivers.

9.3.1. *Measuring the snowpack*

Current observation techniques for characterizing the snowpack use dedicated facilities that provide point measurements or measurements with a very coarse spatial resolution (pixel size). Given the heterogeneity of snowy surfaces (transport by wind, topography), reflectometry appears as an innovative tool with an intermediate spatial resolution that produces continuous recordings.

So far, few studies have demonstrated the possibility of characterizing the internal structure of the snowpack. We will first detail the cGNSS-R by waveform analysis methodology for characterizing the snowpack structure.

We will then present the IPT technique applied to measuring the thickness of the snowpack within the context of a dedicated system, but also in connection with conventional geodetic facilities.

9.3.1.1. *Highlighting snowpack stratification by cGNSS-R measurements*

9.3.1.1.1. Experimental device

In December 2009, an experiment [CAR 12] was used to test the sensitivity of GNSS-R reflectometry in stratification of dry snow masses at the Dome Concordia (Antarctica). The campaign was based on measurements made by the cGNSS-R GOLD-RTR receiver [NOG 07], which was equipped with antennas on a tower over 45 m above the surface. The data were recorded over a period of 20 days.

9.3.1.1.2. Observations

As surface conditions are extremely flat (overall slope < 0.2° and low roughness), one can consider that it is a situation of almost purely specular reflection (incoherent scattering from the ground). However, the observed reflected waveforms are not constant but vary with time. The amplitude oscillations have a similar pattern when the GPS constellation is in an identical configuration, which discards the assumption of variations due to random noise and would be better explained by a contribution from the deep snow layers.

9.3.1.1.3. Modeling the reflections phase

By considering the horizontal reflection planes that correspond to different layers of snow, we can create a model of the reflected signal phase transmitted to the receiver relative to the reference signal using:

$$\phi_m = \frac{\rho}{\lambda} = \frac{2H_m}{\lambda}\sin(e) \qquad [9.7]$$

where ρ is the delay between the direct and reflected signals, λ is the wavelength of the signal, e is the elevation angle of observations and H_m is the distance of the reflector relative to the receiving antenna.

When angular conditions change the phase of the reflected signal, ϕ_m also changes by introducing oscillations in both phase and amplitude.

9.3.1.1.4. Results

A holographic radio technique is used for overall estimation of the depth of snow layers both qualitatively, through simple visualization of fringes that code for distortion (lag-hologram, see Figure 9.6) and quantitatively, by analyzing these fringes. Figure 9.6 clearly shows the signature of several snow layers below the surface in the holographic representation.

The potential vertical resolution of the technique is 5 m and the maximum depth is 250 300 m, which is the depth when the reflected signal becomes too small to be measured.

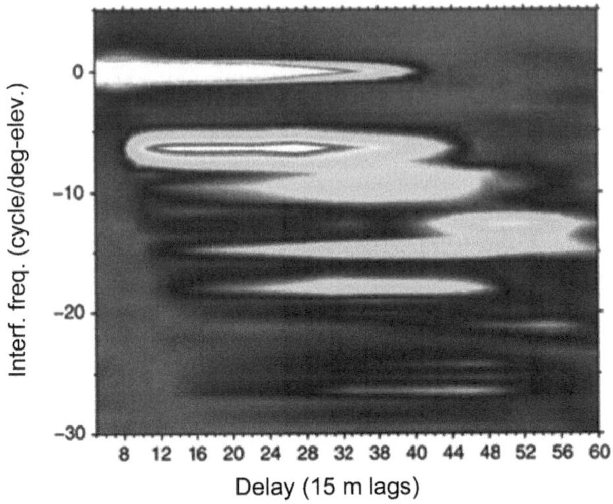

Figure 9.6. *Example of a lag-hologram for a series of observations of 1 s. Frequency 0 corresponds to the reference signal (direct signal) and the negative frequencies illustrate the spread of reflective elements below the air-snow interface. Adapted from [CAR 12]. For a color version of this figure, see www.iste.co.uk/baghdadi/4.zip*

9.3.1.2. *Measuring snow depth by IPT*

The IPT technique simultaneously measures the intereference between the direct and the reflected signals interfere. The strength of the received signal can be expressed in terms of the elevation angle of satellites.

In Figure 9.7(a), one can observe the SNR expressed in terms of sine of the elevation for three depths of snow (0, 0.5 and 1 m). Amplitudes are shifted to better visualize the three signals. The metric linked to the height of SNR oscillations is frequency relative to sine of elevation. A Lomb–Scargle type frequency analysis is used to retrieve the main frequency of these oscillations (Figure 9.7(b)) to then convert it into receiver height.

The advantage of the IPT technique is its ability to provide snow measurements with high temporal, but also spatial, resolution and its ability to monitor temporal and mapping dynamics of accumulation and ablation of the snowpack. Acquiring new data on snow depths at altitudes where few measurements have ever been made is a valuable source of information for hydrometeorological and avalanche models.

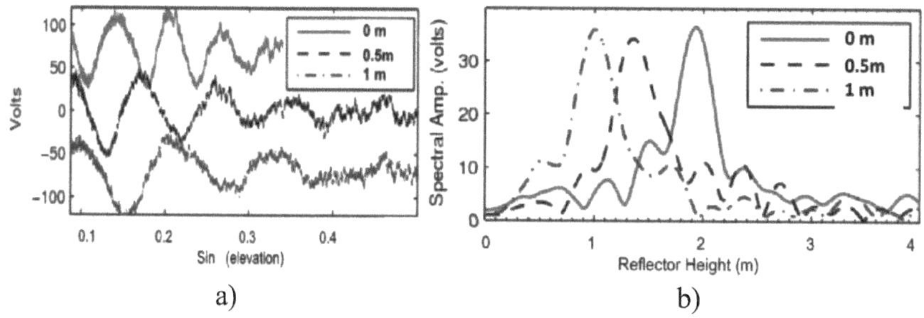

Figure 9.7. *a) Example of SNR according to the sine of the elevation angle for different depths of snow. b) Corresponding frequency analysis where the frequency of SNR oscillation based on the sine of the elevation is directly expressed in antenna height h. For a color version of this figure, see www.iste.co.uk/baghdadi/4.zip*

9.3.1.2.1. IPT method accuracy

Accuracy of the IPT method was evaluated by Gutmann *et al.* [GUT 12] on a site in the American Rocky Mountains (Niwot Ridge) for a hydrological season using several techniques: laser and *in situ* measurements, as well as aerial lidar measurements. A good consensus is achieved on the whole, even if there is a bias of 10 cm (Root Mean Square Error (RMSE) = 13 cm) when the snow is at its thickest. Moreover, Nievinski and Larson [NIE 14] evaluated the performance of the IPT technique for a forest site by performing over a hundred measurements around the antenna. This experiment confirmed the strong correlation (0.99) between *in situ* measurements and GNSS. More systematically, the IPT technique used for snow depth measurements was evaluated on over 100 sites in the American West and were compared to digital simulations from the SNODAS American model [BON 15a]. These comparisons were used to illustrate the use of independent observations of snow depth via the IPT technique.

9.3.1.2.2. Implementation in the Alps

The IPT technique is being implemented in the French Alps to be applied to existing sites of the French National permanent GNSS network (RESIF-RENAG). The original method by Larson *et al.* [LAR 09] initially applied to relatively flat and open areas, so the system needed to be characterized and calibrated for effects of complex and heterogeneous topography in the alpine area. Encouraging results have been recorded at the before site: experimental of the Col du Lac Blanc (CNRM/CEN, Météo France/CNRS) by the ultrasonic sensor (LACB) there (Figure 9.8). Three GNSS traces with three different azimuths were considered (Q2, Q3 and Q4). For traces Q2 and Q3, a good consensus with the ultrasound probe was observed. For the Q4 trace, snow height estimations were about 1.5 m higher relative to the ultrasound probe. This can be explained by the strong heterogeneity of the snowpack and topography surrounding this high-altitude station (2720 m). The trace Q4 crosses for example a zone of enhanced snow accumulation during frequent blowing snow events.

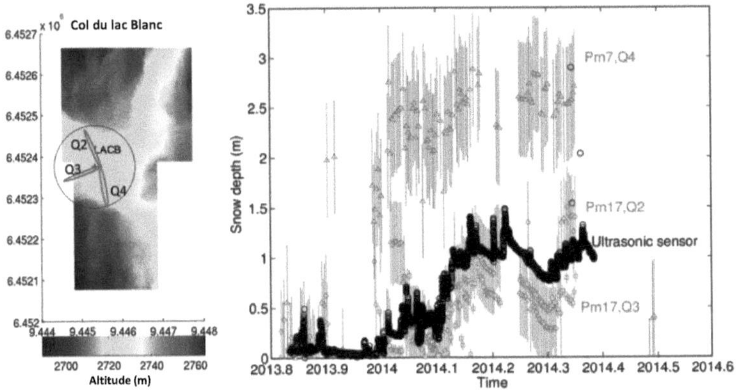

a) b)

Figure 9.8. *a) Topography of the area of Lac Blanc, with the position of the ultrasonic sensor (LACB) and the three considered GNSS traces (Q1, Q2 and Q3); b) snow depth measured by GNSS for two satellite traces (green and purple) and measurements by ultrasonic sensor (black) at the Lac Blanc site [BON 15b]. For a color version of this figure, see www.iste.co.uk/baghdadi/4.zip*

The IPT technique has been validated in different configurations and for different geographical areas. The next step consists of improving satellite track selection to reflect the heterogeneity of the distribution of snow, especially in mountainous areas.

9.3.2. *Measuring the water level*

9.3.2.1. *Altimetry by code delay in cGNSS-R*

9.3.2.1.1. Code delay

The time delay $\Delta\tau$ between the direct correlation peak and the reflected correlation peak observed in waveforms (see section 9.2.1) is directly related to the difference in distance traveled by the reflected signal, depending on receiver height h above the surface and the elevation angle θ of the satellite that is transmitting the signal (see geometric configuration, Figure 9.9). A bias b linked to the difference in position and instrumental delay between direct and reflected antenna is added, and b is considered constant for each estimate:

$$\Delta\tau(t) = 2h \sin\theta(t) + b \quad [9.8]$$

By solving a linear system of several equations for all visible satellites at a time t, it is possible to estimate h and b.

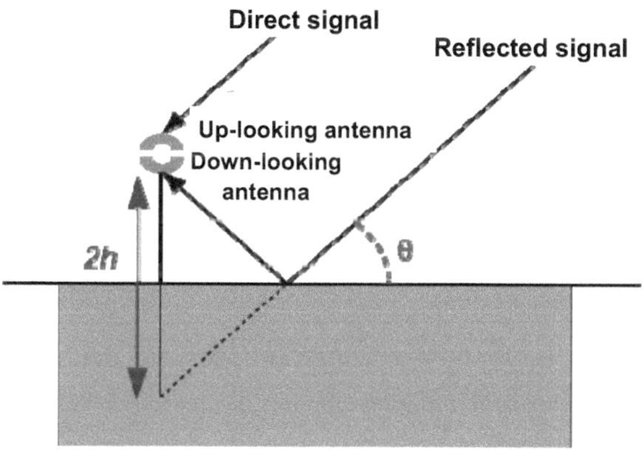

Figure 9.9. *Altimetry measurement configuration*

9.3.2.1.2. Estimation of delay $\Delta\tau$ between direct and reflected signals

Within the context of GNSS-R measurements and for a specular reflection on a planar surface, the delay Δτ between direct and reflected signals can be calculated [HEL 08] in a simplified manner, such as the difference between the position of the correlation maximum in the reflected waveform $\tau_{Yr,max}$ and the position of the correlation maximum in the direct waveform $\tau_{Yd,max}$:

$$\Delta\tau = \tau_{Yr,max} - \tau_{Yd,max} \quad [9.9]$$

The solution Δτ of the delay estimate for the correlation maximum is limited to the inverse of the sampling frequency F_S:

$$\delta\tau = 1/F_S \quad [9.10]$$

This corresponds, for example, to an accuracy of 50 ns (or 15 m) for a receiver operating at 20 M samples/s.

The shape of the autocorrelation function (section 9.2.1) and, therefore, the width of the correlation peak in the time domain are directly related to frequency of the modulating code and the type of modulation. For the L1 C/A GPS signal, as the code is modulated at 1.023 MHz, the width of the correlation peak is equal to a code "chip": $\frac{1}{1.023\ 10^6} = 978$ ns, which more or less corresponds to 300 m traveled by the signal. For L5 GPS or Galileo E5, the frequency of the code is 10.23 MHz, reducing the width of the correlation peak by a factor of 10. The higher the code modulation frequency (and consequently the bandwidth of the wide band signal), and the narrower the correlation peak, the more accurate the altimetric estimate.

Several parameters can degrade the altimetry resolution:

– for a limited bandwidth of the receiver relative to the total bandwidth of the signal, the correlation peak is less pronounced, adding uncertainty to the location of the correlation peak;

– for a low-reflecting or very rough surface, or a remote receiver, the reflected waveforms can be very noisy, adding uncertainty to the location of the correlation peak;

– for satellites at low elevations, the trajectory difference becomes very small between the direct and the reflected signal and the signals suffer the effects of the troposphere (mitigation, curvature), making it difficult to estimate height;

– for satellite receivers, the ionosphere introduces a systematic error in the estimation of the relative delay of the direct and reflected signal.

9.3.2.1.3. Techniques for improving altimetry resolution

In order to overcome the intrinsic limitations of the signal and the receiver mentioned above, several techniques can be used to reduce measurement uncertainty to a decametric level.

One of these methods is to adjust a parabola around the peak of the waveform and estimate where the top is [EGI 10]. This technique overcomes the limited temporal resolution of the receiver and by averaging successive observations, reduces measurement error.

Using a Kalman filter with parameters calculated based on receiver geometry and the phenomenon to be observed (tides, for example) allows us to estimate measurements confined to previous measurements [EGI 10].

The average of observations from different satellites at various elevations also reduces measurement uncertainty.

Finally, a multifrequency receiver (L1, L2, L5) has to be used to overcome the effect of the ionosphere in the case of space-borne receivers.

9.3.2.2. *Altimetry by GNSS-R phase difference*

A GNSS-R receiver observable used for accurate altimetry is the phase $\phi_{ICF}(t)$ of the *ICF(t)* interferometric complex field (see section 9.2). One can get $\phi(t)$ as the angle of the ICF relative to the origin of the complex plane thanks to the atan2 function, which uses the real part $\Re$ and the imaginary part $\Im$ of the complex interferometric field [HEL 08]:

$$\phi_{ICF}(t) = \text{atan2}\left(\Re(ICF(t)), \Im(ICF(t))\right) \quad [9.11]$$

If the surface reflection can be considered as specular (reflective planar surface), the signal will retain its coherence, meaning that the phase properties will be kept. In this case, the measured phase $\phi_{ICF}(t)$ will be directly linked to the additional distance $D(t)$ traveled by the reflected signal relative to the direct signal, to which a noise phase (b_ϕ), which is dependent on the receiver and surface, will be added, as well as an integer N of unknown cycles called integer ambiguity, linked to the periodic nature of the phase:

$$\phi_{ICF}(t) = \frac{2\pi}{\lambda} \cdot D(t) + b_\phi - 2\pi \cdot N \quad [9.12]$$

The distance variation $D(t)$ can have four origins:

change in position of the receiver (zero in the case of a fixed instrument);

– change in position of the transmitter, which can be calculated using satellite orbit parameters;

variation in surface height, with is the observable to be determined;

– variation of the penetration depth of the wave into the ground.

The phase is cyclic and therefore relative, thus the difficulty of the technique comes from:

– solving for the ambiguity in the integer number of cycles N to calibrate the start of the variation to an absolute reference;

– the need to keep a log of phase variations and detect any cycle skip that would result in a change the initial integer ambiguity value.

The ambiguity of the measurement may be resolved in several ways:

– having prior knowledge of the surface height: fixed reference measurement, if the uncertainty does not exceed the signal wavelength (19 cm L1 GPS), or by using an advanced altimetry code method to calibrate the measurement with decametric accuracy and then smooth out the variations with phase measurements. This is called code-phase altimetry;

– double difference methods using the fact that several satellites are being observed and that the phase variation is different for each depending on the vertical height;

– dual-frequency systems (L1 and L2) also theoretically allow the ambiguity to be solved, although this has not been used in GNSS-R.

As phase is relative, each measurement discontinuity requires absolute recalibration. For this reason, long measurements with shorter breaks are preferred for this type of measurement. Cycle breaks can be detected and corrected using phase unwinding methods such as the "unwrap" operator of signal processing software.

These considerations apply to fixed receivers. The implementation of altimetry techniques by phase on aerial platforms is more challenging because of the difficulty in taking changes in positioning and attitude of the aircraft into account. However, recent advances in the field of accurate positioning and real-time attitude estimation should make it possible, in the near future, to estimate height from a plane in conditions of light wind and small waves.

9.3.2.3. *GNSS-R altimetry: estimating the height of lakes*

Many studies have been conducted since the late 90s on the use of code and phase in altimetry. Resolutions in the order of the centimeter, then smaller than the centimeter, were carried over to static configurations for monitoring lake levels [TRE 01].

As an example, we present an operational system for measuring the height of Lake Laja in Chile [EGI 10]. This lake stores meltwater used to produce hydropower. The measurement setup involves fixing a modified version of a commercial GNSS-R reflectometry device, Oceanpal (Figure 9.10(a)), onto the shore of the reservoir. The device consists of a mast above the body of water, equipped with two antennas (RHCP Zenith, LHCP Nadir) linked to a receiver that carries out the cGNSS-R processing.

Direct and reflected waveforms are generated in real time every 20 minutes and are processed using the technique described in sections 9.3.2.1 and 9.3.2.2. The goal is to extract the phase and resolve integer ambiguities through a method of double differences, using the information of all visible satellites. A Kalman filter is used to smooth out the output results.

The service has been running uninterrupted for several months, providing height information every hour. In Figure 9.10(b), we can see the evolution of the estimated height of the lake relative to the geoid between April and September 2009. A comparison to the radar sensor for reference shows an average square error of around 5 cm.

a) b)

Figure 9.10. *a) The installation of the GNSS-R sensor (Oceanpal) on the shore of Lake Laja in Chile. b) Result of measurements and comparison with the radar sensor (DGA sensor) installed on the same lake. From [EGI 10]*

9.3.2.4. *Altimetry by phase difference: estimating the height of lakes*

Several studies have demonstrated the feasibility of monitoring lake hydrosystems using altimetry by phase difference. The information obtained is accurate to a few centimeters. One of the latest and most authoritative studies was conducted with the GNSS Occultation Reflectometry Scatterometry receiver (GORS), which was developed by the *Geo Forschungs Zentrum* (GFZ). It comprises one antenna for detecting direct signals from right polarization (RHCP) and two antennas for detecting reflected signals from right (RHCP) and left (LHCP) polarization. On board a Zeppelin, at an average altitude of 500 m above the lake, it has taken measurements of the height of the surface of Lake Constance (on the border between Germany and Switzerland) for 64 km along the lake.

The GORS receiver outputs the different parameters needed to estimate the height (section 9.2.1.1) at a frequency of 200 Hz. This includes delay adjustment on the code, Doppler frequency, replica of the carrier signal for the direct signal and interferometric complex field for the reflected signal [SEM 13]. The distance variation due to reflection is then estimated from the phase of the interferometric complex field. The height of the lake is then estimated by taking in account: *a priori* height of the lake, the baseline of the antennas, transmitter and receiver position changes, the effects of atmospheric refraction on trajectories and delays. The lake levels thus obtained are accurate to 7 ± 3 cm for the right polarization antenna and 5 ± 4 cm for the left polarization antenna [SEM 14].

9.4. Soil moisture

Estimation of soil moisture by cGNSS-R was designed in the early 2000s [ZAV 00] and validated from an instrumented tower [ZAV 03], and aircraft [KAT 06], but it was only 10 years later that long term [EGI 13] or aerial [EGI 14] experiments have been introduced to extend the study of the technique's scope.

Regarding IPT, [LAR 08] demonstrated first that soil moisture could be estimated from existing GNSS mono-antenna installations, opening up a new potential for measurements through the global presence of extensive GNSS geodetic networks. In France, for example, the IGN's permanent GNSS network (RGP) and its 427 stations could be a source of national humidity data for Météo-France: since 2015, the PRISM[1] program has been studying the

1 PRISM: potentiality of *in situ* and mobile GNSS reflectometry, see www.fondation-stae.net/fr/actions/projets-soutenus/?pg3.

possibility of assimilating GNSS-R soil moisture data into Météo-France's ISBA Dedicated instrument based on the same interferometric technique were also developped to maximize the sensitivity [ROD 11] and demonstrated to retrieve soil moisture and other land parameters.

In this section, we introduce methods for restoring soil moisture, both by cGNSS-R and IPT, and we will illustrate them through concrete examples.

9.4.1. *cGNSS-R methodology (bistatic scatter)*

The dielectric properties of soil are intrinsically linked to the water content. Soil moisture directly affects the reflectivity of the terrain, Γ_{pq}, defined as the ratio of the incident strength in polarization p over the reflected strength in polarization q (section 9.2.1.3).

9.4.1.1. *Apparent reflectivity*

The reflectivity for each polarization can be estimated by a cGNSS-R receiver through the ratio of the amplitude of the reflected correlation peak $Y_{\tau,q}(\Delta\tau,f)$ over the amplitude of the direct correlation peak $Y_{d,p}(0,f)$. This measurement, calculated from the interferometric complex field (section 9.2.1.5) is called the apparent reflectivity Γ'_{pq} [EGI 13]:

$$\Gamma'_{pq} = \left|\left\langle \frac{Y_{\tau,q}(\Delta\tau,f)}{Y_{d,p}(0,f)} \right\rangle\right|^2 \qquad [9.13]$$

where p and q are the polarizations of the transmitter and receiver, respectively, and $\Delta\tau$ is the estimated delay between the direct and the reflected signal.

Γ'_{pq} is the basic observable for measurements based on the scattering properties of soil such as soil moisture or biomass. It includes both coherent and incoherent signal components.

9.4.1.2. *Modeling the impact of soil moisture on reflectivity*

Reflectivity of the ground Γ_{pq} depends on the angle of incidence and the dielectric properties of the surface. These dielectric properties themselves depend on soil type, liquid water content and frequency of observation.

These relationships have been empirically characterized by Hallikainen *et al.* [HAL85] and Dobson *et al.* [DOB 85].

Generally, the relationship between reflectivity and angle of incidence is expressed in terms of vertical and horizontal polarization (Figure 9.11(a)). However, in the case of GNSS, the signals *x(t)* are transmitted in circular polarization, which can be expressed as a combination of vertical and horizontal polarizations:

$$\begin{bmatrix} x_{ll} & x_{rl} \\ x_{rl} & x_{rr} \end{bmatrix} = \frac{1}{2}\begin{bmatrix} x_h + x_v & x_h - x_v \\ x_h - x_v & x_h + x_v \end{bmatrix} \qquad [9.14]$$

Figure 9.11 shows the relationship between reflectivity and angle of incidence for HH and VV polarizations (Figure 9.11(a) and in circular polarization RL and RR (Figure 9.11(b)) for various soil moisture levels. The interesting result for GNSS-R is the close relationship between soil moisture and reflectivity in RL polarization and the relatively constant relationship for angles of incidence below 50°. This allows us to use either of the satellites at different elevations without the need for normalization.

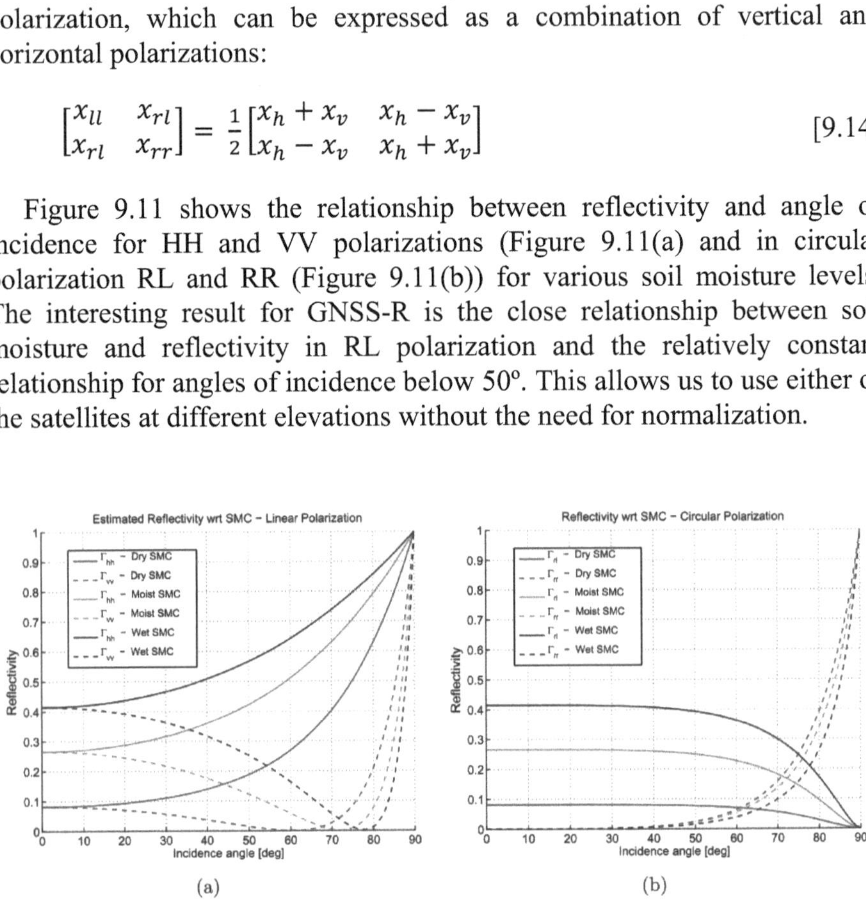

Figure 9.11. *Modeling reflectivity in: a) vertical and horizontal polarization and b) left and right circular polarizations relative to the angle of incidence for different soil moisture levels consisting of 31% sand, 56% silt and 13% clay (Dry SMC = 5 vol. %, Moist SMC = 20 vol. %, Wet SMC = 35 vol. %). From [EGI 13]. For a color version of this figure, see www.iste.co.uk/baghdadi/4.zip*

9.4.1.3. *Bistatic signals (coherent and incoherent components)*

In connection with the geometry of the surface, the signal is reflected in two ways:

– partially in a specular way, in a preferred direction in the case of planar surfaces. This is called a coherent contribution, since the phase coherence is maintained;

– partially in a scattered way, with greater angular dispersion, in the case of rough surfaces, in which case the signal phase becomes random and is called an incoherent contribution.

Conventional electromagnetic scattering models that are typically tested in backscatter (see Chapter 1) can be adapted to describe the influence of surface condition on the distribution of GNSS signals. The Advanced Integral Equation Model (AIEM) was used by Pierdicca *et al.* [PIE 14] in a bistatic configuration to confirm the sensitivity of the GNSS-R to geophysical parameters and their impact on coherent and incoherent components.

Figure 9.12 shows the simulated contributions of coherent and incoherent components for different values of soil moisture when considering a ground roughness value as constant h_{RMS} = 2 cm. For LHCP polarization reflectivity (Figure 9.12(a)), we see that the contribution of the coherent component increases when soil moisture and/or angle of incidence increases. For the incoherent component, we note the same increase for increasing soil moisture. However, the relationship with the angle of incidence is inverted, meaning that the contribution of the coherent component decreases with increasing angle of incidence. In RHCP polarization (Figure 9.12(b)), there is also an increase in reflectivity with increasing soil moisture and the angle of incidence for the coherent component. However, we see very little dependence for the incoherent component.

9.4.1.4. *Effects of roughness on GNSS-R measurements*

Surface roughness (standard deviation of height h_{RMS}), causes a reduction in ground reflectivity by increasing the diffusion of the incident energy, which turns into an omni-directional incoherent scattering component. In Figure 9.13, we see that for low roughness conditions (h_{RMS}= 0.7 cm), the coherent component of the signal is significantly higher than the incoherent part (Figure 9.13(a)) by at least 10 dB (largest difference for high incidences) and by at least 4 dB (most important difference for high incidences) for average roughness (h_{RMS} = 1.5 cm). However, for high roughness levels (h_{RMS} = 3 cm), the incoherent component becomes dominant for incidences below 40°.

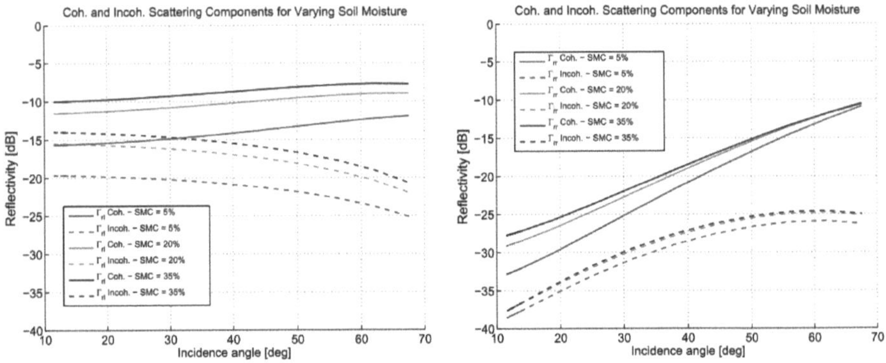

(a) (b)

Figure 9.12. *Modeling reflectivity as a function of incidence angle for different values of soil moisture. a) LHCP polarization, b) RHCP polarization. The solid lines represent the coherent component, the dotted lines represent the incoherent component. Surface roughness is considered constant* h_{RMS} *= 2 cm. From [EGI 13]. For a color version of this figure, see www.iste.co.uk/baghdadi/4.zip*

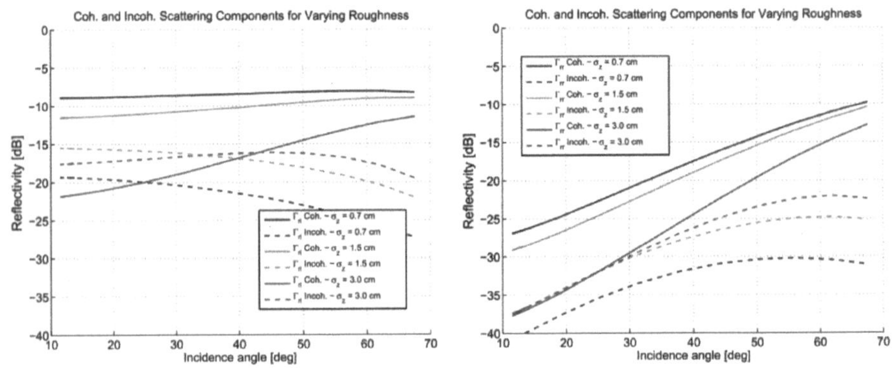

Figure 9.13. *Modeling reflectivity as a function of incidence angle for different soil roughness values. a) LHCP polarization, b) RHCP polarization. The solid lines represent the coherent component, the dotted lines represent the incoherent component. The water content of the soil is considered constant at SMC = 20 vol. %. From [EGI 13]. For a color version of this figure, see www.iste.co.uk/baghdadi/4.zip*

Another observation is that the RHCP component (Figure 9.13(b)) is much more attenuated (–20 dB) than the LHCP component for low incidence angles. However, as the incidence angle increases, the scattering becomes more specular as the effect of the effective surface roughness is less significant.

9.4.2. *Example of monitoring the moisture level of an agricultural plot using cGNSS-R*

The examples of soil moisture restitution presented below come from the LEiMON campaign that took place from March to September 2009 in Italy between the villages of Cerbaia and Montespertoli (43.673°N, 11.126°E). The experimental setup was based on a cGNSS-R receiver for which the antennas were mounted on a 25 m high crane overlooking an agricultural field (Figure 9.14) divided into two halves with different roughness levels. For the period of interest, the eastern part had a slightly higher roughness (h_{RMS} = 2.5 cm) than the western part (h_{RMS} = 2 cm).

a) b)

Figure 9.14. *LEiMON experiment. a) Experimental setup with the 25 m high crane overlooking the studied field. b) The study area separated into two parts (East/West) with two different roughness levels. From [EGI 13]*

Analysis of apparent reflectivity sensitivity measured in LHCP polarization Γ'_{rl} and RHCP polarization Γ'_{rr}, as well as the ratio of apparent reflectivity $\Gamma'_{\mathrm{rl}}/\Gamma'_{\mathrm{rr}}$, showed a linear relationship with soil water content measured by probes for low incidence angles between 5° and 25°. The highest sensitivity was achieved using the reflectivity ratio, as it helped mitigate the effects of roughness when its value was low (hypothesis first proposed by [ZAV 00]). In Figure 9.15(a), one can see the linear relationship between the ratio of apparent reflectivities in left and right polarization and soil moisture, with a measured sensitivity of about 0.2 dB/vol. %.

Regression on these data enabled the establishment of an empirical model for estimating soil moisture. In Figure 9.15(b), moisture estimated by cGNSS-R from the linear model and volumetric moisture measured by sensors is well correlated (r = 0.95) with a mean square error of about 1 vol. %.

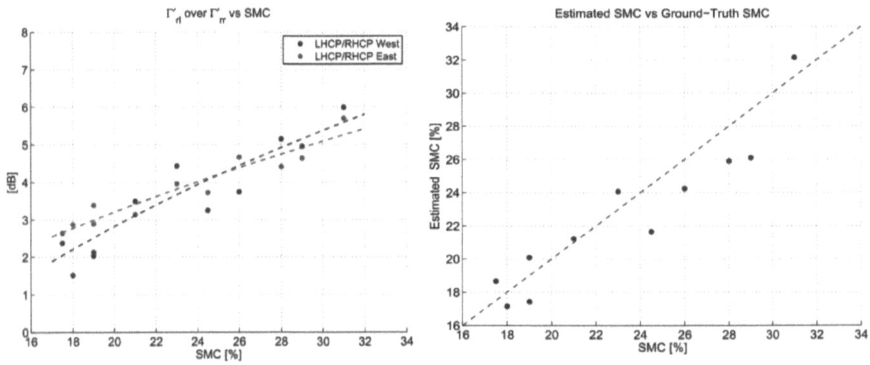

a) b)

Figure 9.15. *Relationships between apparent reflectivity and volumetric moisture: a) calibration line between the ratio of measured reflectivity $\Gamma'_{rl}/\Gamma'_{rr}$and in situ volumetric soil moisture; b) correlation between volumetric soil moisture estimated by GNSS-R and soil moisture measured in situ. From [EGI 13]. For a color version of this figure, see www.iste.co.uk/baghdadi/4.zip*

9.4.3. *Monitoring soil moisture by IPT: methodology*

9.4.3.1. *Amplitude and phase of SNR oscillations due to multipath*

The high frequency part of SNR (denoted by SNR_r), which is mainly due to multipath trajectory, can be expressed as follows [LAR 08]:

$$SNR_r = A_m cos\left(\frac{4\pi h}{\lambda} sin(\theta) + \phi_m\right) \quad [9.15]$$

where h (meters) is the height of the antenna relative to the reflective surface, λ (meters) is the wavelength of the GNSS signal and θ (radians) is the satellite elevation angle. The amplitude A_m and phase ϕ_m are estimated for each satellite passage by solving a matrix through the least squares method.

Chew *et al.* [CHE 14] showed that these two parameters vary depending on soil moisture. This phenomenon was observed *in situ* during several measurement campaigns [LAR 08, LAR 10] and there was usually a better correlation of superficial soil moisture with phase ϕ_m rather than with amplitude A_m, as Chew *et al.* [CHE 14] theoretically demonstrated.

9.4.3.2. *Change in effective height of the antenna relative to the reflective surface*

Furthermore, soil moisture changes the dielectric properties of soil, which has a direct effect on the penetration depth of GNSS waves. Thus, in the example of an agricultural plot in the Toulouse area (see section 9.4.4) with relative permittivity $\epsilon_r = 10.15 - 2.59i$ for a volumetric water content of 23.2 vol. % and $\epsilon_r = 14.61 - 3.59i$ for a volume content of 29.9 vol. % [HAL 85], there will be a theoretical penetration depth of the GNSS electromagnetic (EM) wave ($\lambda_{GPS-L1} = 19.05$ cm) of 3.7 and 3.2 cm respectively. This penetration depth P_d is calculated using the following equation [BEH 06], where $Re(\epsilon_r)$ and $Im(\epsilon_r)$ are the real and imaginary parts of relative permittivity:

$$P_d = \frac{\lambda\sqrt{Re(\epsilon_r)}}{2\pi Im(\epsilon_r)} \qquad [9.16]$$

The effective height h_{eff} (height measured h_s + penetration depth P_d) of the antenna is usually slightly larger than the measured height h_s since the penetration depth of the EM wave is directly dependent on soil moisture. The higher the volume humidity, the less an EM wave can penetrate. We can express the frequency of the high frequency signal present in the SNR as a function of this height h_{eff} [ROU 15, ROU 16]:

$$f = \frac{4\pi}{\lambda}\left(h_{eff} + \dot{h}\frac{\tan(\theta)}{\dot{\theta}}\right) \qquad [9.17]$$

where $\dot{\theta} = \frac{d\theta}{dt}$ is the rate of change of the elevation angle (rad/s) and $\dot{h} = \frac{dh_{eff}}{dt}$ is the effective height variation h_{eff} versus time (m/s).

9.4.4. *Example of monitoring the moisture level of an agricultural field by IPT*

In this section, we illustrate the application of the IPT method for measuring soil moisture. For this, we use a standard GNSS receiver (multi-frequency Leica GR25) and an omni-directional geodetic dual polarization antenna (Leica AR10). The direct signal is picked up by the upper hemisphere of the antenna in RHCP-dominant polarization, just like in conventional GNSS reception, while the reflected signal is captured by the lower hemisphere LHCP-dominant polarization. The one and a half month measurement campaign took place on the site of Lamasquère (France). The antenna (L-band) was installed at a height h_s of 1.69 m from the surface of a bare field. Moisture sensors were located close to each other (distance < 2 m) and at two different depths of 2 and 5 cm. These probes showed variations in superficial volumetric soil moisture of 23.2 vol. % to 29.8 vol. % (Figure 9.16(a)) [ROU 16].

9.4.4.1. *Study on the phase of SNR oscillations*

Figure 9.16(a) shows the evolution of a time series (time resolution of one day) of phase ϕ_m (in °) for the GPS satellite PRN 32. The correlation obtained was 0.97 for moisture measured at 2 cm depth (denoted P_{2cm}), decreasing to 0.95 when measured at 5 cm depth (denoted P_{5cm}) and 0.96 with the surface moisture empirically calculated from the slope formed by P_{2cm} and P_{5cm}. Calculated empirical calibration laws (Figure 9.16(b)) can directly invert phase measurements to get soil moisture.

9.4.4.2. *Study on effective antenna height*

By simultaneously combining frequency measurements f (periodogram, as defined in [9.16]) from different satellites, it is possible to jointly determine h_{eff} and h (see [ROU 15, ROU 16] for more details). Figure 9.17 shows the evolution of the time series $h_{eff}(t)$ obtained during the same measurement campaign as before. A correlation of 0.94 is obtained between GNSS-R restitution and the moisture measurement at 2 cm; it is 0.93 for the moisture measurement at 5 cm. The time resolution here is 10 minutes, like for the soil moisture probes. This time series was obtained by combining all GPS and GLONASS constellation satellites on L1 frequency.

9.4.4.3. *Performance and uses*

The first method, which uses an analysis of amplitude and phase of SNR oscillations due to multipath trajectories (section 9.4.4.1) is used to spatialize the measured data. However, its temporal resolution is limited by the number of visible satellites and repeatability of their orbit. It is not possible to use the GLONASS constellation with this method because of the low level of their short-term orbit repeatability (> 1 week). However, measuring the effective height (section 9.4.4.2) allows us to integrate all satellites of all constellations and thus temporal resolution can be improved significantly.

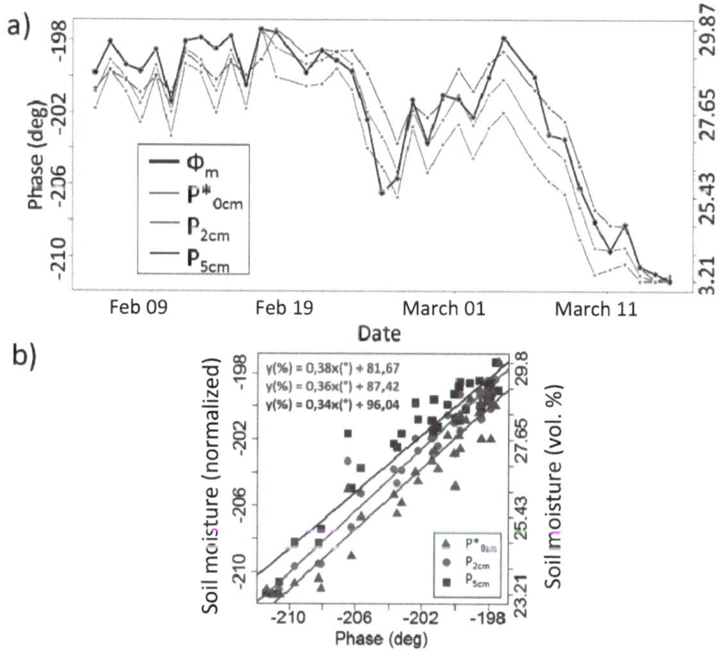

Figure 9.16. *a) Change in volume moisture level based on phase variations ϕ_m of the SNR_r. Time series of phase ϕ_m of the SNR signal recorded on the L_1 of the GPS satellite PRN32 during a measurement campaign on an agricultural plot in 2014 at Lamasquère (France). P_{2cm} and P_{5cm} are measurements from independent moisture probes (Theta probe) installed near the site, respectively 2 and 5 cm deep. $P^*_{0\,cm}$ is the surface moisture level that was empirically calculated from the gradient of P_{2cm} and P_{5cm}; b) changes in soil moisture based on phase ϕ_m. The equations of the regression lines applied to the three obtained point clouds (for P^*_{0cm}, P_{2cm} and P_{5cm} respectively) provide empirical laws to invert the phase (°) to obtain the corresponding soil moisture value (%). From [ROU 16]. For a color version of this figure, see www.iste.co.uk/baghdadi/4.zip*

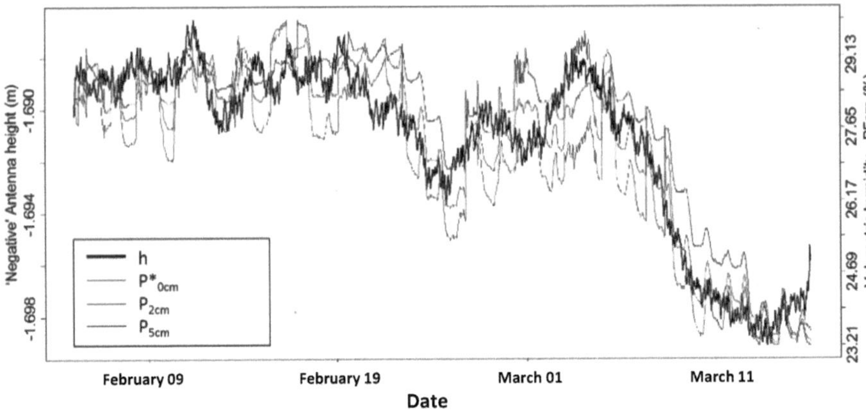

Figure 9.17. *Change in the effective height h_{eff} of the antenna from the ground over time between 5 February and 15 March 2014 (measurements on bare ground). These data are compared to humidity probes (Figure 9.16). From [ROU 16]. For a color version of this figure, see www.iste.co.uk/baghdadi/4.zip*

9.5. Vegetation cover

9.5.1. *Biomass estimation by cGNSS-R: theoretical aspects*

Electromagnetic interactions between microwave signals and vegetation cover come from complex processes, which are difficult to simulate, particularly due to the fact that the L-band wavelengths and vegetation elements are similar in size.

Many approaches have been used to understand the physical processes and simulate electromagnetic response of vegetation cover (see Chapter 1). Three major categories can be found in the literature: semi-empirical, empirical and physical approaches. It is estimated that the latter is best suited to describing vegetation scattering. The main advantage of this type of models is that it takes the scattering properties of each element into account and its ability to combine vegetation scattering theories using accurate geometrical descriptions with advanced ground scattering theories, such as the AIEM. In addition, the input variables of these models are usually linked to quantities that can be measured in the field.

The simulations presented below were carried out with the SAVERS simulator developed by Pierdicca *et al.* [PIE 14]. This tool includes a bare soil scattering model that is coupled to a vegetation model. Figure 9.18 shows simulations for different development stages of sunflower planes (different water content of vegetation). An increase in biomass produces greater attenuation of the coherent contribution of the reflected signal in left polarization (from –10 dB to –15 dB at an incidence of 40° with increasing water content). However, in a less predictable manner, the model predicts an increase in incoherent contribution in right polarization, whereas it remains stable in the left polarization.

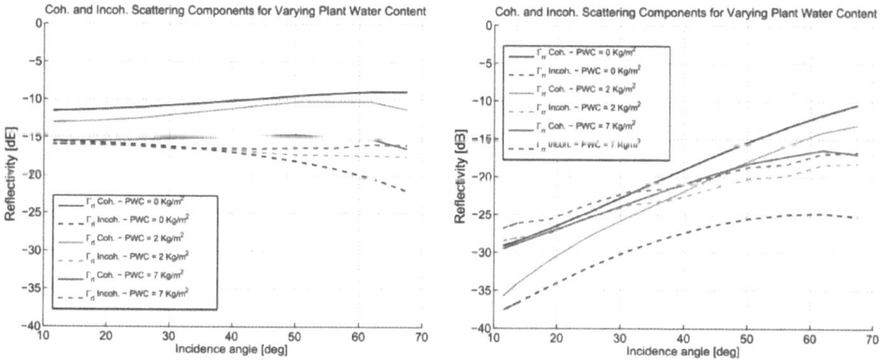

Figure 9.18. *Simulation of the impact of plant water content (PWC) on coherent and incoherent components of reflection coefficients in a) left circular polarization, b) right circular polarization. Soil moisture is constant (20 vol. %), as is its roughness (h_{RMS} = 1.5 cm). From [EGI 13]. For a color version of this figure, see www.iste.co.uk/baghdadi/4.zip*

These simulations show the importance of measuring the reflected signal in both LHCP and RHCP polarizations. They also provide the need for high instrument sensitivity (up to –30 dB relative to the direct signal) in order to measure RHCP polarization at low angles of incidence. These simulations also highlight the importance of the decomposition into coherent and incoherent components to be able to distinguish between different surface contributions.

9.5.2. *Example of forest biomass estimation by cGNSS-R*

In 2011, a series of flights took place in Italy between the villages of Ponte a Elsa and Castelfiorentino near Florence. The aim of the experiment was to test the sensitivity of an airborne polarimetric cGNSS-R receiver for biomass estimation. The plane flew at an altitude of 150 m above the poplars, for which the biomass varied from 4 to 250 t/ha, as well as other dense forests for which the biomass reached 400 t/ha.

The apparent reflectivity in left polarization Γ'_{rl} at 20° incidence was considered to be the most dynamic parameter for the present biomass. Figure 9.19 shows the correlation between measured reflectivity Γ'_{rl} and measured biomass. The measurements show a sensitivity of 1.5 dB/[t/ha] with very good correlation (0.91) [EGI 14]. It is interesting to note that this technique does not exhibit reflectivity saturation for high biomass values, unlike measurements made with a mono-static radar (saturation around 150 t/ta in L-band). So cGNSS-R looks more promising than the mono-static radar for measuring biomass in dense forests (> 400 t/ha).

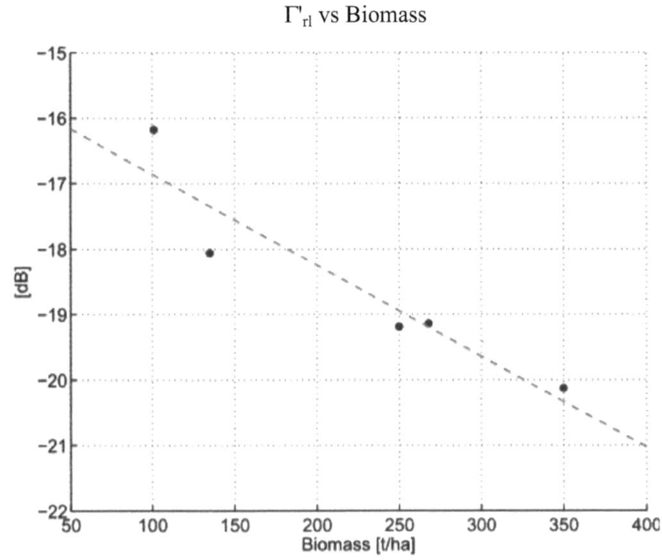

Figure 9.19. *Correlation between left polarization reflectivity Γ'_{rl} at 20° and biomass measured by the cGNSS-R sensor at the L1 GPS frequency (1.575 GHz). From [EGI 13]*

9.5.3. *Monitoring vegetation growth using IPT*

As detailed in section 9.5.1, vegetation has complex interactions with electromagnetic waves due to the presence of leaves and branches (see [VIL 16]). A recent study showed that it is possible to establish a link between vegetation growth and SNR_r thanks to GPS multipath [SMA 10]. Experiments on this site have shown that SNR_r amplitude is inversely proportional to estimated vegetation growth using *in situ* measurements of biomass and vegetation water content. This was done for nine sites covering a surface area of 1,000 m^2 and matching several kinds of vegetation coverage: agricultural plots for growing corn, grasslands and shrublands. Correlations between SNR time series and Normalized Difference Vegetation Index (NDVI) from MODIS[2] compositions over 16 days provided coefficient determination values (r^2) greater that were than 0.8.

9.6. Conclusions and perspectives

Many ground, airborne on balloons, or space GNSS-R campaigns are currently being studied or deployed in operational contexts. Below, we present the topics, missions and instruments that best represent the effort and interest in reflectometry by opportunistic signal at the time of writing of this book.

9.6.1. *IPT network*

Following the success of the IPT method for restitution of soil moisture, snow and vegetation parameters from commercial GNSS receivers, major effort is underway to take advantage of existing GNSS measurement networks for geodesy around the globe. Current research areas include assimilation of information on soil moisture and snow in operational hydrological models in order to improve the estimation of water availability across a territory.

Another key hydrological variable is the water equivalent of snow. This variable is derived from the thickness of snow and density of the snowpack.

2 https://lpdaac.usgs.gov/dataset_discovery/modis/modis_products_table/mod13q1.

Another challenge now is to estimate the density of snow from reflected GNSS signals.

Coastal altimetry with high vertical resolution using IPT is another example of a topic that is of broad and current interest.

9.6.2. *Aerial/balloon measurements*

After two decades that mainly focused on oceanographic applications [GAR 98], several airborne campaigns [EGI 14, MOT 16] are in progress with a particular focus on land applications: altimetry of rivers and lakes, soil moisture, biomass estimation.

Similarly, in 2015, the first campaign aboard a stratospheric balloon [CAR 15] for characterizing the response of the boreal forest to GNSS signals at an altitude of several kilometres took place. The first results showed a sensitivity to biomass variations and the contribution of polarimetry to the estimation of vegetation.

In coastal areas, the study of reflectivity changes related to variations in salinity is a major challenge, which the European Mistrale program is attempting to answer by comparing IPT and iGNSS measurements.

In oceanography, the goal is to show complementarity with radar and radiometric measurements and see if the technology can be used in a spatial context.

Another topic under study is the use of new opportunistic signals, including satellite radio signals in P and S-bands. Lin *et al.* [LIN 15] showed, theoretically, that P-band reflectometry measurements could measure moisture in the root zone (> 20 cm) with a mean square error of 4 vol. %.

9.6.3. *Spaceborne measurements*

Following the success of the GNSS-R instrument aboard the UK-DMC, a new attribute dedicated to GNSS-R was designed and installed on the British technology demonstration satellite TechDemoSat-1 [UNW 11], which was launched in 2014. Subsequent to its launch into orbit, several research projects are now underway, primarily related to oceanography and glaciology, first analysis of signals reflected by land surfaces shows correlation with soil moisture and biomass [CHE16].

Several missions are underway to operationally test viable GNSS-R from space:

– CyGNSS. Led by NASA and the University of Michigan [RUF 12], this space mission is based on a constellation of eight micro-satellites carrying a GNSS-R receiver that is very similar to the one flying on TechDemoSat-1. Deployment is planned for 2016;

– 3Cat-2 is a nano-satellite (< 10 kg) designed by the Polytechnic University of Catalonia [CAR 13]. It carries a dual-frequency, bi-polarization, multi-constellation and multi-modulations receiver, which has already proven its capabilities on fixed or airborne installations and on a stratospheric balloon. The launch is scheduled for mid 2016;

– GEROS-ISS, which was proposed by Martín-Neira *et al.* [MAR 11] and supported by the GFZ [WIC 13]. was designed to be installed on the International Space Station and will use the iGNSS technique (section 9.2.1.2).

These missions provide a chance to gain unprecedented access to coverage including over land surfaces. Early studies suggest a possibility of detecting ice. Sensitivity to land parameters was demonstrated, but further work is needed in order to provide reliable useful products. If it is adequate, it could complement other water resource monitoring missions.

9.7. Key points

– GNSS-R is a technique that dates back to the early 1990s. It is based on the study of navigation signals (GPS, GLONASS, GALILEO) reflected off the surface of the Earth.

– The two main techniques are conventional GNSS-R, or cGNSS-R, which is based on the use of dedicated bi-antennas receivers, and the Interference Pattern Technique, or IPT, which is based on the use of standard mono-antenna receivers such as those used in geodetic networks.

– After a decade during which studies focused on oceanography (altimetry, roughness, currents and winds), attention was paid to the cryosphere and finally to land applications (land altimetry, soil moisture, biomass).

– Altimetry measurements are achieved by estimating the delay between direct and reflected signals, and can reach a decametric resolution. This method can be supplemented by an analysis of variation of phase of the reflected signal, improving accuracy to the centimeter level.

– Measuring soil moisture by GNSS-R is based, like for radar, on the relationship between the ground dielectric constant and the water content. Measuring the relative strength of the reflected signal in right and left circular polarizations allows us to estimate moisture to an accuracy of about 5 vol. %, partially overcoming issues related to soil roughness conditions.

– Estimating biomass for both crops and forest is based on the study of distribution of the reflected signal in right and left polarizations attenuated by the medium. Similar sensitivities to measurements by conventional radars can be achieved.

– Several terrestrial, airborne and space projects are under study and development, including CyGNSS, an operational constellation of 8 microsatellites in low orbit dedicated to GNSS-R for which the launch is planned for 2016.

9.8. Bibliography

[AUB 94] AUBER J.C., BIBAUT A., RIGAL J.M., "Characterization of multipath on land and sea at GPS frequencies", *Proceedings of the 7th International Technical Meeting of the Satellite Division of The Institute of Navigation*, pp. 1155–1171, September 1994.

[BEH 06] BEHARI J., *Microwave Dielectric Behavior of Wet Soils*, Springer, 2006.

[BON 15a] BONIFACE K., BRAUN J.J., MCCREIGHT J.L. *et al.*, "Comparison of SNODAS to GNSS reflectometry snow depth in the Western United States", *Hydrological Processes*, vol. 29, pp. 2425–2437, 2015.

[BON 15b] BONIFACE K., WALPERSDORF A., GUYOMARC'H G. *et al.*, "GNSS reflectometry measurement of snow depth and soil moisture in the French Alps", *International Geoscience and Remote Sensing Symposium (IGARSS)*, pp. 5205–5207, 26-31 July 2015, 2015.

[CAR 12] CARDELLACH E., FABRA F., RIUS A. *et al.*, "Characterization of dry-snow sub-structure using GNSS reflected signals", *Remote Sens. Environ.*, vol. 124, pp. 122–134, 2012.

[CAR 13] CARRENO-LUENGO H., CAMPS A., PEREZ-RAMOS I. *et al.*, "[3]Cat-2: A P(Y) and C/A GNSS-R experimental nano-satellite mission", *International Geoscience and Remote Sensing Symposium (IGARSS),* pp. 843–846, 21-26 July 2013, 2013.

[CAR 15] CARRENO-LUENGO H., AMÈZAGA A., VIDAL D. *et al.*, "First polarimetric gnss-r measurements from a stratospheric flight over boreal forests", *Remote Sens.*, vol. 7, pp. 13120–13138, 2015.

[CHE 14] CHEW C., SMALL E., LARSON K. *et al.*, "Effects of near-surface soil moisture on GPS SNR data: Development of a retrieval algorithm for soil moisture", *IEEE Trans. Geosci. Remote Sens.*, vol. 52, no. 1, pp. 537–543, 2014.

[CHE 16] CHEW C., SHAH R., ZUFFADA C. *et al.*, "Demonstrating soil moisture remote sensing with observations from the UK TechDemoSat-1 satellite mission", *Geophys. Res. Lett.*, March 2016.

[DAR 16] DARROZES J., ROUSSEL N., ZRIBI M., "The reflected global navigation satellite system (GNSS-R): from theory to practice", in BAGHDADI N., ZRIBI M. (eds), *Microwave Remote Sensing of Land Surfaces*, ISTE Ltd, London and Elsevier, Oxford, 2016.

[DOB 85] DOBSON M.C., ULABY F.T., HALLIKAINEN M.T. *et al.*, "Microwave Dielectric Behavior of Wet Soil-Part II: Dielectric Mixing Models", *IEEE Transactions on* in *Geoscience and Remote Sensing*, vol. GE-23, no. 1, pp. 35–46, January 1985.

[EGI 10] EGIDO A., CAPARRINI M., "Monitoring Water Level with GNSS", *GPS World Innovation Column*, available at http://gpsworld.com/natural-resourcesinnovation-friendly-reflections-10470/, September 2010.

[EGI 13] EGIDO A., GNSS reflectometry for land remote sensing applications, PhD thesis, University Politecnica de Catalunya, 2013.

[EGI 14] EGIDO A., PALOSCIA S., MOTTE E. *et al.*, "Airborne GNSS-R polarimetric measurements for soil moisture and above-ground biomass estimation", *Journal of Selected Topics in Applied Earth Observations and Remote Sensing*, vol. 5, pp. 1522–1532, 2014.

[FER 11] FERNÁNDEZ-PRADES C., ARRIBAS J., CLOSAS P. *et al.*, "GNSS-SDR: an open source tool for researchers and developers", *Proceeding of the ION GNSS 2011 Conference*, Portland, Oregon, 19-23September 2011.

[GAR 98] GARRISON J. L., KATZBERG S.J., HILL M.I., "Effect of sea roughness on bistatically scattered range coded signals from the global positioning system," *Geophysical Research Letters*, vol. 25, no. 13. p. 2257, 1998.

[GUT 12] GUTMANN E.D., LARSON K.M., WILLIAMS M.W. *et al.*, "Snow measurement by GPS interferometric reflectometry: an evaluation at Niwot Ridge, Colorado", *Hydrological Processes*, vol. 26, no. 19, pp. 2951–2961, 2012.

[HAL 85] HALLIKAINEN M., ULABY F., DOBSON M. *et al.*, "Microwave dielectric behavior of wet soil–Part 1: Empirical models and experimental observations", *IEEE Trans. Geosci. Remote Sens.*, vol. GE-23, no. 1, pp. 25–34, 1985.

[HAL 88] HALL C.D., CORDEY R.A., "Multistatic Scatterometry", *International Geoscience and Remote Sensing Symposium*, pp. 561–562, 12–16 September 1988.

[HEL 08] HELM A., Ground-based GPS altimetry with the L1 OpenGPS receiver using carrier phase-delay observations of reflected GPS signals, PhD thesis, Deutsches GeoForschungsZentrum GFZ, 2008.

[JIN 14] JIN S., CARDELLACH E., XIE F., *GNSS Remote Sensing*, Springer, 2014.

[KAT 06] KATZBERG S.J., TORRES O., GRANT M.S. *et al*, "Utilizing calibrated GPS reflected signals to estimate soil reflectivity and dielectric constant: Results from SMEX02," *Remote Sens. Environ.*, vol. 100, no. 1, pp. 17–28, January 2006.

[LAR 08] LARSON K., SMALL E., GUTMANN E. *et al.*, "Using GPS multipath to measure soil moisture fluctuations: Initial results", *GPS Solutions*, vol. 12, no. 3, pp. 173–177, 2008.

[LAR 09] LARSON K., GUTMANN E., ZAVOROTNY V. *et al.*, "Can we measure snow depth with GNSS receivers?", *Geophys. Res. Lett.*, vol. 36, p. L17502, 2009.

[LAR 10] LARSON K., BRAUN J., SMALL E. *et al.*, "GPS multipath and its relation to near-surface soil moisture content", *IEEE J. Sel. Topics Appl. Earth Observ. Remote Sens.*, vol. 3, no. 1, pp. 91–99, 2010.

[LAR 13] LARSON K., NIEVINSKI F.G., "GPS snow sensing: results from the EarthScope Plate Boundary Observatory", *GPS solutions*, vol. 17, no. 1, pp. 41–52, 2013.

[LES 16] LESTARQUIT L., PEYREZABES M., DARROZES J. *et al.*, "Reflectometry with an open-source software GNSS receiver: use case with carrier phase altimetry", *Journal of Selected Topics in Applied Earth Observations and Remote Sensing*, 2016.

[LIN 15] LIN Y.C., STIENNE G., GARRISON J. *et al.*, Remote Sensing of Soil moisture with Signals of Opportunity, GNSS+R 2015 Workshop, GFZ, Potsdam, Germany, 2015.

[LOW 02] LOWE S.T., "First spaceborne observation of an Earth-reflected GPS signal", *Radio Sci.*, vol. 37, pp. 1–28, 2002.

[MAR 93] MARTÍN-NEIRA M., "A Passive Reflectometry and Interferometry System (PARIS)- Application to ocean altimetry", *ESA J.*, vol. 17, no. 4, pp. 331–355, 1993.

[MAR 11] MARTÍN-NEIRA M., D'ADDIO S., BUCK C. *et al.*, "The PARIS ocean altimeter in-Orbit demonstrator", *IEEE Trans. Geosci. Remote Sens.*, vol. 49, pp. 2209–2237, 2011.

[MAS 04] MASTERS D., AXELRAD P., KATZBERG S., "Initial results of land-reflected GPS bistatic radar measurements in SMEX02", *Remote Sensing of Environment*, vol. 92, pp. 507–520, 2004.

[MOT 16] MOTTE E., ZRIBI M., FANISE P. *et al.*, "GLORI: A GNSS-R dual polarization airborne instrument for land surface monitoring", *Sensors*, vol. 16, no. 5, p. 732, May 2016.

[NIE 14] NIEVINSKI F.G, LARSON K.M., "An open source GNSS multipath simulator in Matlab/Octave", *GNSS Solut*, 2014.

[NOG 07] NOGUÉS-CORREIG O., GALÍ E.C., CAMPDERRÓS J.S. *et al.*, "A GPS-reflections receiver that computes Doppler/delay maps in real time", *IEEE Transactions on Geoscience and Remote Sensing*, vol. 45, no. 1, pp. 156–174, 2007.

[PIE 14] PIERDICCA N., GUERRIERO L., GIUSTO R. *et al.*, "SAVERS: A simulator of GNSS reflections from bare and vegetated soils", *IEEE Trans. Geosci. Remote Sens.*, vol. 52, no. 10, pp. 6542–6554, 2014.

[ROD 11] RODRIGUEZ-ALVAREZ N., CAMPS A., VALL-LLOSSERA X. *et al*, "Land geophysical parameters retrieval using the interference pattern GNSS-R technique", *IEEE Trans. Geosci. Remote Sens.*, vol. 49, no. 1 PART 1, pp. 71–84, 2011.

[ROU 14] ROUSSEL N., FRAPPART F., RAMILLIEN G. "Simulations of direct and reflected wave trajectories for ground-based GNSS R experiments", *Geosci. Model Dev.*, vol. 7, no. 5, pp. 2261–2279, 2014.

[ROU 15] ROUSSEL N., RAMILLIEN G., FRAPPART F. *et al.*, "Sea level monitoring and sea state estimate using a single geodetic receiver", *Remote Sensing of Environment*, vol. 171, pp. 261–277, 2015.

[ROU 16] ROUSSEL N., FRAPPART F., RAMILLIEN G. *et al.*, "Detection of soil moisture variations using GPS and GLONASS SNR data for elevation angles ranging from 2° to 70°", *IEEE Journal of Selected Topics in Applied Earth Observations and Remote Sensing*, 2016.

[RUF 12] RUF C.S., GLEASON S., JELENAK Z. *et al.*, "The CYGNSS nanosatellite constellation hurricane mission", *International Geoscience and Remote Sensing Symposium (IGARSS)*, pp. 214–216, 22-27 July 2012.

[SEM 13] SEMMLING A.M., WICKERT J., SCHÖN S. *et al.*, "A zeppelin experiment to study airborne altimetry using specular Global Navigation Satellite System reflections", *Radio Sci.*, vol. 48, pp. 427–440, 2013.

[SEM 14] SEMMLING M., BEYERLE G., BECKHEINRICH J. *et al.*, "Airborne GNSS reflectometry using crossover reference points for carrier phase altimetry", *IEEE International Geoscience and Remote Sensing Symposium*, Quebec City, Canada, pp. 3786–3789, 2014.

[SMA 10] SMALL E., LARSON K.M., BRAUN J.J., "Sensing vegetation growth with reflected GPS signals", *Geophysical Research Letters*, vol. 37, p. L12401, 2010.

[TRE 01] TREUHAFT R.N, LOWE S.T., ZUFFADA C. *et al.*, "2-cm GPS altimetry over Crater Lake", *Geophys. Res. Lett.*, vol. 28, no. 23, pp. 4343–4346, 2001.

[UNW 11] UNWIN M., BLUNT P., DE VOS VAN STEENWIJK R. *et al.*, "GNSS Remote Sensing and Technology Demonstration on TechDemoSat-1", *Proceedings of the 24th International Technical Meeting of The Satellite Division of the Institute of Navigation (ION GNSS)*, Portland, OR, pp. 2970–2975, September 2011.

[VIL 16] VILLARD L., LE TOAN T., HO TONG MINH D. *et al.*, "Forest biomass from radar remote sensing", in BAGHDADI N., ZRIBI M. (eds), *Land Surface Remote Sensing in Agriculture and Forest*, ISTE, London and Elsevier, Oxford, 2016.

[WIC 13] WICKERT J., ANDERSEN O., BEYERLE G. *et al.*, "GEROS-ISS: Innovative GNSS reflectometry/occultation payload onboard the International Space Station for the Global Geodetic Observing System", *American Geophysical Union, Fall Meeting, Abstract G51A-0871*, 2013.

[ZAV 00] ZAVOROTNY V.U., VORONOVICH A.G., "Bistatic GPS signal reflections at various polarizations from rough land surface with moisture content", In *Proceedings of the 2000 IEEE International Geoscience and Remote Sensing*, IGARSS 2000, Honolulu, Hawaii, USA, pp. 2852–2854, 24-28 July 2000.

[ZAV 03] ZAVOROTNY V., MASTERS D., GASIEWSKI A. *et al.*, "Seasonal polarimetric measurements of soil moisture using tower-based GPS bistatic radar", *International Geoscience and Remote Sensing (IGARSS),* pp. 781–783, 21–25 July 2003.

10

Energy Balance of Continental Surfaces and the Use of Surface Temperature

10.1. Introduction

The transfer of energy and water, as well as gaseous components, in the soil–plant–atmosphere system plays a crucial role in many processes involved in climate change. The surface energy balance drives both evapotranspiration (closely linked to the CO_2 flux) and sensible heat flux. By providing direct information on heat dissipation efficiency, surface temperature appears to be a key variable. In particular, a lot of work has been conducted on estimating evapotranspiration from thermal infrared (TIR) remote sensing measurements, in order to derive information relevant to agriculture (monitoring plant growth, detection of water stress, crop yield forecasting, etc.) or hydrology (water cycle monitoring, catchment water budget, etc.). This chapter first very succinctly presents the relationship between surface temperature and the surface energy balance, providing the reader with reference publications for further reading. This is followed by a review of the data available in the thermal infrared (TIR) domain and an analysis of various sources of uncertainty affecting surface temperature measurements and their impact on the final accuracy. The different methods for estimating and spatializing actual evapotranspiration (AET) are then discussed. Finally, practical applications of TIR information other than AET are briefly mentioned.

Chapter written by Jean-Pierre LAGOUARDE and Gilles BOULET

10.2. Energy budget and surface temperature

Energy balance results from the energy conservation law within a given system. Several assumptions are usually made when it is applied to vegetated continental surfaces (artificial surfaces, such as urban areas, are not discussed in this chapter). The surface is seen here as a mixture of vegetation and substrate (usually soil) in varying proportions ranging from partial coverage (sparse vegetation in arid areas or during early crop growing stages, for example) to dense and continuous canopies that completely cover the ground. The canopy is assumed to have a large enough horizontal extension so that horizontal exchanges can be ignored (advective exchanges between neighboring plots, for example), and only transfers in a vertical direction considered. In addition, the canopy is identified with a "thin" surface so that energy storage in vegetation can be ignored. This assumption is valid for crops with a limited vertical extension. However, for canopies such as forests, this assumption is only appropriate within a time frame of 24 hours for which the storage term budget is practically zero. For shorter time steps, a storage term should be included into the surface energy balance for rigor.

Energy exchanges at the canopy level are then reduced to radiation exchanges, convective exchanges (within the atmosphere) and conduction exchanges (within the ground). Other phenomena (photosynthesis, precipitation heat input, etc.) could be added, but they only involve low energy amounts and are therefore ignored. The surface energy budget reflects the balance of all energy exchanges that affect the surface. The energy budget is usually expressed as follows, by separating the radiation budget from other fluxes:

$$Rn = H + LE + G \qquad [10.1]$$

where Rn denotes net radiation, H and LE are the sensible heat flux and latent heat flux, respectively, and G is the conduction flux in the soil. Each of these terms is further detailed below. Energy balance in the form of equation [10.1] assumes that fluxes are counted positively during daytime, the radiative incoming Rn energy towards the surface being compensated by energy losses under convective form (in the atmosphere) and conductive form (in the ground).

10.2.1. *Radiative budget*

Radiation exchanges mainly occur at "short" (solar domain between 0.3 and about 5.0 μm) and "large" (thermal infrared TIR domain between about 4 and 50 μm) wavelengths, for which the atmosphere has relatively large transmission coefficients. The solar radiation (global radiation, Rg) that reaches the surface results from the contributions of a direct component from the Sun itself and a diffuse component (scattering by atmospheric gases, clouds, aerosols). It is partially reflected at the surface with a coefficient, the albedo a, that depends on the surface type (type of vegetation, canopy structure, soil type, presence of snow, etc.), so that the surface only absorbs Rg (1-a) in the solar domain.

In the TIR domain, the ground emits a radiation $L\uparrow = \varepsilon\sigma T_s^4$ according to the Stefan–Boltzmann law, where T_s denotes the surface temperature (here in K), ε the emissivity and σ the Stefan–Boltzmann constant ($\sigma = 5.67\ 10^{-8}\ Wm^{-2}K^{-4}$), L↑ being expressed in Wm^{-2}. The surface receives a downward radiation L↓ (resulting from the contribution in the TIR of all particles – air, water molecules, CO_2, aerosols, etc. – present in the atmosphere). A portion of this radiation is reflected by the surface with a coefficient (1 - ε). The budget in the TIR domain is ultimately $\varepsilon\ (L\downarrow - \sigma T_s^4)$. The surface temperature T_s in the expression of the net radiation below [10.2] is a surface "radiative" temperature. It results from the integration of the TIR upward emission of each of the canopy elements.

The radiation budget mixing all wavelengths represents the overall radiative energy available at the surface, and corresponds to the net radiation Rn:

$$Rn = Rg\,(1 - a) + \varepsilon\,(L\downarrow - \sigma T_S^4) \qquad [10.2]$$

During daytime, Rn is positive: the surface therefore receives energy in radiative form, inducing both its warming and the development of a temperature profile in the air with increasing temperature when getting closer to the surface. Conversely, Rn is negative at night because radiative exchanges are reduced to long wavelengths terms: the surface loses energy and cools down and the temperature profile in the air is inverted.

Rn can be estimated fairly accurately from albedo and surface temperature obtained from remote sensing [BOE 99, KUS 94]. Mira *et al.* [MIR 16] showed that when using remotely sensed data to estimate the different components of Rn, the final accuracy is mainly limited by the accuracy of the downward incident radiation.

10.2.2. *Convective fluxes*

These are so called because they are closely related to the turbulent nature of air flow, which induces, via eddies, transfers of different elements: water vapor, heat, gas, etc.

Several convective regimes naturally coexist: forced convection corresponds to turbulent movements driven by dynamic effects only (wind effect), while free convection corresponds to vertical air movements generated by differences in density (buoyancy) linked to temperature differences (along vertical profiles). The general case is that of a mixed convection that combines both processes. For daytime conditions, convection is triggered by the presence at ground level of warm air layers, which are lighter than the air above them (the air moisture that results from vegetation evaporation acts in the same way and contributes to enhancing this phenomenon). The atmosphere in these conditions is thus "unstable". At night, however, the stratification of the air in the lower atmosphere is "stable", that is cold air layers near the surface tend to dampen the turbulence of air flow.

In general, convective fluxes F_c are expressed, by analogy with the laws of diffusion, as the product of a transfer coefficient K_c and a vertical concentration gradient dC/dz of a scalar variable:

$$F_c = - K_c \frac{dC}{dz} \quad [10.3]$$

This equation is strictly valid in the atmosphere, but it has been extended between the surface itself and a reference level above. It can thus be shown that the sensible heat flux H in the case of a uniform surface can be written as:

$$H = \rho c_p \frac{T_0 - T_a}{r_a} \quad [10.4]$$

where ρ and c_p respectively represent the density (kg m^{-3}) and specific heat (J kg^{-1} C^{-1}) of air, and T_a the air temperature at the reference level as measured by meteorological stations, or estimated by climate models from the first level of the atmospheric profile. r_a is the aerodynamic resistance (s m^{-1}). H is expressed in W m^{-2}. Resistance r_a will not be discussed here (we refer the reader to [BRU 82]); it depends on the canopy structure (characterized by parameters such as the displacement height and the mechanical and thermal roughness) and on the degree of stability/instability of the atmosphere (characterized by the Monin–Obukhov length).

The temperature T_0, which appears in the expression of the convective flux [10.4], is called the "aerodynamic" temperature. It is equivalent to an air temperature within the canopy, at a level determined from the study of turbulent transfers within the canopy, and it results from flux-profile relationships.

Similarly, we show that for a sufficiently dense vegetation cover, for which we can identify the whole canopy to a single-leaf stomatal cavity (big-leaf model [MON 81] generalized [RAU 95, DAI 04]), the latent heat flux LE (expressed in W m^{-2}, E being the mass evaporation in kg m^{-2} $s^{-1}$and L the latent vaporization heat in J kg^{-1}) can be written as:

$$LE = \frac{\rho c_p}{\gamma} \frac{e_s(T_0) - e_a}{r_a + r_c} \qquad [10.5]$$

where $e_s(T)$ denotes the saturation water vapor pressure curve (Pa) as a function of temperature T, e_a is the air vapor pressure (Pa) at the selected reference level above the canopy and γ is the psychrometric constant around 66 Pa C^{-1}. The resistance to water vapor extraction r_c is added to the aerodynamic resistance and reflects the greater or lesser leaf stomatal closure in response to the conditions to which the vegetation is subjected. r_c depends on both:

– the demand exerted by the meteorological conditions (global solar radiation Rg, air temperature T_a, dryness level of air expressed through the air vapor pressure deficit $e_s(T_a) - e_a$);

– the water status of the plant and the availability of soil water, which constrain the possibilities of water extraction by the roots.

10.2.3. Soil *conductive flux*

The heat flux in the soil depends on the temperature gradient within the top centimeters of soil and on the thermal properties (capacity, conductivity) of the soil surface layer – hence on the texture, the density, the organic matter and surface water contents. In the absence of sufficiently accurate spatial information on these variables, G is usually scaled to the net radiation available at the ground. Since this essentially depends on the fraction of solar radiation intercepted by vegetation, the G flux is conventionally expressed as a fraction of total net radiation, which can easily be estimated by remote sensing. If an estimation of G at the same time each day is needed, to analyze surface temperatures provided by a sun-synchronous satellite sensor for instance, the ratio $\xi = G/Rn$ can be expressed as a simple semi-empirical function of radiative surface properties [SUN 13], in particular the fraction of radiation intercepted within the canopy:

$$\xi = f_c A_c + (1 - f_c) A_s \qquad [10.6]$$

where f_c is the fraction of vegetation cover (usually deduced from NDVI), and A_c and A_s are the typical ratios for pure bare soil and fully vegetated surfaces (respectively).

If one wants to estimate G at different times of the day, the diurnal cycle of the G/Rn ratio must be taken into account because of the phase shift between the two fluxes (indeed, Rn reaches its maximum around noon whereas G is at its maximum in mid-morning [GEN 10]). The diurnal evolution of the G/Rn ratio is a sinusoidal time function calibrated against measurements performed for different canopies displaying a contrasted leaf area index [SAN 03]:

$$\xi = A \cos[2\pi(t + 10800)/B] \qquad [10.7]$$

where A is of the order of 0.2 for dense canopy and 0.4 for bare soil; B is between 8.0 10^4 s and 10^5 s. In order to obtain a more realistic and general expression accounting for both temporal and surface characteristics, it is recommended to combine [10.6] and [10.7] and to replace A in [10.7] by [10.6].

10.2.4. *Energy budget equation*

By replacing each of the terms in [10.1] with expressions [10.2], [10.4], [10.5] and [10.6], the energy budget equation takes the form:

$$(1-\xi)[Rg\,(1-a)+\varepsilon(L\downarrow-\sigma T_s^4)] = \rho c_p \frac{T_s - T_a}{r_a} + \frac{\rho c_p}{\gamma}\frac{e_s(T_s) - e_a}{r_a + r_c} \qquad [10.8]$$

On the left hand side part of equation [10.8], temperature T_s is a radiative temperature (see [10.2]), whereas on the right hand side, it substitutes the aerodynamic temperature T_0 (see [10.4] and [10.5]). Expressing [10.8], considering a single temperature T_s only, requires an assumption to be made on the relationship between the aerodynamic temperature T_0 and the radiative temperature. Much research is still underway to establish relationships between these two temperatures and ensure consistency of the equations of flux and energy budget [KUS 09]. This is further discussed in section 10.4.1. T_0 and T_s are generally close in the case of bare soils or very dense homogeneous covers. Not taking their differences into account can result – especially for partial cover – in significant errors in the turbulent terms of the energy budget and ultimately, evapotranspiration.

Surface temperature T_s appears as the variable solution of [10.8]. As it is linked to each term of the energy budget, it provides information on all surface fluxes and should be considered as the "signature" of the energy balance. Since, by definition, this equation must remain true at all times, the surface temperature value is constantly adjusting in order to satisfy the energy balance closure. Thus, a variation of Rg under the influence of passing clouds or a variation of the aerodynamic resistance under the influence of wind whirls for instance, will result in instantaneous changes in T_s.

Similarly, the various characteristics of the surface (geometry, water status, etc.) will directly translate into surface temperature variations driven by the corresponding flux differences.

The energy budget equation has been established on the assumption of a uniform surface. However, it can be applied at spatial scale (leaf, soil, plant, canopy, plot, region, etc.). In the case of composite surfaces however (such as a fragmented agricultural landscape), the parameters introduced into the budget equation must be adapted and correspond to "effective" values that integrate the effect of spatial heterogeneity. This is especially important when working at a regional scale with images of low spatial resolution, for which pixels (kilometer size or even more for instance) generally include a large number of different surface types.

Finally, let us remember that the albedo and emissivity parameters in equation [10.8] are integrated on a wide portion of spectrum (broadband) and should not be confused with spectral measurements from satellites.

10.2.5. *Significance of surface temperature*

The energy budget equation [10.8] has been developed assuming a "thin", flat and continuous surface. In reality, vegetation covers grow up to a certain height. The elements inside the canopy (leaves, stems, soil) have different temperatures depending on their positions within the canopy and whether they are directly illuminated by the sun or shaded (Figure 10.1). The surface temperatures of each of these elements are determined by energy budgets at their own level and are therefore governed by a complex system of coupled energy and radiative transfers within the canopy. By conditioning both air flow within the canopy (which generates a wind profile) and penetration of radiation into the layers of vegetation, the canopy structure (height and spacing of plants, leaf density, possible row effects, etc.) plays a leading role in the spatial distribution of temperatures of elementary facets.

Integration of emissions from each soil or vegetation elementary surface in a given direction provides a directional radiative temperature. The temperature measured by a radiometer that observes the surface within a field of view (FOV) and in a given direction is a so-called "brightness temperature", which incorporates both the emission of the surface (the radiative temperature) and the reflection of the downward long wave radiation by the surface. Estimating radiative surface temperature from a measurement of the brightness temperature requires knowledge of the emissivity of the surface. These issues will not be discussed here, but the reader is referred to [BRI 16, NOR 95] for terminology clarification.

Figure 10.1. *Example of thermography of a corn plot (FLIR SC2000 camera, 15/06/2012, 11:16:05.26 GMT, Lamasquère site in South-west France) depicting temperature variability within a vegetation canopy: the sunlit ground is above 40°C and about 32°C if shaded, while leaf temperature varies within a range of about 5°C (between 25 and 30°C). For a color version of this figure, see www.iste.co.uk/baghdadi/4.zip*

10.3. Surface temperature data

Surface temperature measurements from space are affected by various sources of error and uncertainty. Correction of atmospheric effects in the thermal infrared has been described in [BRI 16]. Furthermore, the measurement carried out at the surface by a sensor provides a "brightness" temperature T_b that results from:

– the proper emission from the surface at the radiative temperature T_s weighted by emissivity ε;

the reflection, with a coefficient $(1 - \varepsilon)$ of the downward atmospheric longwave radiation L↓ (section 1.4.2 in [BRI 16]). In the same chapter, methods for separating temperature and emissivity from a brightness temperature measurement are described.

10.3.1. *Directional effects in thermal infrared*

Measuring surface temperature in the TIR domain is prone to significant directional anisotropy effects. This anisotropy (difference between temperatures observed in oblique and nadir viewing angles) is induced by the fact that the structure of any surface (vegetation, urban canopy, etc.) determines both the temperature variability of the various individual surface elements (leaves, etc.) within the canopy through radiative and turbulent coupled exchanges, and the "visibility" of these elements for a sensor located in a given viewing direction. Comprehensive experimental work on directional anisotropy has been conducted since the 1960s [PAW 92]. Recent experimental work based on airborne thermal imaging camera measurements allowed us to illustrate directional effects in the TIR domain for all azimuth viewing directions and on a range of zenith viewing angles up to 60° on forest and urban canopies [LAG 00, LAG 10]. They have highlighted significant hotspot effects (Figure 10.2). These are explained by the fact that when the surface is viewed exactly in the direction of the Sun, the sensor only sees sunlit elements, leading to a peak in temperature. In the immediate vicinity of this specific direction, the measurements are increasingly affected by shaded elements.

In the case of discontinuous canopies, directional anisotropy effects can be exacerbated due to the stronger contribution of soil, especially if the latter is dry. Figure 10.3 shows the impact of row directions in a vineyard on the directional effects that may here exceed 15°C [LAG 14].

Angular effects can significantly affect satellite measurements for various reasons. For orbiting satellites with a polar orbit and a scan close to an East–West direction, Figure 10.2 allows us to understand that the effects are tightly affected by the overpass time. Very large cross track scanning angles achieved when using MODIS or VIIRS sensors (about ±55°) or the obliqueness of acquisitions at the edge of the observation disk when using geostationary satellites are also the source of angular effects that must be corrected.

Modeling efforts (not detailed here) are conducted to simulate directional anisotropy through various approaches, ranging from simple geometric models to 3D models coupled with radiative and energy transfer (a brief review of this can be found in Verhoef *et al.* [VER 07]).

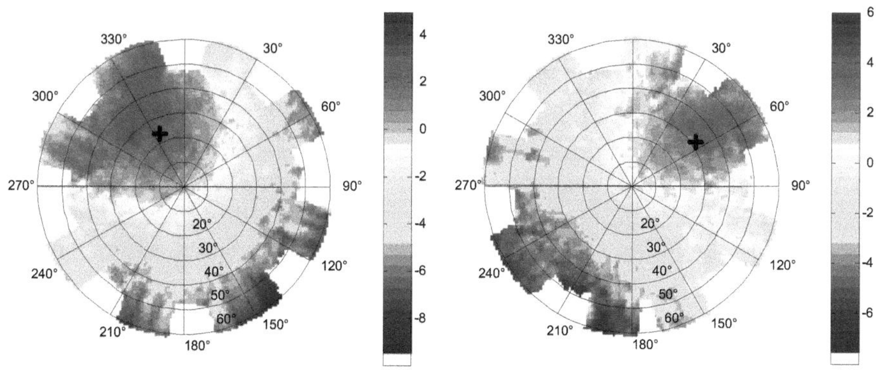

Figure 10.2. *Polar representation of TIR directional anisotropy measured above an urban environment (Toulouse, France) on 15 July 2004 around 11:30 a.m. (left) and 2:00 p.m. UT (right). The difference (in °C) between measurements acquired obliquely and at nadir is color-coded. The radii indicate the azimuthal viewing directions and the concentric circles show the zenith viewing angles. The cross indicates the direction of direct illumination from the Sun. We observe that it corresponds to the position of the hot spot and that it moves during the day according to the Sun's movement [LAG 10]. For a color version of this figure, see www.iste.co.uk/baghdadi/4.zip*

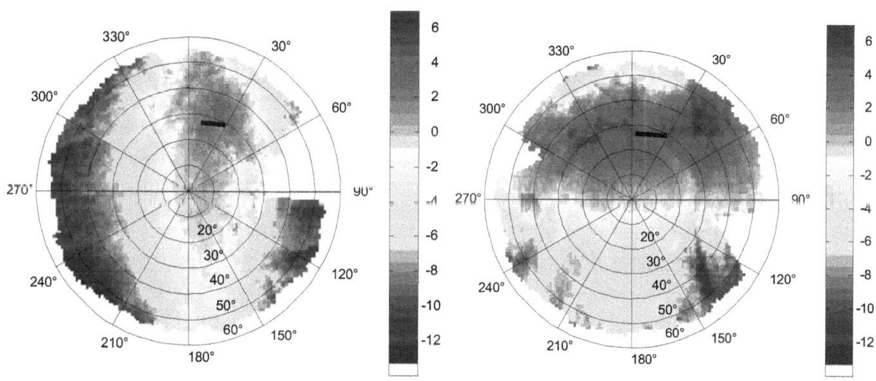

Figure 10.3. *Effect of row direction (North–South, 12:29 to 1:06 p.m. UT on the left, East–West, 12:15 to 1:06 p.m. on the right) on the TIR directional anisotropy of a vineyard (Médoc region, France, 3 August 1996). The anisotropy representation system is the same as in Figure 10.2. The direct illumination from the Sun varies with its position during the acquisition of measurements and is indicated by the black segment. For a color version of this figure, see www.iste.co.uk/baghdadi/4.zip*

10.3.2. *Impact of atmospheric turbulence on measurement uncertainty*

Surface temperature (T_S) is prone to significant temporal fluctuations linked to the turbulence of air flow near the surface [KAT 98]. The intensity and frequency of these fluctuations depend on the atmospheric turbulent structures that generate them, and their impact on the estimation of T_S depends on the spatial resolution of sensors. High frequency structures in the surface boundary layer correspond to typical scales of a few meters linked to the size of the surface roughness elements. Fluctuations in T_S associated with these structures are similar in the mechanical domain to the "waves" or "honamis", which propagate over wheat fields under the influence of wind [DUP 10]. These temporal fluctuations are smoothed out for decametric pixels (50-100 m) such as what is envisaged for future HyspIRI or THIRSTY missions at high resolution in the TIR. In contrast, low frequency structures within the planetary boundary layer have typical scales of several hundred meters and can trigger significant fluctuations both in time and in amplitude on pixels of smaller size. Figure 10.4 illustrates the passing of thermal turbulent structures over a plot of corn and provides an example of the temporal evolution of surface temperature over twenty minutes at a resolution of 50 m. It also provides a simulation of the expected errors depending on the spatial resolution at which a satellite measurement is performed.

The "remote sensing" community still remains largely unaware of the existence of these fluctuations and of the "turbulent" character of surface temperature. However, they have significant consequences for the interpretation of remotely sensed T_S measurements, as well as for the definition of specifications and the operation of future TIR missions:

– they induce uncertainty in surface temperature measurements; the satellite speed (7-8 km s^{-1}) is high enough so that a measurement performed on a pixel can be considered as an instantaneous sample of a random signal;

– the estimation of surface flux is affected. Indeed, since models and retrieval algorithms assume stationary conditions over time steps of the order of 15 to 30 minutes, any deviation between an instantaneous measurement of T_S and its average value over the model time step is a source of flux error;

– it should also be considered in calibration/validation activities of satellite products, both satellite and ground reference measurements being affected by measurement errors and uncertainty;

– finally, the knowledge of this uncertainty is necessary for optimizing space mission specifications. The uncertainty inherent to T_s makes it unnecessary – at least for the continental biosphere – to have a very low sensitivity requirement (this sensitivity corresponds to the ability to discriminate between two neighboring temperatures, or NedT – Noise Equivalent Delta Temperature). It is also a strong argument in favor of higher revisit frequencies, which is the only way, through the multiplication of observations each affected by an error, to force more constraint on T_s models. Ultimately, it can constrain the optimum spatial resolution, the latter resulting from a trade-off between the typical size of the plots to observe on the ground and an acceptable error measurement (sharpness of the resolution being accompanied by an increase in measurement uncertainty).

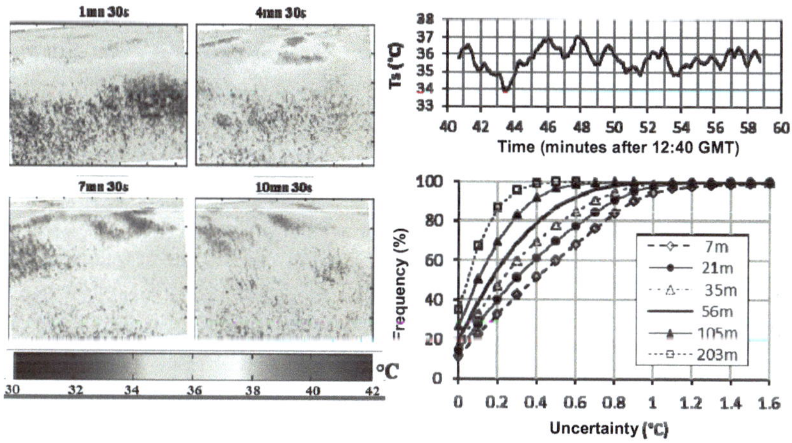

Figure 10.4. *Samples from a sequence of TIR images acquired at high frequency (10 Hz) on a corn plot: the spatial-temporal variability of the surface temperature structure is clearly visible in the thumbnail images extracted from a 20-minute series (left) while the time trace of the surface temperature aggregated at a resolution of 50 m (top right) shows a variation of 3°C. Bottom right: cumulative uncertainty histograms (in absolute value) expected at various spatial resolutions for a maritime pine cover from Large Eddy Simulations; the measurement error decreases inversely with spatial resolution (for example, the measurement error – regarded here as being the deviation of instantaneous values from the average of a time sequence of a few tens of minutes – is less than ±0.4°C for 98 % of measurements at a resolution of 203 m, but for only 80 % and 60 % of them at resolutions of 56 and 21 m) [LAC 15]. For a color version of this figure, see www.iste.co.uk/baghdadi/4.zip*

10.3.3. *Current data and their limitations: a need for new missions in the TIR domain*

The local scale (for example, the plot) is the key scale where the complexity of a large number of physical, biological, social or economic processes that interact in ongoing global changes is expressed. This scale is also that of the management and implementation of policies for adapting to change. It is therefore important to have high spatial resolution data in order to access very small scales, such as that of the agricultural plot. Furthermore, the temperature acquired in TIR reacts to energy exchanges at the surface (under the influence of meteorological forcing and events such as rainfall and irrigations) and is a much more relevant information than, say, the sole reflectance in the visible-near infrared domain, since it reacts much faster to water stress for instance. The quality of information in TIR therefore depends on the ability to grasp the high intrinsic temporal variability of surface temperatures, which in turn requires (keeping in mind the limitations imposed by cloudiness) a high revisit capability.

Researchers and end users, however, are now facing a severe shortage in supply of data in TIR and have to conciliate between spatial resolution and revisiting capacity. On the one hand, systems like VIIRS/SUOMI-NPP, MODIS/TERRA and AQUA provide daily observations (and even hourly with geostationary satellites) with a spatial resolution of a km^2 at best. On the other hand, high spatial resolution is only available with Landsat and ASTER systems at the cost of approximately 2-week revisits, which is completely inconsistent with the monitoring objectives of vegetation. The various attempts to disaggregate TIR data acquired at a kilometric resolution by recreating thermal contrasts based on VIS/PIR data with high spatial resolution [AGA 07, INA 08, MER 10, ZHA 13] have been unsuccessful and cannot be used as an alternative to higher resolution TIR data. Indeed, it lacks crucial information on water status that is required to obtain a reliable disaggregation. Several space mission projects have recently been proposed or are underway to overcome this obstacle, such as MISTIGRI [LAG 13], HyspIRI (http://hyspiri.jpl.nasa.gov/) or THIRSTY [CRE 14], which aim at combining a resolution between 50 and 100 m with a revisit capacity between 1 and 5 days.

It is generally accepted [CRO 08] that an absolute accuracy of 1°C for surface temperature is acceptable to obtain a reasonably reliable evapotranspiration estimate. There are many sources of uncertainty that affect the measurement of T_s by satellite (atmospheric corrections, directional effects, impacts of atmospheric turbulence, accuracy of sensors), yet this 1°C target remains an unreached goal for most space systems. This value of 1°C, however, is not an absolute threshold and must be considered with caution: on the one hand, it does not have the same impact on flux depending on the values of $(T_s - T_a)$, and on the other hand, the requirements in terms of accuracy of evapotranspiration may depend on a greater or lesser sensitivity of the flux to $(T_s - T_a)$ for some phenological stages of vegetation. Ultimately, the requirements in terms of accuracy of measurement of T_s may also depend on the methodologies implemented for calculating evapotranspiration (presented below: residual or contextual methods, assimilation).

10.4. Estimating evapotranspiration

Evapotranspiration (AET = LE/L where L is the latent heat of vaporization) establishes the link between water and energy budgets of land surfaces. Since evaporation requires a large amount of energy (L is about $2.26\ 10^6$ J kg^{-1}), actual evapotranspiration AET – or its equivalent, the total latent heat flux LE – is a major determinant of the surface energy budget. Evapotranspiration depends on three factors: weather, soil water availability and vegetation cover. These three factors are highly variable in time and space. Depending on the application, an estimation of AET is required at hourly (weather applications), daily (hydrology, agronomy) or monthly (surface water-groundwater interactions) time steps.

The main methods for estimating AET are based on solving the energy balance [10.1] in order to provide an explicit link between the temperature T_s measured by satellite and LE. Available energy (Rn-G) is partitioned according to the surface water status between the two turbulent fluxes H and LE. When water is present in the soil in sufficient quantity, available energy is mostly converted into evapotranspiration, while under water stress, this distribution is instead in favor of the sensible heat flux.

Remote sensing is the most suitable means for providing a spatial estimation of AET [COU 05, KAL 08]. Remote sensing provides key variables for forcing the energy budget. The surface temperature acquired in the thermal infrared reflects the intensity of energy exchange and the distribution between the various transfer processes. Information on vegetation (albedo, amount of vegetation, LAI) is provided by measurements in the visible and the near infrared, which is complementary to the TIR.

Often, the information from TIR remote sensing is only available with large temporal gaps between two successive image acquisitions (due to a satellite revisit frequency that is too low, presence of clouds, etc.), while a quasi-daily revisit is required to monitor satisfactorily the succession of wettings and dry downs for instance. One therefore needs to develop methods to produce seamless AET timeseries from infrequent TIR data (i.e. interpolation, gap-filling and other data assimilation methods [OLI 05, PEL 02]).

Two types of methods are currently used to compute LE: the so-called "residual" methods use information from each pixel independently of any other pixel in the image, while the "contextual" methods take advantage of thermal contrasts in the image. They provide instantaneous estimates of latent heat flux at the time of the satellite overpass.

10.4.1. *Residual methods*

These methods estimate LE as the residual term of the energy budget, i.e. the difference between available energy (Rn- G) and H:

$$LE = Rn - G - H \tag{10.9}$$

Equation [10.9] has enabled us to establish the first empirical residual methods relating directly the difference (T_s – T_a) and the daily AET [SEG 94]. Improvements in the description of the various components of the land surface models during the 1990s have offered a physical basis to relate T_s to the various energy budget components beyond these empirical methods. Indeed, for these early approaches, energy balance [10.8] is expressed in a simple way for a hypothetical continuous uniform surface (bare soil or dense vegetation) for which the aerodynamic temperature T_0 and the radiative surface temperature T_s are almost identical. For intermediate vegetation cover fractions, however, the aerodynamic temperature depends, on the one hand, on air temperature and on the other, on the surface temperature of each of the individual foliage elements and the underlying soil surface. Equation [10.9] is more generally expressed from equations [10.2] and [10.4] and becomes:

$$LE = (1 - \xi)[Rg\ (1 - a) + \varepsilon(L \downarrow - \sigma T_s^4)] - \rho c_p \frac{T_0 - T_a}{r_a} \qquad [10.10]$$

For less dense but homogeneous cover, it is possible to define an average scalar profile (wind, temperature) within the canopy. Relationships were therefore established to link the aerodynamic temperature with an average radiative surface temperature. The latter can, as previously shown, be obtained from TIR remote sensing, and is generally higher than the aerodynamic temperature because it integrates all surface elements, particularly the warmer bare ground illuminated by direct sunlight (section 10.2.5). Substituting T_s for T_0 in [10.10] means either weighting H with a correction factor α, or adding an additional resistance r_e ([10.11] and [MAT 05]). Many models provide a way to parameterize α and r_e depending in particular on the leaf area index (e.g. the SEBS model [SU 02]).

$$H = \rho c_p \alpha \frac{T_s - T_a}{r_a} = \rho c_p \frac{T_s - T_a}{r_a + r_e} \qquad [10.11]$$

In the case of sparse vegetation, for which a large fraction of bare soil is directly illuminated, the characterization of an average profile is an issue [BOU 12] and often temperature end members (of the distribution of the foliage and individual soil elements) have to be explicitly considered. The latter often correspond to these illuminated soil and partly shaded vegetation elements, respectively. In addition, these "single-source" models (because they consider the area as a whole) only compute total AET, while for many eco-hydrological and agricultural applications, a separate evaluation of the plant transpiration and soil evaporation components is required.

"Two-source" models solve separate energy budgets for the soil and the vegetation and thus allow us to estimate these two components (Figure 10.5). They derive average equilibrium temperatures for the soil and the vegetation as a whole, which are considered the main sources of heat. In the absence of directional temperature data integrated over a very short time (see ATSR sensor on ENVISAT [JIA 03]), it is not possible to separately estimate these two temperatures. Solving these two unknown temperatures in both energy budgets with a single piece of information (the average radiative temperature) leads to a problem of indeterminacy. We must therefore add an assumption to reduce the number of unknowns. The two-source energy balance model (TSEB [KUS 99, NOR 95]) offers a pragmatic solution based on a realistic assumption, namely that the impact of soil moisture reduction on evaporation (which can extract the water available in the shallow topsoil only) generally precedes water stress of the plant (which has access via its deep roots to a larger amount of water).

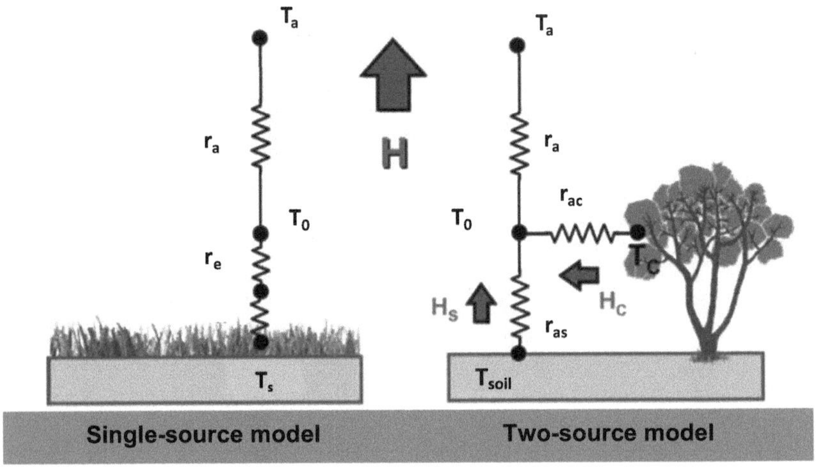

Figure 10.5. *Single- and two-source resistance schemes [KUS 09]: R_{as} and Rac are the soil–air and vegetation–air aerodynamic resistances, respectively. T_{soil} and T_C are ground and canopy temperatures, respectively. Hs and Hc are the sensible heat fluxes from the ground and the canopy, respectively*

The performance of residual methods strongly depends on the accuracy of the climate forcing, in particular air temperature and wind speed data. To overcome this difficulty, it is possible to estimate the temporal evolution of air temperature through a simple model of a boundary layer fueled with geostationary satellite data (ALEXI models DiSALEXI, [AND 11a]) at the same resolution as those satellites, that is between 3 and 5 km. By estimating air and surface temperatures from the same data, the difference $T_s - T_a$ used in the calculation of H gains in consistency and therefore in robustness.

Despite these improvements, residual methods accumulate various instantaneous estimation errors for each of the components of the surface energy budget, particularly all fluxes that involve surface temperature. They are therefore particularly sensitive to the absolute error associated with the brightness temperature or the "surface temperature" product. The estimation of the total aerodynamic resistance involved in the calculation of H is also questionable and requires prior knowledge of the average (or "effective") height of the vegetation for each pixel. This value can be derived from NDVI for relatively dense non-woodland cover, for example for agricultural lands, while for bare soil, sparse cover and woody vegetation, such as shrubs or savannah, other remote sensing variables should be used.

10.4.2. *Contextual methods*

To overcome these limitations, and if one assumes that a large part of the total estimation error on the surface temperature product is relatively constant (i.e. systematic) over a given thermal image, it is possible to establish a simple scaling relationship between the different temperatures in a given landscape and the corresponding distribution of water stress within the image (methods known as "contextual"). These methods assume the existence on a given image of extreme water and plant conditions, with the warmest pixels corresponding to the driest points and the coolest pixels the wettest points in the scene, respectively. For a given water status, the impact on surface temperature will also depend on other surface properties, which are largely governed by the vegetation characteristics (LAI, roughness, albedo, etc.). Scaling between the temperature end members requires at least a second surface variable that is accessible through remote sensing: a vegetation index (like the NDVI or the SAVI in Moran *et al.* [MOR 94]), the albedo (S-SEBI [ROE 00]) or both [MER 14]. Various schematics shown in Figure 10.6 illustrate the most commonly used methods and contextual relationships that are applied to derive aerial estimates of latent heat flux.

Most contextual methods do not use much auxiliary data, mainly just the available energy that can be determined in a rather robust way through remote sensing [MIR 16]. They are less sensitive to the precision in the meteorological variables. However, they are based on many simplifying assumptions that tend to restrict their conditions of applicability. Firstly, scaling of the water status between the hottest pixels that are considered dry and the coldest ones that are therefore close to saturation is based on the area's average available energy and climatic conditions, which are thus considered homogeneous over the entire image. Then, the assumption of a simultaneous presence within a given image of sufficiently contrasted pure pixels (homogeneous in terms of vegetation and water status) is often no longer valid when the image is acquired at the very beginning or the very end of a dry down after rainfall. The selection of temperature end members must be done carefully and within an area that is not too big so that average climate conditions can be representative, but large enough so that extremes can be selected from a sufficiently exhaustive sample of pixels. These limit conditions are therefore difficult to identify in homogeneous areas such as natural vegetation or rain-fed agricultural landscapes, and for low resolution data that encompass very different plots without revealing average extreme situations, in terms of moisture levels, and/or vegetation extent. In this case, residual methods may be better suited than contextual methods.

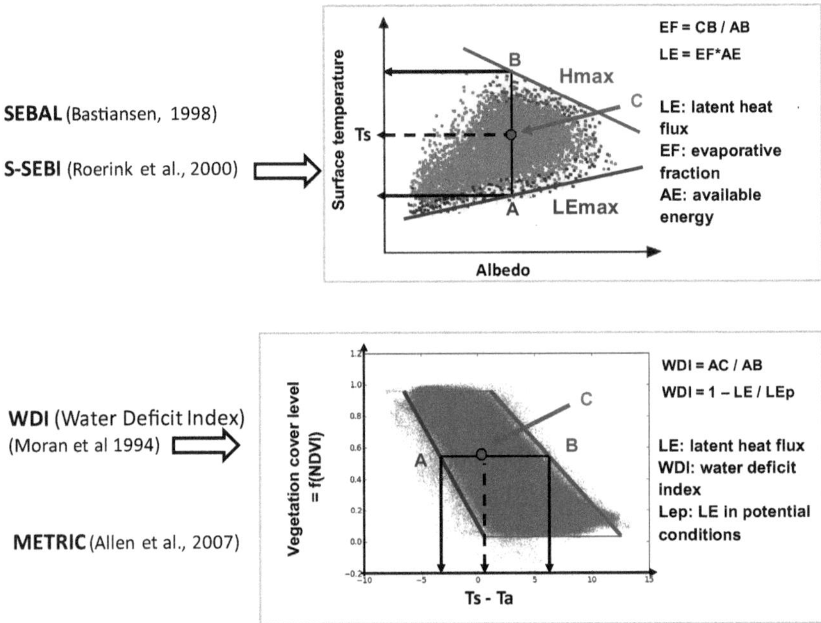

Figure 10.6. *Traditional contextual methods: top, graphical illustration of S-SEBI, which calculates an evaporative fraction at any point C depending on the temperature limits for each albedo class A and B; bottom, graphical illustration of the WDI model (Water Deficit Index) that estimates a stress factor for a given point C according to temperature limits A and B, which is identified by range of vegetation coverage deduced from solar domain data*

Methods used to solve the surface energy budget therefore mainly differ in the way the available energy is partitioned between the two turbulent fluxes. Both methods can be combined to produce complex "contextual" models (SEBAL [BAS 98], METRIC [ALL 07]). The EVASPA platform (evapotranspiration assessment from space [GAL 13]) brings all these algorithms together to provide an ensemble simulation, which allows deriving an idea about the model uncertainty: the simulation average obtained by combining the different models provides an optimal estimate of AET and the ensemble spread of an associated standard deviation. For now, the platform in version 2.0 enables us to operationally process MODIS and Landsat data (Figure 10.7). EVASPA uses different reflectance inputs that allow us to compute emissivity, LAI and albedo, as well as brightness temperature, but can also use the corresponding products when they are available (such as MODIS).

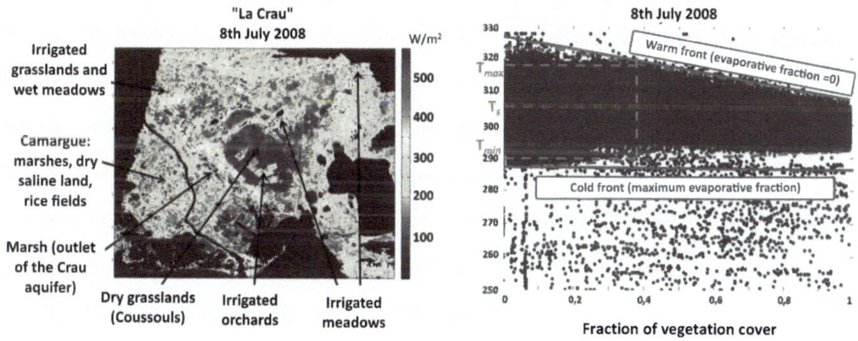

Figure 10.7. *Map of latent heat flux (LE) on 8 July 2008 (Landsat-7 data) and the corresponding T_S-vegetation fraction cover diagram obtained for the Crau–Camargue region (France) using EVASPA. For a color version of this figure, see www.iste.co.uk/baghdadi/4.zip*

10.4.3. *Assimilation of TIR data*

The methods presented in the previous section allow us to estimate AET for dates when spatial data are available (dates resulting from an intersection of revisit capacity and absence of clouds). Prior knowledge of the water status of the surface and the history of water supply is unnecessary here.

Another approach is based on assimilation methods (see [OTT 16]) that are commonly used in weather forecasting. It requires the implementation of a soil–vegetation–atmosphere transfer (SVAT) model that solves the coupled water and energy budget equations and derives the evolution over time of turbulent fluxes as well as a surface temperature equilibrium. Besides numerous surface biophysical parameters, the SVAT model requires a time series of meteorological and water supply (rain or irrigation) input data. Assimilation means adjusting the state variables or the parameters of the SVAT model in order to minimize the difference between the remotely sensed biophysical variable at the satellite overpass time and the corresponding value simulated by the SVAT model [OTT 94].

Radiative surface temperature can be directly assimilated, but one can also use the various instantaneous AET estimates obtained discretely over time by either methods (residual or contextual) described above in order to provide continuous AET time series.

Assimilation methods are used to adjust the parameters that are the most difficult to specify (minimum stomatal resistance, hydrodynamic soil properties, see [COU 07]), the initial conditions (for example, sowing date) or the system state variables (usually soil moisture). The most common methods use observations when they are available (sequential methods [CRO 08]) or integrate them over a given interval (smoothing, variational methods [JAR 14]). Their use is still limited, not just because TIR data at a satisfactory resolution (agricultural land) are only available for intervals that in practice often exceed the average duration between two successive rainfalls, but also because it is often difficult to accurately quantify the various errors and covariance errors associated with the model or with observations. The quantification of these uncertainties is the limiting factor for the performance of assimilation platforms.

10.4.4. *Operational methods adapted to the availability of satellite data*

Water stress and evapotranspiration estimated through the residual and contextual methods are instantaneous values at the time of the satellite overpass. The reconstruction of daily or seasonal evapotranspiration therefore requires the use of extrapolation (reconstitution of the diurnal cycle) and interpolation (estimation between two successive images,

[DEL 12]) methods. Apart from the assimilation methods mentioned above, which allow us to continuously simulate AET over time according to the water supply forcing data but are relatively complex to implement, different simple reconstitution methods of temporal variations of AET are proposed. Water stress (see WDI in Figure 10.6) or an equivalent variable such as the evaporative fraction EF (the ratio between latent heat flux and available energy Rn-G, see Figure 10.6) have the advantage of being dimensionless and relatively stable during the day, and are therefore good candidates for these reconstruction techniques. The evaporative fraction is relatively constant around the solar noon, especially during periods of water stress [GEN 11]).

Daily AET and the diurnal cycle of LE can be estimated by multiplying the instantaneous value of EF or the complementary part to one of the WDI by the daily total or the diurnal cycle of the hourly available energy or potential evapotranspiration rate (respectively). The estimate between two successive dates for which an image has been acquired can be obtained through a simple interpolation of the EF or the WDI between these dates, since the available energy or potential evapotranspiration are known throughout the interpolation interval.

10.5. Other applications

One of the most important applications of TIR is to access the water cycle, and to follow rapid changes of water status during dry down periods (after each rainfall or irrigation input). The main application, which is described in the previous section, is the estimation of evapotranspiration. Through it, applications of TIR extend to other variables of interest for continental surfaces, since evapotranspiration directly reflects the satisfaction or not of plant water needs, CO_2 absorption and, ultimately, the growth of vegetation and its production. Water transfers in the soil–plant–atmosphere system strongly influence the transport of various mineral compounds and chemical elements in solution, making evapotranspiration a major determinant of biogeochemical cycles. Applications of TIR are finally expected in the field of hydrology, as the evapotranspiration is governing soil moisture at the surface and in the root zone through water extraction by plants, with significant consequences for infiltration, runoff and, ultimately, for the whole catchment water budget.

10.5.1. *Agriculture and forestry*

10.5.1.1. *Estimating production and its limiting factors*

Photosynthesis processes govern biomass production. Production thus depends on the amount of radiation absorbed by plants (f_{APAR}, fraction of absorbed photosynthetically active radiation) and then on their ability to convert it into biomass. This was formalized in the Kumar and Monteith model [KUM 81], which connects dry matter (DM) production during a time interval [t_1,t_2] to the global solar incident radiation R_g in the following manner:

$$DM = \int_{t1}^{t2} \varepsilon_b f_{APAR} \varepsilon_c R_g dt \qquad [10.12]$$

where ε_c is the climatic efficiency or part of incident solar radiation that is useful for photosynthesis ($\varepsilon_c \sim 0.42\text{-}0.50$) and ε_b is the conversion efficiency into dry biomass.

In practice, measuring the useful fraction of radiation absorbed by plants is difficult and an approximation consisting of equating absorbed radiation with intercepted radiation is often carried out. As the incident radiation is split between an intercepted part and a part reflected by the plant canopy, the fraction of intercepted radiation (sometimes referred to as interception efficiency ε_i) can be deduced from the measurement of the reflected radiation by remote sensing. Many studies have been conducted [REM 13] to statistically link ε_i or f_{APAR} to various vegetation indices derived from satellite measurements in the visible and near infrared. The conversion efficiency ε_b can be prescribed provided the vegetation type is known. However, as CO_2 and water transfers are conditioned by the same stomatal regulation processes, the conversion efficiency is very closely linked to the level of satisfaction of the plant water needs and, ultimately, to their transpiration. TIR information therefore appears as complementary to that in the solar domain, by providing the means to derive ε_b from water stress indices (Figure 10.8) and to take the temporal evolution of water conditions into account. Insufficient spatial resolution (at best, the km^2) of TIR satellite data, as mentioned previously, undoubtedly slowed works in this direction (aside from "primary production" MODIS products that were based on a simplified Monteith approach). These should be stimulated in the future through access to the agricultural plot scale, expected from TIR mission projects, combining high spatial resolution with high revisit capacities [LAG 13, CRE 14].

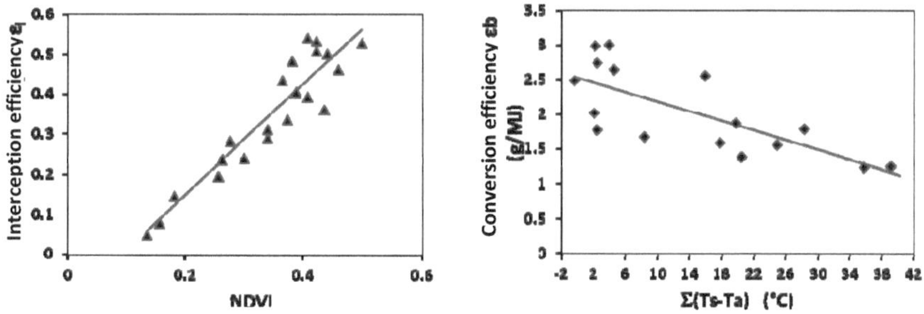

Figure 10.8. *Relationships between the efficiency of interception and vegetation index NDVI [BEG 93] on the left, and between the efficiency of conversion at crop cycle scale and stress degree day calculated by summing (T_s – T_a) over the same period [OUA 94] on the right. Case of Mil in Niger*

In many regions, tree crops and orchards have been converted from rainfed to irrigated agriculture. Estimating evapotranspiration in these systems still remains difficult because of their complexity (heterogeneity of canopies and root systems) and because of the size of the fields, which are often small [BER 09].

In woodlands, TIR data help detect the effects of prolonged drought and trigger many other applications such as monitoring tree growth and timber production, but also sensitivity to fire risks or parasitic attacks that may affect the economic potential of the forest [JAN 06].

Ultimately, we can include mapping of frost risk or detection of radiative frost, which can significantly affect the level of agricultural production [KER 96, TAI 03].

10.5.1.2. *Detecting water stress through stress indices*

Identification of water stress has significant applications for rainfed agriculture (vulnerability of eco-agrosystems to drought, ecosystem services, etc.) and for irrigated agriculture (dynamic mapping of irrigated areas, irrigation management, etc.). Evapotranspiration is the major dissipation flux of available energy in the absence of stress, which helps maintain a surface temperature close to an equilibrium computed in potential (unstressed) conditions. This temperature can be estimated from remote sensing data in the visible-near infrared range (NDVI, albedo) and from meteorological data by solving the energy balance at the surface [BOU 07]. When vegetation reduces its level of transpiration during a water stress episode, the surface temperature deviates from the temperature in potential conditions. Thus, the difference between these two temperatures is a good water stress proxy and is accessible by remote sensing.

The algorithms discussed in section 10.4 provide an estimate of surface water stress. It is expressed similarly to the WDI according to temperature end members that correspond to limit conditions (potential evapotranspiration or zero evapotranspiration) that are observed on an image (contextual methods) or calculated by an energy balance model (residual methods).

The result is a dimensionless index that allows water stress to be mapped and can also be converted into relative humidity in the root zone, for example [HAI 09]. At regional and continental scales, due to the availability of kilometric data for almost two decades, drought areas can be monitored. An illustration for the contiguous United States (CONUS) zone is presented by Anderson *et al.* [AND 12].

10.5.1.3. *Irrigation*

Water use in agriculture now represents 70 % of global resources, making sustainable irrigation an important issue. Applications of TIR can here be found for each spatial scale.

At the regional scale, automatic detection and mapping of irrigated farmland area is vital for many services in charge of water management, both administratively and in terms of equipment (for network design purposes, for example). In that respect, TIR brings, in addition to visible and near infrared, key information on irrigated areas for which the development of vegetation reaches maximum values during the seasonal cycle and which simultaneously most often display lower surface temperatures.

At the scale of the agricultural plot, the expected constraints on water resources under global change today impose an implementation of more efficient irrigation practices. Methods for accurate control of water supplies are now essential (whether in deficit irrigation or supplementary irrigation), for example by setting automatic water distribution rules based on a level of critical stress that triggers an average irrigation dose [SIM 09]. The decrease of moisture within the soil after water supply can be evaluated from the surface moisture estimated by radar (see Chapter 1), but observations in the TIR remain better-suited to monitoring vegetation water stress by taking advantage of the methods described previously to continuously monitor evapotranspiration and water balance or to monitor stress indices. The intervention scale is the agricultural plot (Figure 10.9), and irrigation monitoring must be adapted to the proper needs of each of the cultures at a short time scale in a given area. This is why it is important to develop new satellite observation systems in TIR that conciliate spatial resolution and high revisit capabilities. Although they have been used successfully in the USA to evaluate water budgets on the seasonal scale and to evaluate allocation between irrigation activities and hydrology of groundwater and rivers (see the work of Allen from the University of Idaho, for example), Landsat data remain – because of low revisit rates – in most regions unsuited to the objectives of water status monitoring in real time and of irrigation management, which are essential for optimizing production.

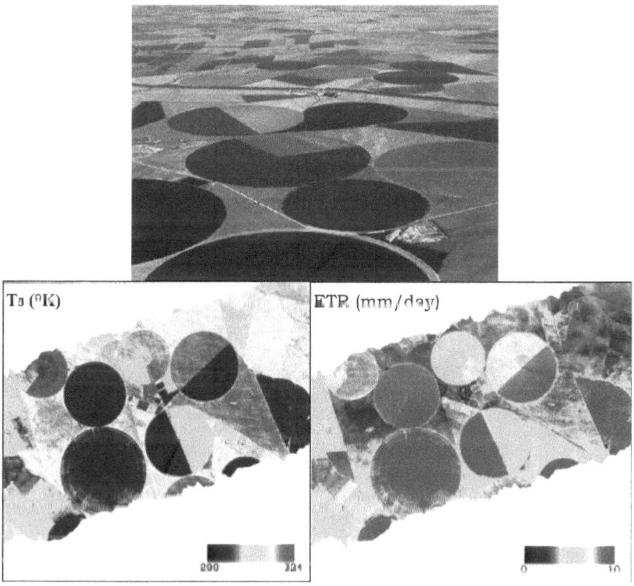

Figure 10.9. *Surface temperature (T_S, scale 290-324°K) and daily evapotranspiration (ETR, scale 0–10 mm per day) from an airborne hyperspectral scanner data (AHS) and the S-SEBI model over the agricultural area of Barrax (Spain) [SOB 08]. For a color version of this figure, see www.iste.co.uk/baghdadi/4.zip*

TIR can also provide information on the maintenance of irrigation systems: the use of aerial data allows us to detect leaks in piping networks (manifested by colder areas in daytime) or operating anomalies in the sprinklers (for example, clogged nozzles on pivots that cause warmer circular structures to appear). Many commercial companies are now developing such services, particularly in the United States.

10.5.2. *Climate monitoring and health of ecosystems*

For agricultural areas, as for natural vegetation, surface temperature allows us to monitor large-scale functioning of ecosystems, and their long-term evolution in relation to climate change. In addition to indices that are only based on reflectance in the solar domain [REM 13], indices that incorporate TIR satellite data have also been proposed to characterize the extent and intensity of droughts [MU 13, AND 11b]. In order to anticipate production for food security, early warning or insurance objectives, we can also mention the VHI (vegetation health index [ROJ 11]) and the Agricultural Stress Index System, set up by the FAO (Figure 10.10). Anomaly products of vegetation, rain or evaporation (the latter being based on surface temperature – see for example, the FEWS-net network, http://www.fews.net/fr) delivered by various organizations for the same purpose are also added to this list.

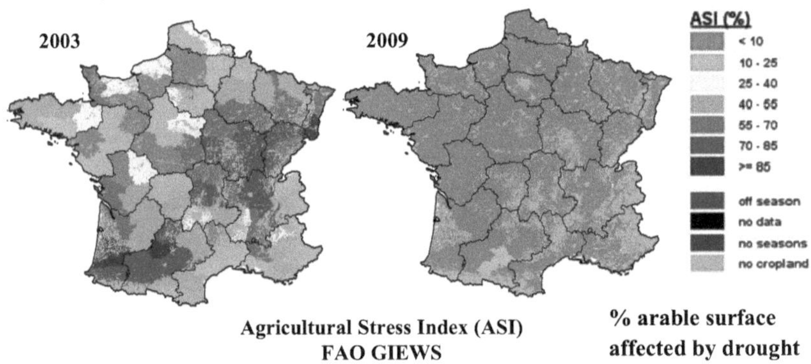

Figure 10.10. *Comparison of the annual ASI over France in 2003 (affected by a drought) and 2009 (without any marked stress episodes). ASI indicates the percentage of arable land affected by drought (according to the FAO,* Global Information and Early *Warning* System on food and *Agriculture GIEWS). Taken from: http://www.fao.org/giews/earthobservation/country/index.jsp?lang=fr&code=FRA#. For a color version of this figure, see www.iste.co.uk/baghdadi/4.zip*

TIR data also contribute to the monitoring of ecosystems that are vulnerable to climate change. In northern regions, detection and mapping of seasonally frozen soils (which are indicators of the global warming process [COM 03, HAC 09]) have become critical for economic reasons (infrastructure damage) but also because of the potential negative feedback effects on climate. On a smaller scale, monitoring riparian forests [AND 12] and wetlands should help authorities manage water supplies (see for example the case of the Yellow River Delta in China [JIA 09]).

10.5.3. *Biogeochemical cycles and soil pollution*

Water is the vector of nutrient transport but also of many contaminants (residues of pesticides and fertilizers, in particular) into soil and water systems. An environmental application that is becoming important is the evaluation of degradation of the quality of water and soil as a result of intensive use of agricultural inputs to increase production. Better control of water cycle and irrigation practices, particularly through the development of robust methods for estimating evapotranspiration, is now essential in order to better adjust water supplies to crop requirements and to improve and maintain water and soil quality. The carbon cycle is of paramount importance with respect to climate change, as is the transfer of other volatile compounds such as nitrogen oxides NO_x, which are also directly linked to transpiration and stomatal closure [COR 04]. Finally, but prospectively, applications to processes of degradation and mobilization of inputs, which are under the joint dependency of moisture and soil temperature [ALL 06, LOU 07], could bc sought.

10.5.4. *Hydrology*

At the continental scale, about 60% of annual precipitation goes back into the atmosphere by evapotranspiration. For mountainous watersheds, the percentage is lower given the significant amount of runoff and subsurface lateral flow.

Continuous monitoring of AET during a succession of dry downs allows us to estimate soil moisture and the amount of remaining water in the root zone and at the surface before a new rainfall event. Those conditions impact runoff mechanisms and subsurface transfers during floods. Hydrological models that simulate flows during a flood require initialization of the soil water content profile. Accurate information on water in the first centimeters of soil allows us to better simulate runoff by excess infiltration, while deep soil moisture is important for monitoring the development of saturated surface areas to know the amount of water present at the beginning of a flood in a soil column above the impermeable layer.

Estimating AET as accurately as possible is also essential for determining the least known water budget terms of these watersheds, including deep transfers that are difficult to measure, except by using expensive geochemical analyses. These deep fluxes can be quantified in a residual manner from rainfall, AET and the observed flow rate. Slow subsurface flows maintain river base flows and ensure the water supply of orchards during summers with high water depletion rates. They also provide important environmental services in the riparian zone.

In large irrigated plains, accurate knowledge of AET allows us to quantify groundwater extraction including deep aquifers with low renewal rates [SAA 15]. In this case, AET maps from low resolution TIR data (for example, MODIS) are particularly useful. At a smaller scale, Anderson *et al.* [AND 12] provide examples of AET maps from Landsat thermal data applied to the evaluation of water access rights for farmers.

Beyond AET, TIR data have other applications in hydrology. These include the mapping and monitoring of surface flows in wadi beds in arid areas using thermal contrasts. However, aerial or UAV monitoring is necessary to obtain an adequate resolution [BUR 08]. Similarly, despite the difficulty of projecting remote sensing data onto surface topography, temperature data can also be used in the spatial monitoring of snow cover. Negative surface temperatures indicate where cooling of the snowpack takes place, while the 0° isotherm can locate areas that are potentially melting. Surface temperature is also a good proxy for climatic conditions and can, under certain conditions, help derive a rough estimate of the impact of the lapse rate and exposure effects on air temperature and hence the rate of melting [BOU 11]. The dependence on water supply for many people living in South America or Asia is very reliant on the amounts of water stored as snow during the spring in the Andes or the Himalayan mountains.

10.6. Prospects

Estimation of AET methods from TIR data enables a number of functional applications, from water management at plot scale to drought mapping up to continental areas [AND 12]. These applications will be more numerous and stronger when we have acquired data with high revisit frequency (ideally one day), and with a resolution that is compatible with agricultural landscapes that are usually very fragmented. Space mission projects such as MISTIGRI, or more recently THIRSTY, aim to fill this gap.

Given the number of sources of error involved in measuring surface temperature, estimates of AET from the instantaneous energy budget are tainted with an inevitable degree of uncertainty. Several ways of improvement are worth exploring. First of all, it is important to continue developing methodologies to control the measurement error itself (improving atmospheric correction algorithms, taking directional anisotropy effects into account, etc.). Another way to reduce the range of uncertainty associated with AET products is to add information from other wavelengths, such as radar data, which are used directly to quantify the soil water status or by implementing assimilation techniques to better constrain surface models (SVAT models). This simultaneously requires continuous improvement of the latter.

10.7. Key points

– There is a close connection between the average radiative temperature, actual evapotranspiration (AET) and the energy budget of continental surfaces. Several methodologies have been developed to estimate AET and characterize water stress using remote sensing data that combine thermal infrared data with information obtained in other spectral ranges.

– Unlike measurements performed in the visible-near infrared domain, a particular feature of surface temperature is to present rapid fluctuations at various time scales (daily cycle, sensitivity to atmospheric turbulence flow), which requires implementing interpretation methods (assimilation, temporal interpolation, etc.) or adapting the conditions of satellite data acquisition (revisit, overpass time, etc.). Progress on methods for estimating AET, as well as the potential availability of TIR data from future satellite missions that combine high spatial resolution and high revisit capabilities, allow us to consider building operational AET products which include an associated estimation error.

10.8. Bibliography

[AGA 07] AGAM N., KUSTAS W.P., ANDERSON M.C. *et al.*, "A vegetation index based technique for spatial sharpening of thermal imagery", *Remote Sensing of Environment*, vol. 107, no. 4, pp. 545–558, 2007.

[ALL 06] ALLETTO L., COQUET Y., BENOIT P. *et al.*, "Effects of temperature and water content on degradation of isoproturon in three soil profiles", *Chemosphere*, vol. 64, pp. 1053–1061, 2006.

[ALL 07] ALLEN R.G., TASUMI M., TREZZA R., "Satellite-based energy balance for mapping evapotranspiration with internalized calibration (METRIC) – Model", *Journal of Irrigation and Drainage Engineering*, vol. 133, no. 4, pp. 380–394, 2007.

[AND 11a] ANDERSON M.C., KUSTAS W.P., NORMAN J.M. *et al.*, "Mapping daily evapotranspiration at field to continental scales using geostationary and polar orbiting satellite imagery", *Hydrology and Earth System Sciences*, vol. 15, no. 1, pp. 223–239, 2011.

[AND 11b] ANDERSON M.C., HAIN C., WARDLOW B. *et al.*, "Evaluation of drought indices based on thermal remote sensing of evapotranspiration over the continental United States", *Journal of Climate*, vol. 24, no. 8, pp. 2025–2044, 2011.

[AND 12] ANDERSON M.C., ALLEN R.G., MORSE A. *et al.*, "Use of Landsat thermal imagery in monitoring evapotranspiration and managing water resources", *Remote Sensing of Environment*, vol. 122, pp. 50–65, 2012.

[BAS 98] BASTIAANSSEN W.G.M., MENENTI M., FEDDES R.A. *et al.*, "A remote sensing surface energy balance algorithm for land (SEBAL). 1. Formulation", *Journal of Hydrology*, vol. 212-213, pp. 198–212, 1998.

[BEG 93] BÉGUÉ A., "Leaf area index, intercepted photosynthetically active radiation, and spectral vegetation indices: sensitivity analysis for regular-clumped canopies", *Remote Sensing of Environment*, vol. 46, pp. 45–59, 1993.

[BER 09] BERNI J.A.J., ZARCO-TEJADA P.J., SEPULCRE-CANTÓ G. *et al.*,"Mapping canopy conductance and CWSI in olive orchards using high resolution thermal remote sensing imagery", *Remote Sensing of Environment*, vol. 113, pp. 2380–2388, 2009.

[BOE 99] BOEGH E., SOEGAARD H., HANAN N. *et al.*, "A remote sensing study of the NDVI-Ts relationship and the transpiration from sparse vegetation in the Sahel based on high-resolution satellite data", *Remote Sensing of Environment*, vol. 69, no. 3, pp. 224–240, 1999.

[BOU 07] BOULET G., CHEHBOUNI A., GENTINE P. *et al.*, "Monitoring water stress using time series of observed to unstressed surface temperature difference", *Agricultural and Forest Meteorology*, vol. 146, nos. 3–4, pp. 159–172, 2007.

[BOU 11] BOUDHAR A., DUCHEMIN B., HANICH L. *et al.*, "Spatial distribution of the air temperature in mountainous areas using satellite thermal infra-red data", *Comptes Rendus Geoscience*, vol. 343, no. 1, pp. 32–42, 2011.

[BOU 12] BOULET G., OLIOSO A., CESCHIA E. *et al.*, "An empirical expression to relate aerodynamic and surface temperatures for use within single-source energy balance models", *Agricultural and Forest Meteorology*, vol. 161, pp. 148–155, 2012.

[BRI 16] BRIOTTET X., "Radiometry in the optical domain", in BAGHDADI N., ZRIBI M. (eds), *Optical Remote Sensing of Land Surfaces*, ISTE Ltd, London and Elsevier Ltd, Oxford, 2016.

[BRU 82] BRUTSAERT W., *Evaporation into the Atmosphere: Theory, History and Applications*, Springer, 1982.

[BUR 08] BURKHOLDER B.K., GRANT G.E., HAGGERTY R. *et al.*, "Influence of hyporheic flow and geomorphology on temperature of a large, gravel-bed river, Clackamas River, Oregon, USA", *Hydrological Processes*, vol. 22, no. 7, pp. 941–953, 2008.

[COM 03] COMISO J.C., "Warming trends in the Arctic from clear sky satellite observations", *Journal of Climate*, vol. 16, pp. 3498–3510, 2003.

[COR 04] CORTINOVIS J., SOLMON F., SERÇA D. *et al.*, "A simple modeling approach to study the regional impact of a Mediterranean forest isoprene emission on anthropogenic plumes", *Atmospheric Chemistry and Physics Discussion*, vol. 4, pp. 7691–7724, 2004.

[COU 05] COURAULT D., SEGUIN B., OLIOSO A., "Review on estimation of evapotranspiration from remote sensing data: From empirical to numerical modeling approaches", *Irrigation and Drainage Systems*, vol. 19, nos. 3–4, pp. 223–249, 2005.

[COU 07] COUDERT B., OTTLE C., "An improved SVAT model calibration strategy based on the optimisation of surface temperature temporal dynamics", *Geophysical Research Letters*, vol. 34, p. L04402, 2007.

[CRE 14] CREBASSOL P., LAGOUARDE J.P., HOOK S., "THIRSTY Thermal InfraRed SpaTial System", *IEEE Geoscience and Remote Sensing Symposium* (IGARSS '14), Québec, Canada, available at http://dx.doi.org/10.1109/IGARSS.2014.6947113, pp. 3021–3024, 13-18 July 2014.

[CRO 08] CROW W.T., KUSTAS W.P., PRUEGER J.H., "Monitoring root-zone soil moisture through the assimilation of a thermal remote sensing-based soil moisture proxy into a water balance model", *Remote Sensing of Environment*, vol. 112, no. 4, pp. 1268–1281, 2008.

[DAI 04] DAI Y.J., DICKINSON R.E., WANG Y.P., "A two-big-leaf model for canopy temperature, photosynthesis, and stomatal conductance", *Journal of Climate*, vol. 17, no. 12, pp. 2281–2299, 2004.

[DEL 12] DELOGU E., BOULET G., OLIOSO A. *et al.*, "Reconstruction of temporal variations of evapotranspiration using instantaneous estimates at the time of satellite overpass", *Hydrology and Earth System Sciences*, vol. 16, no. 8, pp. 2995–3010, 2012.

[DUP 10] DUPONT S., GOSSELIN F., PY C. *et al.*, "Modelling waving crops using large-eddy simulation: comparison with experiments and a linear stability analysis", *Journal of Fluid Mechanics*, vol. 652, pp. 5–44, 2010.

[GAL 13] GALLEGO-ELVIRA B., OLIOSO A., MIRA M. *et al.*, "EVASPA (EVapotranspiration Assessment from SPAce) Tool: An overview", *Procedia Environmental Sciences*, vol. 19, pp. 303–310, 2013.

[GEN 10] GENTINE P., ENTEKHABI D., POLCHER J., "Spectral behaviour of a coupled land-surface and boundary-layer system", *Boundary-Layer Meteorology*, vol. 134, no. 1, pp. 157–180, 2010.

[GEN 11] GENTINE P., ENTEKHABI D., POLCHER J., "The diurnal behavior of evaporative fraction in the soil–vegetation–atmospheric boundary layer continuum", *Journal of Hydrometeorology*, vol. 12, no. 6, pp. 1530–1546, 2011.

[HAC 09] HACHEM S., ALLARD M., DUGUAY C.R., "Using the MODIS land surface temperature product for mapping permafrost: an application to Northern Quebec and Labrador, Canada", *Permafrost and Periglacial Processes*, vol. 20, pp. 407–416, 2009.

[HAI 09] HAIN C.R., MECIKALSKI J.R., ANDERSON M.C., "Retrieval of an available water-based soil moisture proxy from thermal infrared remote sensing. Part I: methodology and validation", *Journal of Hydrometeorology*, vol. 10, no. 3, pp. 665–683, 2009.

[INA 08] INAMDAR S., BOVOLO F., BRUZZONE L. *et al.*, "Multidimensional probability density function matching for preprocessing of multitemporal remote sensing images", *IEEE Transactions on Geoscience and Remote Sensing*, vol. 46, no. 4, pp. 1243–1252, 2008.

[JAN 06] JANG J.D., VIAU A.A., ANCTIL F., "Thermal-water stress index from satellite images", *International Journal of Remote Sensing*, vol. 27, no. 8, pp. 1619–1639, 2006,

[JAR 14] JARLAN L., BOULET G., "Data assimilation for the monitoring of continental surfaces", in TUPIN F., INGLADA J., NICOLAS J-M. (eds), *Remote Sensing Imagery*, ISTE Ltd, London and John Wiley & Sons, New York, 2014.

[JIA 03] JIA L., LI Z.L., MENENTI M. *et al.*, "A practical algorithm to infer soil and foliage component temperatures from bi-angular ATSR-2 data", *International Journal of Remote Sensing*, vol. 24, no. 23, pp. 4739–4760, 2003.

[JIA 09] JIA L., XI G., LIU S. *et al.*, "Regional estimation of daily to annual regional evapotranspiration with MODIS data in the Yellow River Delta wetland", *Hydrology and Earth System Sciences*, vol. 13, pp. 1775–1787, 2009.

[KAL 08] KALMA J.D., MCVICAR T.R., MCCABE M.F., "Estimating land surface evaporation: a review of methods using remotely sensed surface temperature data", *Surveys in Geophysics*, vol. 29, nos. 4-5, pp. 421–469, 2008.

[KAT 98] KATUL G.G., SCHIELDGE J., HSIEH C.I. *et al.*, "Skin temperature perturbations induced by surface layer turbulence above a grass surface", *Water Resources Research*, vol. 34, no. 5, pp. 1265–1274, 1998.

[KER 96] KERDILES, H., GRONDONA M., RODRIGUEZ R. *et al.*, "Frost mapping using NOAA AVHRR data in the Pampean region, Argentina", *Agricultural and Forest Meteorology*, vol. 79, pp. 157–182, 1996.

[KUM 81] KUMAR M., MONTEITH J.L., "Remote sensing of crop growth", in SMITH H. (ed), *Plants and the Daylight Spectrum*, Academic Press, London, 1981.

[KUS 94] KUSTAS W.P., PINKER R.T., SCHMUGGE T.J. *et al.*, "Daytime net radiation estimated from semiarid rangeland basin from remotely-sensed data", *Agricultural and Forest Meteorology*, vol. 71, nos. 3-4, pp. 337–357, 1994.

[KUS 99] KUSTAS W.P., NORMAN J.M., "Evaluation of soil and vegetation heat flux predictions using a simple two-source model with radiometric temperatures for partial canopy cover", *Agricultural and Forest Meteorology*, vol. 94, no. 1, pp. 13–29, 1999.

[KUS 09] KUSTAS W.P., ANDERSON M., "Advances in thermal infrared remote sensing for land surface modeling", *Agricultural and Forest Meteorology*, vol. 149, no. 12, pp. 2071–2081, 2009.

[LAG 00] LAGOUARDE J.P., BALLANS H., MOREAU P. *et al.*, "Experimental study of brightness surface temperature angular variations of maritime pine", *Remote Sensing of Environment*, vol. 72, pp. 17–34, 2000.

[LAG 10] LAGOUARDE J.P., HÉNON A., KURZ B. *et al.*, "Modelling daytime thermal infrared directional anisotropy over Toulouse city centre", *Remote Sensing of Environment*, vol. 114, pp. 87–105, 2010.

[LAG 13] LAGOUARDE J.P., BACH M., SOBRINO J.A. *et al.*, "The MISTIGRI thermal infrared project: scientific objectives and mission specifications", *International Journal of Remote Sensing*, vol. 34, nos. 9-10, pp. 3437–3466, 2013.

[LAG 14] LAGOUARDE J.P., DAYAU S., MOREAU P. *et al.*, "Directional Anisotropy of Brightness Surface Temperature Over Vineyards: Case Study Over the Medoc Region (SW France)", *IEEE Geoscience and Remote Sensing Letters*, vol. 11, no. 2, pp. 574–578, 2014.

[LAG 15] LAGOUARDE J.P., IRVINE M., DUPONT S., "Atmospheric turbulence induced errors on measurements of surface temperature from space", *Remote Sensing of Environment*, vol. 168, pp. 40–53, 2015.

[LOU 07] LOUCHART X., VOLTZ M., "Aging effects on the availability of herbicides to runoff transfer", *Environmental Science & Technology*, vol. 41, pp. 1137–1141, 2007.

[MAT 05] MATSUSHIMA D., "Relations between aerodynamic parameters of heat transfer and thermal-infrared thermometry in the bulk surface formulation", *Journal of the Meteorological Society of Japan*, vol. 83, pp. 373–389, 2005.

[MER 10] MERLIN O., DUCHEMIN B., HAGOLLE O. *et al.*, "Disaggregation of MODIS surface temperature over an agricultural area using a time series of Formosat-2 images", *Remote Sensing of Environment*, vol. 114, no. 11, pp. 2500–2512, 2010.

[MER 14] MERLIN O., CHIROUZE J., OLIOSO A. *et al.*, "An image-based four-source surface energy balance model to estimate crop evapotranspiration from solar reflectance/thermal emission data (SEB-4S)", *Agricultural and Forest Meteorology*, vol. 184, pp. 188–203, 2014.

[MIR 16] MIRA M., OLIOSO A., GALLEGO-ELVIRA B. *et al.*, "Uncertainty assessment of surface net radiation derived from Landsat images", *Remote Sensing of Environment*, submitted, 2016.

[MON 81] MONTEITH J.L., "Evaporation and surface-temperature", *Quarterly Journal of the Royal Meteorological Society*, vol. 107, no. 451, pp. 1–27, 1981.

[MOR 94] MORAN M.S., CLARKE T.R., INOUE Y. *et al.*, "Estimating crop water deficit using the relation between surface-air temperature and spectral vegetation index", *Remote Sensing of Environment*, vol. 49, no. 3, pp. 246–263, 1994.

[MU 13] MU Q., ZHAO M., KIMBALL J.S. *et al.*, "A remotely sensed global terrestrial drought severity index", *Bulletin of the American Meteorological Society*, pp. 83–98, doi: 10.1175/BAMS-D-11-00213.1, 2013.

[NOR 95a] NORMAN J.M., BECKER F.,"Terminology in thermal infrared remote sensing of natural surfaces", *Agricultural and Forest Meteorology*, 77, p. 153-166, 1995.

[NOR 95b] NORMAN J.M., KUSTAS W.P., HUMES K.S., "Source approach for estimating soil and vegetation energy fluxes in observations of directional radiometric surface temperature", *Agricultural and Forest Meteorology*, vol. 77, nos. 3-4, pp. 263–293, 1995.

[OLI 05] OLIOSO A., INOUE Y., ORTEGA-FARIAS S. *et al.*, "Future directions for advanced evapotranspiration modeling: Assimilation of remote sensing data into crop simulation models and SVAT models", *Irrigation and Drainage Systems*, vol. 19, no. 3, pp. 377–412, 2005.

[OTT 94] OTTLÉ C., VIDAL-MADJAR D., "Assimilation of soil-moisture inferred from infrared remot-sensing in a hydrological model over the HAPEX-MOBILHY region", *Journal of Hydrology*, vol. 158, nos. 3-4, pp. 241–264, 1994.

[OTT 16] OTTLÉ C., MAHFOUF J.-F., "Data assimilation of observations from space", in BAGHDADI N., ZRIBI M. (eds), *Microwave Remote Sensing of Land Surfaces*, ISTE Ltd, London, Elsevier Ltd, Oxford, 2016.

[OUA 94] OUAIDRARI H., Utilisation des données NOAA-AVHRR pour modéliser la production primaire du mil dans l'écosystème sahélien (Niger), PhD Thesis, University of Paris, France, 1994.

[PAW 92] PAW U.K.T., "Development of models for thermal infrared radiation above and within plant canopies", *ISPRS Journal of Photogrammetry and Remote Sensing*, vol. 47, pp. 189–203, 1992.

[PEL 02] PELLENQ J., BOULET G., "A methodology to test the pertinence of remote-sensing data assimilation into vegetation models for water and energy exchange at the land surface", *Agronomie*, vol. 24, no. 4, pp. 197–204, 2002.

[RAU 95] RAUPACH M.R., "Vegetation atmosphere interaction and surface conductance at leaf, canopy and regional scales", *Agricultural and Forest Meteorology*, vol. 73, nos. 3-4, pp. 151–179, 1995.

[REM 13] REMBOLD F., ATZBERGER C., SAVIN I. *et al.*, "Using low resolution satellite imagery for yield prediction and yield anomaly detection", *Remote Sensing*, vol. 5, pp. 1704–1733, 2013.

[ROE 00] ROERINK G.J., SU Z., MENENTI M., "S-SEBI: A simple remote sensing algorithm to estimate the surface energy balance", *Physics and Chemistry of the Earth Part B-Hydrology Oceans and Atmosphere,* vol. 25, no. 2, pp. 147–157, 2000.

[ROJ 11] ROJAS O., VRIELING A., REMBOLD F., "Assessing drought probability for agricultural areas in Africa with coarse resolution remote sensing imagery", *Remote Sensing of Environment*, vol. 115, pp. 343–352, 2011.

[SAA 15] SAADI S., SIMONNEAUX V., BOULET G. *et al.*, "Monitoring irrigation consumption using high resolution NDVI image time series. Calibration and validation in the Kairouan plain (Tunisia)", *Remote Sensing*, vol. 7, no. 10, pp. 13005–13028, 2015.

[SAN 03] SANTANELLO J.A., FRIEDL M.A., "Diurnal covariation in soil heat flux and net radiation", *Journal of Applied Meteorology*, vol. 42, no. 6, pp. 851–862, 2003.

[SEG 94] SEGUIN B., COURAULT D., GUERIF M., "Surface temperature and evapotranspiration: Application of local scale methods to regional scales using satellite data", *Remote Sensing of Environment*, vol. 49, no. 3, pp. 287–295, 1994.

[SIM 09] SIMONNEAUX V., LEPAGE M., HELSON D. *et al.*, "Estimation spatialisée de l'Evapotranspiration des cultures irriguées par télédétection. Application à la gestion de l'Irrigation dans la plaine du Haouz (Marrakech, Maroc)", *Sécheresse,* numéro spécial eau et zone arides, vol. 20, no. 1, pp. 123–130, 2009.

[SOB 08] SOBRINO J.A., JIMÉNEZ-MUÑOZ J.C., SÒRIA G. *et al.*, "Thermal remote sensing in the framework of the SEN2FLEX Project: field measurements, airborne data and applications", *International Journal of Remote Sensing*, vol. 29, nos. 17-18, pp. 4961–4991, 2008.

[SU 02] SU Z., "The Surface Energy Balance System (SEBS) for estimation of turbulent heat fluxes", *Hydrology and Earth System Sciences*, vol. 6, no. 1, pp. 85–99, 2002.

[SUN 13] SUN Z., GEBREMICHAEL M., WANG Q., "Evaluation of empirical remote sensing-based equations for estimating soil heat flux", *Journal of the Meteorological Society of Japan*, vol. 91, no. 5, pp. 627–638, 2013.

[TAI 03] TAIT A., ZHENG X. ,"Mapping frost occurrence using satellite data", *Journal of Applied Meteorology*, vol. 42, no. 2, pp. 193–203, 2003.

[VER 07] VERHOEF W., JIA L., XIAO Q. *et al.*, "Unified optical-thermal four-stream radiative transfer theory for homogeneous vegetation canopies", *IEEE Transactions in Geoscience and Remote Sensing*, vol. 45, no. 6, pp. 1808–1822, 2007.

[ZHA 13] ZHAN W.F., CHEN Y.H., ZHOU J. *et al.*, "Disaggregation of remotely sensed land surface temperature: Literature survey, taxonomy, issues, and caveats", *Remote Sensing of Environment*, vol. 131, pp. 119–139, 2013.

11

Remote Sensing Data Assimilation: Applications to Catchment Hydrology

11.1. Introduction

The modeling of water and energy transfer at the interface between the surface and the atmosphere is not only essential for a good representation of hydrological phenomena, but also for anticipating their response to climatic and anthropogenic stress. In this context, several Soil–Vegetation–Atmospheric Transfer (SVAT) models have been developed since the late 1980s, in order to represent the processes operating in the first decimeters of soil, exclusively emphasizing the vertical interactions between water and heat. For certain hydrological applications, SVAT models may lack horizontal spatial redistribution since the lateral flows of water are not modeled [WOO 91]. The simplifications present in these models can cause errors in the estimation of the surface soil moisture, and therefore in the spatial redistribution of the water.

In order to improve the modeling of hydrological processes, some hydrological models take into account the horizontal redistribution of the water, which allows modeling of the lateral exchanges of water in addition to a river network implementation. It is therefore possible to more accurately model the surface runoff and the lateral subsurface flow until it is intercepted by a watercourse.

Chapter written by Delphine LEROUX and Thierry PELLARIN.

If the assimilation of satellite observations has been strongly developed for SVAT models, initiatives were less common for hydrological models. The goal of this chapter is to provide an overview of their assimilation into hydrological models, in particular for a better representation of the discharge at the outlet of watersheds. The assimilation of satellite observations can play an important role during extreme hydrological events, for example during major floods.

Many examples of assimilation will be presented in this chapter: soil moisture, river water depth and water storage. A case study is also presented at the end of the chapter with an assimilation of the surface soil moisture for a West African watershed.

11.2. Hydrological models

Hydrological models are designed to represent water and energy transfer within a region or a watershed. These models are essential to study and model hydrological processes, as well as the impact of modern anthropogenic factors such as changing land use, deforestation, or even the temperature increase on the water cycle. Mass and energy conservation laws are used to describe spatial and temporal variations in a hydrological system.

There are two main families of hydrological models: conceptual (or empirical) and physically-based models. The simplest conceptual models consider the watershed as a single entity and connect precipitation with the outlet discharge. The model parameters then represent spatial averages of the basin. The advantage of this type of model is the simplicity and the low computing cost. In contrast, physically-based models define equations, sometimes complex ones, representing the spatial variability of the different hydrological processes. This physical description allows a more precise study of the spatial distribution of different variables, such as the presence of snow, the run-off, the volume of groundwater, etc., as well as their impact on the hydrological cycle of the basin. We also find models that fall between those two approaches, which focus on replicating the dominant processes (runoff, evaporation, karstic flow, etc.) while conceptualizing others: we refer to them as conceptual physically-based models or semi-conceptual models.

Through the example of the distributed hydrological model Distributed Hydrology Soil-Vegetation Model (DHSVM) [WIG 94] which will be used for the application at the end of the chapter, Figure 11.1 schematically shows the key processes generating flows of water and energy within a watershed, and the way in which they are represented in a distributed hydrological model with a physical basis. Forced by meteorological conditions (precipitation, radiation, humidity and air temperature, wind, pressure) which represent the model inputs, different water fluxes are described according to the geometrical characteristics of the watershed (topography, slopes), soil types (texture, organic matter content) and the presence of vegetation (forests, herbaceous vegetation, crops). All of these characteristics represent the set of parameters that need to be calibrated according to the study area (using samples collected on the field, for example). This way, the model can estimate the amount of runoff and drainage (water that the plant can use or that will feed the deeper aquifers). The model will also calculate the evapotranspiration rate, and *ultimately*, the water discharge. All of these different variables simulated by the model show the situation of the basin studied and are usually called state variables. Depending on the study, some state variables will be more emphasized than others, even if they are all linked in the model.

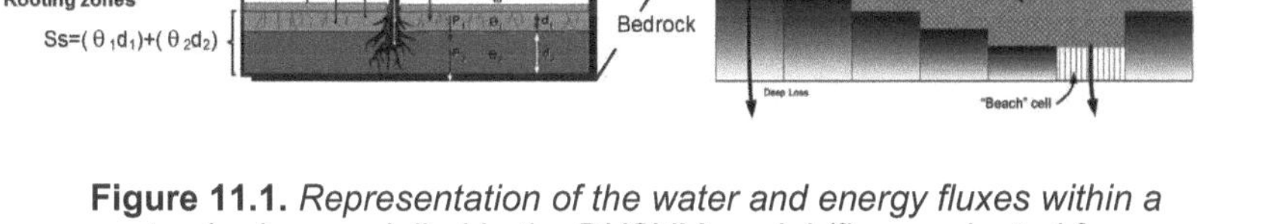

Figure 11.1. *Representation of the water and energy fluxes within a watershed as modelled in the DHSVM model (figure adapted from: www.hydro.washington.edu/Lettenmaier/Models/DHSVM/overview.shtml)*

11.3. Satellite data available for assimilation

Useful observations for hydrology at the watershed scale can be divided into two complementary categories: field measurements or *in situ* measurements allowing continuous monitoring of a particular point, and remote sensing observations (satellite or airborne) offering a larger spatial coverage. In terms of spatial and temporal resolutions, the needs are not the same if the study focuses on a watershed of a few km^2 or one of several thousand km^2. In spite of the global coverage that remote sensing observation provides, a fine spatial resolution is often necessary to accurately represent the heterogeneity of the study zone. The temporal resolution is also an important element which is required to correctly describe the vegetation dynamics. Continuous improvements in the quality and resolution of the satellite products during the last few years have made it easier to integrate this information into hydrological models and opened new research perspectives.

Adapted from Lahoz and Lannoy [LAH 14], Table 11.1 details some of the current space missions (in 2015) relevant to the observation of the hydrological cycle. Space missions dedicated to the monitoring of vegetation such as MODIS, Landsat or even PROBA-V could be added to this list due to their connection to the water cycle *via* transpiration.

Hydrological variable	Observation technique	Time resolution	Spatial resolution	Observation conditions	Samples of space missions
Precipitation	Thermal infrared	hour	4 km	Tropical convective clouds only	GOES, MODIS
		1 day	1 km		AVHRR
		15 days	60 m		Landsat, ASTER
	Passive microwave	3 h	10 km	Land calibration problems	TRMM, SSM/I, AMSR-E, GPM
	Active microwave	1 day	10 m		TRMM, GPM
Surface soil moisture	Passive microwave	1-3 days	25-100 km	Limited to sparse vegetation, low topographic relief	AMSR-E, SMOS, Aquarius, SMAP
	Active microwave	3 days	3-25 km	Significant noise from vegetation and surface roughness	ERS, ASCAT
		30 days	10 m		Sentinel-1, Radarsat, PALSAR, ASAR
Surface skin temperature	Thermal infrared	1 h	4 km	Cloud contamination, soil and vegetation average	GOES, MODIS
		1 day	1 km		AVHRR
		15 days	60 m		Landsat, ASTER

Snow cover	Visible/ thermal infrared	1 h	4 km	Cloud contamination, vegetation masking, bright soil problems	GOES
		1 day	500 m – 1 km		MODIS, AVHRR
		15 days	30-60 m		Landsat, ASTER
SWE (*Snow Water Equivalent*)	Passive microwave	1-3 days	10 km	Limited penetration depth	AMSR-E
	Active microwave	30 days	100 m	Limited spatial coverage	SCLP, Cryosat-2
Water level	Laser	10 days	100 m	Cloud contamination	ICESAT, ICESAT-2, DESDynl
	Radar	7-30 days	100 m – 1 km	Limited tolarge rivers	TOPEX/POSEIDON, SWOT
Total water storage changes	Change in Earth's gravity	30 days	1 000 km	Change of water storage at large scale	GRACE
Evapotranspiration	Visible/ thermal infrared	1 h	4 km	Significant assumptions	GOES, MODIS
		1 day	500 m – 1 km		AVHRR
		15 days	30-60 m		Landsat, ASTER

Table 11.1. *Non exhaustive list of space missions for the monitoring of the water cycle (adapted from Lahoz and Lannoy [LAH 14])*

11.4. Description of data assimilation

The choice of the assimilation method and its implementation is a whole study area and cannot be summarized in this chapter. The publication *Data Assimilation Making sense of observations*, Lahoz *et al.* [LAH 10] provides a very good overview of the different methods of assimilation as well as the various applications at multiple scales in Earth science (Ottlé and Mahfouf, Chapter 8, Volume 2). Here, the most common methods are described without explicitly explaining all of the mathematical developments and all the assumptions applied.

To predict change in a system (for example, the development of a flood), we start from a theoretical state of the system (called background associated with an error $\varepsilon^b = x_b - x_{true}$) which is then refined based on available observations. This refined state (called analysis x_a, with an error

$\mathcal{E}^a = x_a - x_{true}$) is therefore the background to which a correction has been added. This can be mathematically formalized as follows:

$$\begin{cases} x_a = x_b + \text{correction} \\ \text{correction} = K\big(z - \mathcal{H}(x_b)\big) \end{cases} \quad [11.1]$$

where z represents the observations and $\mathcal{H}(x_b)$ the model simulations interpolated on the observation grid. When the observed variable is not one of the model variables (i.e.: observation of a radiance for the soil temperature), the operator $\mathcal{H}$ is not only one interpolator allowing to go from one grid to another, but it is also a model (or *forward model*) transforming a model state into an observation variable (for example, a radiative transfer model). $\big(z - \mathcal{H}(x_b)\big)$ is therefore the difference between the observations and the simulations, also called the innovation, which is weighted by a coefficient or gain K, to balance the confidence between the simulations and the observations. If a single variable is observed and only one point of the model grid is concerned, the gain could be reduced to a simple ratio:

$$x_a = x_b + \frac{\sigma_{model}}{\sigma_{model} + \sigma_{obs}}\big(z - \mathcal{H}(x_b)\big) \quad [11.2]$$

with σ being the model or the observation uncertainty. If the observation is very accurate, then σ_{obs} tends to 0 and K tends to 1, replacing the simulation x_b by the observation z. On the contrary, if the model is very accurate compared to the observation, the gain K tends to 0 and the simulation $x_b$is almost not adjusted.

When several observations are available and where several grid points of the model are involved, we talk about state vectors x and gain matrix K. The simple case where model uncertainties and observations are assumed to be known is called optimal interpolation (OI), which is formalized as follows:

$$\begin{cases} x_a = x_b + K\big(z - \mathcal{H}(x_b)\big) \\ K = BH^T(HBH^T + R)^{-1} \\ A = (I - KH)B \end{cases} \quad [11.3]$$

with B and R being the error covariance matrices of the model state and of the observations (square matrices whose sizes depend on the number of variables in the state vector x_b and on the number of observations in vector z). Since the operator $\mathcal{H}$ is not supposed to be linear (which is often the case for observation models), it is not possible to formalize it as a matrix, and we use instead its linearized form H, its Jacobian (partial derivatives, $H = \partial\mathcal{H}/\partial x$, rectangular matrix allowing to go from the "model" space to the "observation" space). If the state variable is directly observed (example: water level), then $\mathcal{H}$ and $H$are merged, assuming the interpolation is linear as well. These formulas for x_a and K are obtained through the BLUE methodology (*Best Linear Unbiased Estimator*), in order to minimize the variance of the analysis ($tr(A)$). It is also assumed that the observation and the model errors are Gaussian and unbiased.

In the OI case, the error covariances of the model state and of the observations (B and R) are supposed to be known. In reality, it is impossible to determine exactly the errors due to the complexity of the models (errors in the climate forcing, modeling errors, parameterization errors, etc.) and of the observation system (instrumental noise, calibration error, etc.).

The most commonly used assimilation scheme is the Kalman filter (KF, [KAL 60]) where the matrix of covariance error of the model is propagated through the model in the same way as the state variables. It is therefore updated every time step k, according to the previous iteration and estimated model accuracy:

$$\begin{cases} \text{Prediction from } (\mathrm{k}-1) \text{ to } (\mathrm{k}): \\ x_b^k = F^k x_a^{k-1} \\ \text{Propagation error from} (\mathrm{k}-1) \text{ to } (\mathrm{k}): \\ B^k = F^k A^{k-1} \left(F^k\right)^T + Q^k \\ \text{Update at } (\mathrm{k}): \\ K^k = B^k \left(H^k\right)^T \left(H^k B^k \left(H^k\right)^T + R^k\right)^{-1} \\ x_a^k = x_b^k + K^k\left(z^k - H^k x_b^k\right) \\ A^k = \left(I - K^k H^k\right) B^k \end{cases} \qquad [11.4]$$

with F^k being the transition model going from time step (k-1) to (k), Q and R being the model and the observations errors. The KF was originally developed for linear models F, which can therefore be formed as matrices in the equations above. In the same way as for the observation operator $\mathcal{H}$, the transition model can be non-linear, and the Jacobians of these models are used instead ($F = \partial\mathcal{F}/\partial x$ and $H = \partial\mathcal{H}/\partial x$). This is referred to as the Extended Kalman Filter (EKF). This method provides the ability to tackle the condition of nonlinearity, provided that the model is not strongly non-linear between two observations. It is still important to note that the EKF is very costly in terms of computation time, especially because of the number of operations required to calculate the Jacobian.

Finally, it is possible to evaluate these error covariance matrices (instead of estimating them until now) by generating assemblies that evolve in time using the model $\mathcal{F}$. This method is called the Ensemble Kalman Filter (EnKF, [EVE 94]). The error covariance matrix$B$is then evaluated for each time interval (indices omitted in the following equations) using the statistical dispersion of the set composed of$N$members:

$$\begin{cases} B = \frac{1}{N-1}\sum_{i=1}^{N}\left(x_b^i - \langle x_b\rangle\right)\left(x_b^i - \langle x_b\rangle\right)^T \\ BH^T = \frac{1}{N-1}\sum_{i=1}^{N}\left(x_b^i - \langle x_b\rangle\right)\left(\mathcal{H}(x_b^i) - \langle\mathcal{H}(x_b)\rangle\right)^T \\ HBH^T = \frac{1}{N-1}\sum_{i=1}^{N}\left(\mathcal{H}(x_b^i) - \langle\mathcal{H}(x_b)\rangle\right)\left(\mathcal{H}(x_b^i) - \langle\mathcal{H}(x_b)\rangle\right)^T \\ K = BH^T(HBH^T + R)^{-1} \\ A = (I - KH)B \end{cases} \quad [11.5]$$

The number of elements in the ensemble must be large enough so that the calculations have statistical meaning. Most applications use between 10 and 100 members, which multiplies the computing time by the same number. Nonetheless, this remains reasonable compared to the EKF. In the case of the EnKF, no linearization or Jacobian estimation is needed for the calculation of the Kalman gain K, which makes it one of the most widely used methods.

Here are a few lines to summarize some of the assimilation methods mentioned so far:

– Optimal interpolation (OI): very simple method, requires little calculation time, no linearity assumption for model but needs to specify the covariance error matrices B and R, optimal method only when these matrices are perfectly defined (never the case, unless the observations are synthetic and the model is perfectly known);

– Kalman Filter (KF): B matrix changes with the model at each time step in order to better estimate its error, linearity assumption for models $\mathcal{F}$ and $\mathcal{H}$, optimal method only in cases where noises are white (Gaussian distribution with a zero mean value);

– Extended Kalman Filter (EKF): sub-optimal method with linearization of models $\mathcal{F}$ and $\mathcal{H}$ involving intensive partial differential calculations (can be a good alternative to KF as long as the observations are not too far apart in time, so that the hypothesis of linearity between two observations is valid);

– Ensemble Kalman Filter (EnKF): sub-optimal method to generate a set of noisy states used to calculate the error of the model B. No linearization is necessary, less time-consuming in computing time than the EKF, but it is necessary to control the ensemble dispersion (if it is too small, the model will face too much weight towards observations, if there is an excessive dispersion, the model can diverge). A variant of the widely used EnKF is the particule filter, where members of the ensemble do not have the same weight; their weight depends on their location in the distribution function.

Assimilation can be applied sequentially or in a variational method (Ottlé and Mahfouf, Chapter 8, Volume 2). In the sequential method, the model runs until an observation is available. The Kalman gain is therefore calculated based on the model and observation uncertainties in order to update the model variables. The model starts again using the adjustments and runs until the next available observation (Figure 11.2). For the variational method, a temporal window is defined, and may contain several observations. The model is stopped before the beginning of each temporal window in order to adjust the model variables. This provides the optimum match between the simulations and the observations (weighted according to their uncertainties).

The counterpart of the BLUE analysis by variational optimization can be formalized as follows:

$$\begin{cases} x^a = \arg\min \mathcal{J}(x) \\ \mathcal{J}(x) = (x - x^b)^T B^{-1}(x - x^b) + (z - \mathcal{H}(x))^T R^{-1}(z - \mathcal{H}(x)) \end{cases} \quad [11.6]$$

where $\mathcal{J}$ is the cost function that we aim to minimize by varying the analysis x^a. Once the cost function is well defined, only the minimization method remains to be chosen (gradient method, Newton method, conjugate gradient, etc.). When the time variable is not involved (as in OI, matrices B and R are fixed), we talk about the 3D-Var method. When a time dimension is involved, we instead use a method called 4D-Var.

As a comparison, the KF and 4D-Var methods are equivalent (they provide the same results) if the models $\mathcal{F}$ and $\mathcal{H}$ are linear. In the case of a small nonlinearity, we would rather choose the EKF or the incremental 4D-Var (not detailed here) using the Jacobian forms of models $\mathcal{F}$ and $\mathcal{H}$.

Assimilation can be used for two reasons:

– the calibration of model parameters;

– the update of model state variables.

This method only makes sense if a connection can be established between what is observed and what is simulated by the model. A model sensitivity study should also be performed before and to ensure that the model is sensitive to the parameters/variables we wish to adjust. If there is not, the assimilation will provide no additional information. However, if the model is sensitive, the assimilation can improve the simulations and the forecasts.

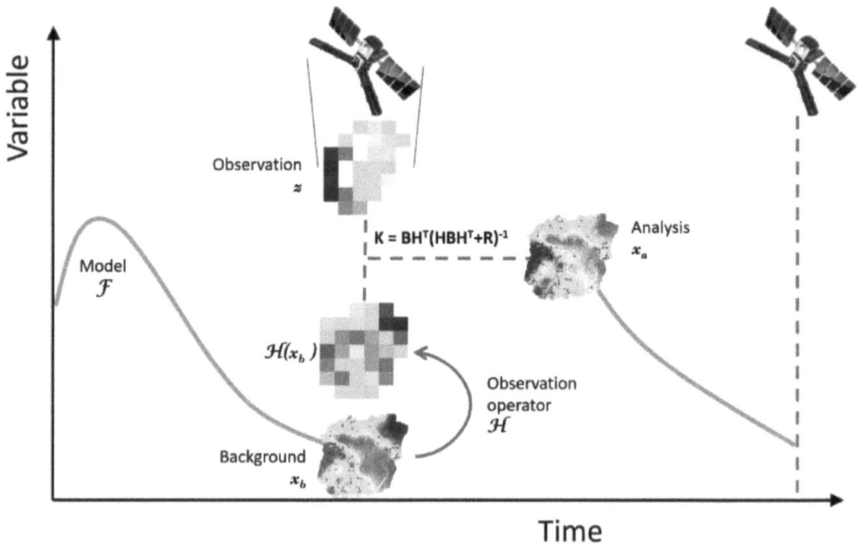

Figure 11.2. *Diagram of sequential assimilation of a spatial observation with adjustment of the observed state variable*

11.5. Examples of assimilation in hydrological models

Assimilation in hydrological models has experienced a growing development in the last two decades. Numerous studies assimilate a variety of variables observed from space: soil moisture, snow cover, snow water equivalent (SWE), soil temperature, evapotranspiration, leaf area index (LAI), free water level, water storage. Some examples are presented later in the course of this chapter, focusing especially on the assimilation of soil humidity, water depth of rivers and the global water stock in soils. The last section presents a more detailed example of the assimilation of humidity products, SMOS, in a hydrological model on a watershed in Benin, West Africa.

11.5.1. *Assimilation of surface soil moisture*

The surface soil moisture plays an important role in the processes that govern the water and energy balance, since it is located at the intersection of the surface and the atmosphere. It is also involved in partitioning the incoming solar and atmospheric radiations into latent, sensible and ground heat fluxes (energy balance), as well as in the partitioning of precipitation into surface runoff and infiltration towards deeper soil layers (water balance).

Accurate simulations of soil moisture would theoretically allow better modeling of the various hydrological processes. Nonetheless, better forecasting has not yet been formally demonstrated by the assimilation of soil moisture in hydrological models, due to certain aspects that are still not well understood in terms of data assimilation. Among these difficulties, we can mention the quantification of model errors and observation errors (matrices B and R), the difference in scale between the observations and the model, the choice of the assimilation method, or the high dependence of the assimilation performance on soil and climatic conditions [BRO 12]. Today, the assimilation is implemented on a case-by-case basis and is intrinsically linked to a model with performances that differ with the study area. For this reason, the results presented in this chapter can have very different conclusions that are specific to the choices and assumptions of the authors.

A number of soil moisture assimilation studies were conducted before the 2000s. Ottlé and Vidal-Madjar [OTT 94] were the first to assimilate soil moisture, retrieved from measurements in the thermal infrared, in order to improve the modeling of evaporation fluxes for a watershed in southwestern France. Thereafter, Houser *et al.* [HOU 98] assimilated retrieved soil moisture from airborne radar observations over the Walnut Gulch watershed in the United States using the TOPLATS model. Galantowicz *et al.* [GAL 99] assimilated soil moisture derived from passive microwave measurements at Beltsville in a soil propagation model. From the 2000s onwards, several studies have shown the benefit of soil moisture assimilation in the simulation of streamflow at various watershed outlets. By modifying the water content of the soil layers, the surface runoff and the infiltration of the water into the deeper layers are also modified. Ultimately, the volume of water going to the river is impacted and the stream flow is modified. We can also mention the studies of Pauwels *et al.* [PAU 01, 02], where the soil moisture retrieved from radar observations of satellites ERS-1 and ERS-2 have been assimilated into the TOPLATS model. Their works found that it was sufficient to assimilate the mean and the standard deviation of soil moisture (instead of the observed values) in order to improve the estimation of the streamflow in a non-distributed model, and that the assimilation in the distributed version of the model does not improve the performances significantly. However, these results are very specific to the experiment and to the chosen model. We will later see that some studies prefer the assimilation of observed values in more complex assimilation schemes, which will provide improved streamflow simulations.

As it has been shown by Reichle *et al.* [REI 08], assimilation of soil moisture in a hydrological model can only have a positive impact if the model and the observation uncertainties (matrices B and R) are correctly defined. It is essentially because of this reason that many studies use the EnKF method, provided that they have the necessary computing resources to generate an ensemble of sufficient size [REI 02]. Sufficient size is defined in a statistical sense, which means that a fixed number cannot be given, since it is the user who will define the "confidence" that needs to be achieved for these results: the larger the ensemble, the more confident the results (for example: an average does not make sense if a large number of elements are are not used). It should still be noted that most of studies require at least

10 members. In the case that the ensemble is not sufficiently populated, the results of the assimilation can even degrade the performance of the model.

Another application of soil moisture assimilation concerns flood simulation, which requires accurate modeling of surface runoff. Brocca *et al.* [BRO 10] have assimilated ASCAT/METOP water content indices in the rainfall–runoff model MISDc *(Modello Idrologico Semi Distribuito in continuo)* over several sub-basins of the upper part of the Tiber watershed in Italy using the method of *nudging*, which makes it possible to manually and artificially push the simulations towards the observations (already used in [PAU 01]). The results show an improvement in the simulation of streamflow, which demonstrates the benefit of having spatial information even at low spatial resolution. In a later study [BRO 12], ASCAT surface and root zone soil moisture were assimilated by EnKF in the same model (MISDc) over the Niccone sub-basin of the Tiber watershed. The results showed a clear improvement of streamflow, thanks to the root zone soil moisture assimilation, while the assimilation of the surface soil moisture had very little impact in this case.

An interesting study was conducted by Han *et al.* [HAN 12] on the Upper Cedar Creek watershed, Indiana, United States of America. In order to quantify the impact of surface soil moisture assimilation on other hydrological components, the authors intentionally perturbed the precipitation in the semi-distributed hydrological Soil and Water Assessment Tool (SWAT) [ARN 05] model. This study showed that the assimilation of soil moisture, only at the surface level, allowed connection of the perturbed precipitation data. Soil moisture at the surface, as well as at deeper layers, have a real impact on the other hydrological variables such as the evapotranspiration, the groundwater level, or its amount. However, little impact has been observed on the simulation of streamflow (Chen *et al.* [CHE 11]) using the same model on the Cobb Creek basin in Oklahoma, United States. This result can be explained by the nature of the SWAT model based on a rainfall–runoff mechanism, with low sensitivity of the runoff to soil moisture. In previous studies, it had been already demonstrated that soil moisture assimilation could only correct for perturbation of the model inputs (precipitation and evapotranspiration) lower than 10% [FRA 03].

Many available soil moisture products usually have a spatial resolution of around 25–40 km (AMSR-E, SMOS, SMAP, ASCAT). Although, it should be noted that the SMAP disaggregated soil moisture product would have been distributed on a 9 km grid without the radar failure which occurred a few weeks after launch. Yet, to properly model the hydrological processes within the watersheds, hydrologic models typically run at much finer spatial resolutions and this can raise scale issues. Several solutions of assimilation at different scales were studied in De Lannoy *et al.* [DEL 10]. Assimilation can be performed at the model resolution or at the observation resolution, pixel by pixel or using an influence radius (pixels closer to the observation are impacted with more weight in matrix H than those that are further). The authors concluded that the most effective method in their case study (assimilation of a snow water equivalent product into a surface model) was the method that used an influence radius because it provided a better transition between the observation pixels (*"edge effect"* is decreased) and paid more respect to the finer model spatial variability (Figure 11.3). Sahoo *et al.* [SAH 13] confirmed that the method using an influence radius was the most. It is not necessarily more efficient in terms of statistical scores, but it preserves the spatial distribution provided by the model (right panel of Figure 11.3).

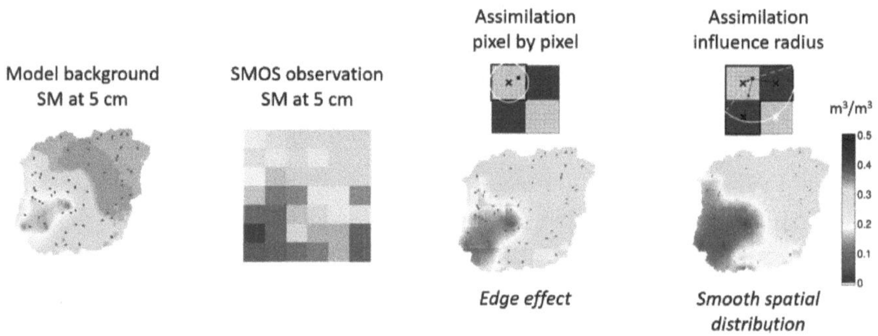

Figure 11.3. *Scale choice to perform the assimilation of the observations: pixel by pixel with an edge effect, or using an influence to keep the spatial distribution provided by the model (black and white diagrams adapted from [DEL 10]). For a color version of this figure, see www.iste.co.uk/baghdadi/4.zip*

Ridler *et al.* [RID 14] addressed the issue of a possible bias between the observations and the model simulations. In this study, the authors compare different techniques of matching the averages of the observations and the model simulations, their standard deviations, their Cumulative Density

Function (CDF) or using copulas. The authors showed that the bias in SMOS soil moisture products (compared to their model simulations) depended on the type of land use or *land cover* (LC, determined by Landsat thermal image at higher resolution) and that it could be taken into consideration at the assimilation step. The simulated soil moisture at 5 cm depth after assimilation was closer to the field measurements (30 stations distributed over the entire basin in this study), but the simulated streamflow was overcorrected by the assimilation. A clear protocol for bias detection and removal has not yet been established and remains a research subject to this day.

At larger scales, several other studies demonstrate the benefit of surface soil moisture assimilation for better simulation at deeper soil layers, in particular in the root zone [DRA 12, REN 14]. Soil moisture assimilation provides the opportunity to change the water content of the whole soil column, which then modifies the groundwater stock, the runoff and the river flows. This positive impact on deeper layers can also lead to other benefits such as a better estimate of drought areas [KUM 14] and agricultural yield [CHA 14].

Data assimilation is not necessarily limited to observations of a single variable; it is also possible to assimilate several variables. In order to illustrate this principle, Aubert *et al.* [AUB 03] proposed the assimilation of soil moisture and streamflow measurements for a sub-basin of the Seine watershed in France with a special focus on flood events. The assimilation method chosen by Aubert *et al.* [AUB 03] was the EKF and the conceptual hydrological model used was the GR4J (*Génie Rural 4 paramètres Jouenalier*, [PER 03]). This study showed that only the assimilation of streamflow is effective in normal hydrological conditions, while the assimilation of only the soil moisture was more effective in extreme hydrological conditions encountered during flood events. These results appear natural since in the case of low precipitations, the streamflow is essentially determined by what comes from upstream and the assimilation of this variable is rather efficient, while in the case of heavy precipitations, the assimilation of soil moisture is a good tool to control the amount of water entering the system. Therefore, it was demonstrated that the combined assimilation of soil moisture and streamflow was very efficient at forecasting streamflow. Although this study was very local within a relatively short period of time using a specific model, it showed that the joint assimilation of several hydrological variables (whether they are distributed or not), can allow better monitoring, better simulation and better prediction of the hydrological cycle in a watershed.

11.5.2. *Assimilation of river water level*

Over the past few decades, measurements of altimetry satellites such as TOPEX and JASON have successfully been used to estimate the water depth variations of the largest rivers on Earth (width > 1 km). With the recent Saral-AltiKa and the preparation of the SWOT mission (*Surface Water and Ocean Topography*, joint mission between JPL-NASA and CNES), the technical progress in measuring will allow extending this measure to all rivers whose width exceeds 100 m. In hydrology, the estimation of river flows is critical but often limited due to a lack of field observations. The establishment of an *in situ* measurement network on a global scale is very expensive; radar altimetry is a promising technique, facilitated by the development of radar interferometry.

Manning's Equation allows connecting the water depth to a flow rate, but it requires perfect knowledge of the morphology of the river as well as its friction coefficient that affects flow velocity:

$$Q = \frac{AR^{2/3}\sqrt{S}}{n} \qquad [11.7]$$

with Q the flow rate [m^3/s], A the area of the submerged section [m^2], R the hydraulic radius (equal to A/P where P is the perimeter of the submerged area, [m]), S is the slope of the river [m/m], and n is the roughness coefficient [$s/m^{1/3}$]. It is possible to estimate, to some extent, the geometry of the riverbed with the use of a very accurate DTM, but it is still difficult to determine the river topography (bathymetry). The friction coefficient or Manning's coefficient is usually a major source of uncertainty for hydraulic or hydrological models. In addition, the geometry of the river and the friction coefficients are subject to change over time: after heavy rainfall or after extreme rainfall events, for example. In this context of variable geometry, hydraulic models are often preferred to more traditional models with fixed geometries. Most studies use a hydrological model to estimate the water volume entering the waterbed and this input is used to force the hydraulic model. Some key findings on the assimilation of water levels are presented here.

In contrast to soil moisture, which is directly affected by the rainfall amount, the water depth or flow of a river is a consequence of numerous processes following the rainfall event, such as surface runoff, evaporation and infiltration. In a sense, water depth is a more integrated variable in space and time than soil moisture.

Let's start with some examples of model calibration through assimilation.

Pedinotti *et al.* [PED 14] assimilate synthetic water levels of the future SWOT satellite to optimize Manning's coefficient n in the ISBA–TRIP model (*Interaction Soil–Biosphere–Atmosphere*, [NOI 89] *–Total Runoff Integrating Pathways* [OKI 98]) by EKF on the Niger catchment. This study shows the possibility of using future SWOT observations in order to obtain a better model parameterization. They show that after 18 months, Manning's coefficient converges to a value that reduces the relative bias on water depth by 30%. The impact on the flow simulations is less important (reduction of the relative bias of 7%) because the amount of water supplied by the ISBA model to the Niger river is not corrected by assimilation. Schumann *et al.* [SCH 07] have already shown interest in spatially optimizing the friction coefficient (not assuming the same value along the entire length of the river) in the river Alzette in Luxembourg, using the SAR observations of satellites ERS-2 and ENVISAT during the major flood that took place from January 2 to January 5, 2003. In this study, the friction coefficient was optimized iteratively not using an assimilation scheme.

Over the same Luxembourg watershed, Montanari *et al.* [MON 09] suggest calibrating another parameter of the hydrological model, which is the ratio between the water flowing to the river and the rainfall amount, called the *storm flow coefficient* C. This coefficient C represents the water status of the basin, with a more important runoff if the soil is closed to the saturation, and therefore it needs to be updated using the ERS-2 and ENVISAT SAR observations (it is updated iteratively here). The other parameters are assumed to be fixed and are calibrated with field data.

At a larger spatial scale, Durand *et al.* [DUR 08] study the contribution of assimilation of future SWOT water depth measurements in the hydrodynamic model LISFLOOD-FP [BAT 00] on the Amazon basin. Assuming that the bathymetric depth and slope do not evolve over time, these two parameters are estimated by EnKF (20 members are used in this study). They were also able to observe that the slope was estimated with an error of 3 mm/km and that the absolute error of the bathymetric depth at the outlet was 56 cm, which represents a decrease of 84% compared to the simulation without assimilation.

We can continue with other examples of model variable correction using assimilation schemes.

In Giustarini *et al.* [GIU 11], which follows the studies of Montanari *et al.* [MON 09] and of Matgen *et al.* [MAT 10], a hydrological model and a hydraulic model are coupled. The hydrological model CLM 2.0 (*Community Land Model* [DAI 03]), forced by meteorological forcing, is coupled with the hydraulic model HEC-RAS 4.0 (*Hydrologic Engineering Center River Analysis System*, http://www.hec.usace.army.mil/software/hec-ras/index.html). The particle filter assimilation method (variant of the EnKF, where each member of the ensemble is assigned a weight depending on the statistical distribution of the ensemble) is used with SAR observations of ERS-2 and ENVISAT satellites in a perfectly calibrated model with the objective of improving water depths along river Alzette in Luxembourg, and thereby to improve flood simulations. This study shows the benefit of satellite observations in providing information throughout the entire river length and not on a single site. However, this creates additional difficulties, particularly due to the irregular temporal sampling as well as in the error estimation (hence the use of an ensemble).

On a broader spatial scale, the Ohio River in the United States has also been studied within the framework of the future SWOT mission by Andreadis *et al.* [AND 07]. Virtual observations of water levels were assimilated by EnKF (20 members) in the hydrodynamic model LISFLOOD-FP. Unlike the above mentioned studies, the roughness coefficients are assumed to be uniform and are set beforehand. Water inflows are supplied by the surface model VIC (*Variable Infiltration Capacity* [LIA 94]). Through assimilation, the adjusted state variables are water depth and flow.

After assimilation, the root-mean-square error on the water height decreases from 56 cm to 22 cm, and that of the flow decreases from 162 m^3/s (relative error of 23%) to 76 m^3/s (relative error of 10%), thus demonstrating the positive impact of assimilation. Note that synthetic observations are used and not real observations.

Biancamaria *et al.* [BIA 11] studied the bottom portion of the river Ob in the Arctic. Similarly, for this study, synthetic observations of the SWOT satellite are assimilated by LEnKS (*Local EnKF Smoother*, variant of the Kalman ensemble with sliding time interval) in the ISBA-LISFLOOD-FP coupled model. The assimilation operates in the hydraulic portion and it is assumed that the roughness coefficient is uniform and does not vary in time. In order to account for the water velocity in the river (measuring more than 1,100 km), a temporal sliding window of a few days is introduced into the method (hence the name *Smoother* in the assimilation method). The size of this window is optimal for a three-day duration, while the water takes 10 days to travel from the study area. After assimilation, the spatial root-mean-square error for water depth decreases from 80 cm to 33 cm, and the average error in time from 111 cm to 38 cm. This is an additional example of the positive impact of assimilation of water depths in coupled hydrological/hydraulic models. Nonetheless, it is important to emphasize once again that the errors were estimated and that an incorrect estimation can lead to biased results.

11.5.3. *Soil water storage assimilation*

Since the launch of the GRACE mission (*Gravity Recovery and Climate Experiment*, a joint mission between NASA and DLR) in March 2002, Earth's gravity variations can be estimated (Ramilien *et al.*, Chapter 6, Volume 2). Using two identical satellites located in the same polar orbit at an altitude of 500 km, separated by a distance of 220 km, a map of gravitational changes is generated monthly with a spatial resolution of 300 km. Once the main contributions were removed (ocean currents, atmospheric pressure and polar caps), the change in continental water storage remains. This improves by a factor of 1,000 the precision of existing Earth gravity maps. Gravity variations allowed generation of the Terrestrial Water Storage (TWS) [ROW 05] product which describes the water content change within the entire observed soil column (snow, soil moisture of the vadose zone and groundwater). Despite their low resolution, this data is very valuable for the evaluation of the hydrological balance of large watersheds such as that of Mississippi or the Amazon. As reported by Li *et al.* [LI 12], the surface water and groundwater, representing a large portion of the TWS, are very interdependent. This connection means that an accurate estimate of the groundwater amount could greatly improve flood forecasts.

However, the assimilation of TWS products in a hydrological or surface-atmosphere model raises several questions [ZAI 08]. First, the assimilation technique must take into account the difference in resolution between the TWS product and the model. In general, a TWS pixel covers several cells of the model and the assimilation must redistribute the GRACE data on a smaller scale (horizontal disaggregation). Second, the TWS represents the water content of the entire soil column and therefore represents several "layers" of the model: snow cover (if present), unsaturated and saturated soil layers which are sometimes not accounted in hydrological models (vertical disaggregation).

In this context, the authors have assimilated the TWS products in the hydrological model Catchment Land Surface Model (CLSM) on the Mississippi catchment, using the EnKS assimilation method, which is very suitable for this case since the TWS products are monthly averaged. They studied numerous hydrological variables and their results show a decrease of the error, an improvement on the correlation between observed and simulated groundwater variations in the basin, an improvement on the simulated hydrological variability, and a slight improvement on the correlation between observed and simulated runoff in the majority of sub-basins. One of the advantages of the GRACE observations is the estimation of the "ground water" component, to which other satellite missions have no access. As a conclusion of this study, the authors suggest use of these measurements to improve drought forecasting.

Following this idea, the work of Li *et al.* [LI 12] investigates the potential contribution of GRACE-TWS concerning the monitoring of droughts in Europe, and shows that the assimilation of TWS products improves the stream flows, but does not improve superficial soil moistures (probably due to the lack of ground stations to cover the field of view of GRACE of

300 km, according to the authors). In contrast, the assimilation of TWS has a significant impact on the volume of groundwater and on its temporal dynamics, which is essential for good water management and for monitoring droughts. Due to a lack of field measurements, this impact could not be validated, but improved streamflow simulations suggest, according to the authors, that the volume of groundwater is better estimated and therefore drought monitoring is more accurate.

Lo *et al.* [LO 10] suggest using GRACE's TWS products together with streamflow measurements to better calibrate the parameters of the hydrological model CLMGW (*Community Land Model 3.0 with a Ground Water parameterization*). Despite its low resolution, the TWS product allows better constraint of the model parameterization, and it results in a better estimate of the water table depth for the three basins of the study located in Illinois, United States of America.

Investigating the snow pack properties is another application of the GRACE assimilation. Forman *et al.* [FOR 12] assimilated the TWS products in the hydrological model CLSM on the Mackenzie River catchment in northwestern Canada. The authors show that the simulations after assimilation have improved the water content of the snowpack and the stream flows.

It is also possible to combine GRACE's monthly water storage products with other satellite observations available at a finer temporal resolution. In order to improve the modeling of snow cover, Su *et al.* suggest coupling the TWS products to the SCF products (*Snow Cover Fraction*) from MODIS in the CLM hydrological model. Their study area includes eight watersheds in North America. Due to the different time resolutions of the two products, assimilation is performed by EnKF and by EnKS. The authors show that for the regions where the assimilation of MODIS by itself provides little information (SCF saturated during the snow accumulation period on boreal forests and tundra), the joint assimilation of GRACE's TWS products improves the simulations of snow cover.

It is important to remember that data assimilation is specific to each experiment conducted. The choice of the method depends on the model's structure and the knowledge on the observations. The definition of errors is a very important step that may advantage or disadvantage the results of assimilation. Whether the goal is to calibrate a model or adjust its state, assimilation is, to date, one of the most used methods for merging model simulations and observations.

11.6. Application example: assimilation of SMOS's soil moisture in the DHSVM hydrological model, on the Ouémé catchment, Benin

Data assimilation in hydrology is also interesting when compensating a lack (or uncertainty) on the model's forcing. This is frequently the case in arid or semi-arid regions where observation networks are often less developed than in temperate regions. In this final part, a study shows the importance of soil moisture measurements to constrain hydrological simulations performed using uncertain satellite rainfall estimates.

11.6.1. *Study area and hydrological model*

Ouémé's catchment is part of the AMMA-CATCH observatory [LEB 09] and benefits from strong instrumentation installed during the AMMA program (African Monsoon Multidisciplinary Analysis [RED 06]). This observatory is dedicated to the monitoring of the water cycle and the vegetation dynamics in the West African region. The Ouémé site has several meteorological and ground-based soil moisture stations, many rain gauges spread throughout the basin, flux measurements, as well as some streamflow stations (some of them are identified on Figure 11.4). The watershed is covered mostly with savannah, forests, and crops for an area of 15,000 km^2. The rainy season runs from April to October for an annual precipitation amount of about 1,200 mm.

Observed ground-based hydrological processes show a significant contribution of subsurface lateral flows in streamflow generation. For this reason, the choice of the DHSVM (*Distributed Hydrology Soil and Vegetation Model* [WIG 94]) hydrological model was made. The model allows, at the same time, generation of surface runoff (during heavy rainfall intensities), subsurface and ground water lateral flow (see Figure 11.1).

DHSVM is used at a spatial resolution of 1 km with a time step of one hour, the ground is discretized into four layers located at 1 cm, 5 cm, 40 cm and 80 cm depth. The first layer is very close to the surface and only has numerical purposes, the second is used for the assimilation of SMOS soil moisture and the two deepest layers as validation with *in situ* measurements.

The numerous rain gauges located throughout the watershed provide precise mapping of precipitation *through* the use of an elaborate interpolation method (Kriging Lagrangian [VIS 11]). In this study, this is a great opportunity to have access to *in situ* spatialized rainfall. In most cases in West Africa, there are not enough rain gauge stations to perform an accurate interpolation: derived precipitation products of satellite observations are used. However, current satellite precipitation products contain numerous sources of uncertainty, and the comparisons with field observations show wide disparities between regions and products [KID 11].

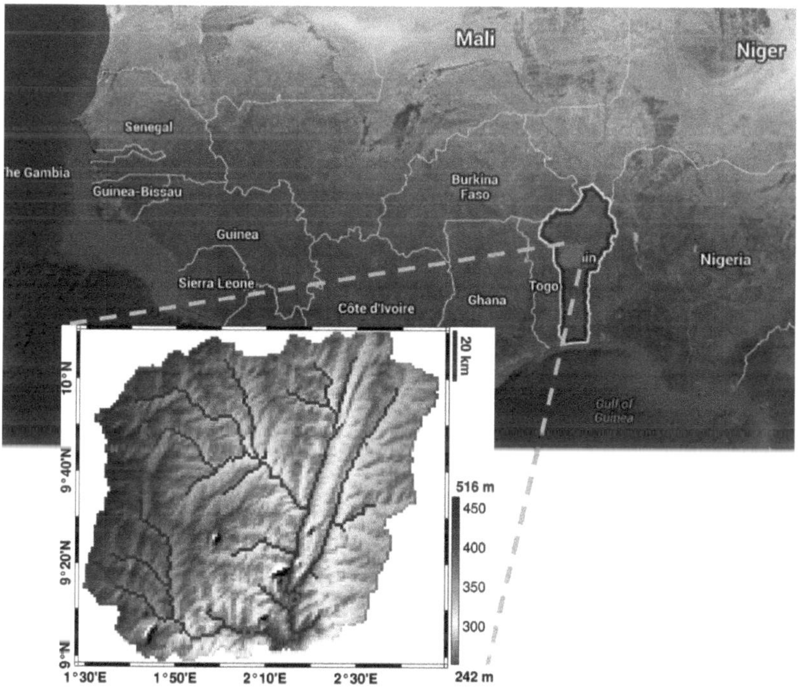

Figure 11.4. *Ouémé's watershed, located in the North of Benin. Meteorological stations with soil moisture sensors are indicated with red crosses, and two streamflow stations are indicated with red circles (outlets of the two sub-watersheds). The topography of the basin is indicated with a gradient of colors (altitude varies between 242 m and 516 m). For a color version of this figure, see www.iste.co.uk/baghdadi/4.zip*

A calibration of the model was performed using the year 2010 based on the ground-based precipitation mapping. The model outputs were evaluated with streamflow observations (two sites) as well as the soil moisture measurements and evapotranspiration. The results show very good representation of streamflows (see Figure 11.5) with a Nash coefficient of around 0.9 and a correlation of 94%. Then, the idea consists of substituting observed rainfalls with satellite rainfalls and performing a twin experiment, where the model's results with and without assimilation of SMOS surface soil moisture will be evaluated.

As emphasized by Bitew *et al.* [BIT 11], calibrating a model with uncertain satellite precipitations can compensate for the rain errors, but the balance of overall water will suffer and other hydrological processes will be disrupted. For this reason, only the calibration based on *in situ* rainfall (which gives a good rain partitioning between infiltration and runoff) is used. As it is not expected for the performance of satellite rainfall simulations to be as good as those *in situ*, only the change in performance between before and after assimilation will be analyzed.

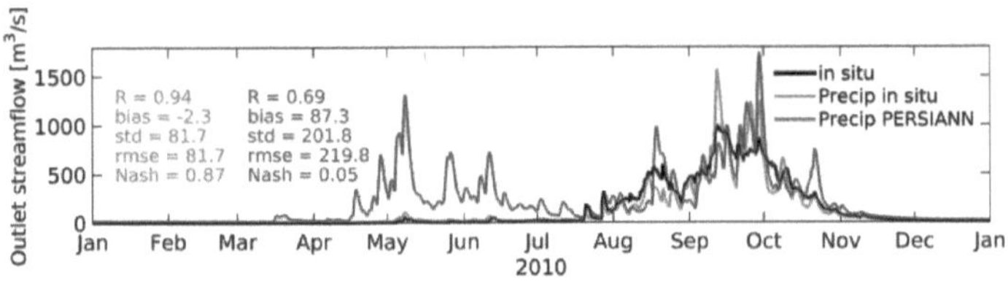

Figure 11.5. *Observed (black) and simulated streamflow (using DHSVM) using* in situ *precipitations (green) and satellite product PERSIANN (blue). For a color version of this figure, see www.iste.co.uk/baghdadi/4.zip*

When *in situ* precipitations are replaced by a product of satellite precipitations (PERSIANN in this example), the quality of the simulated flows logically decrease (Figure 11.5). The cause of this deterioration is related to the difference, sometimes significant, between the satellite rainfall products and field measurements. In this example, PERSIANN precipitations are at a spatial resolution of 0.25° and every three hours. The product significantly overestimates precipitations, especially at the beginning of the rainy season, between May and June. The lack of precision in the amount and spatial distribution of rainfall plays a very important role in the simulation of hydrological processes within the watershed. The correlation is strongly reduced, the error is multiplied by 3 and the Nash coefficient decreases from 0.9 to 0.05.

11.6.2. *Assimilation of SMOS surface soil moisture in the hydrological model*

The hypothesis tested in this study is to evaluate the contribution of the assimilation of SMOS soil moisture products on the simulation of hydrological processes on the Ouémé basin to compensate for quantitative uncertainty on rainfalls. Assimilation of raw SMOS observations, called brightness temperatures, is possible but it requires a radiative transfer model and the observer H should include this model in addition to the interpolation grids. This option puts additional complexity and introduces new unknown errors. For this study, SMOS soil moisture products are directly assimilated [RAM 16].

The SMOS products used are Level-3 data, RE02, produced over 1 day (both 6 h and 18 h were used), available on a grid sampled at 25 km. The first challenge lies in the difference of spatial resolution between the model at 1 km and the SMOS products (on a grid of 25 km but showing an observation of about 40 km). The chosen method of assimilation is the optimal interpolation [DEL 10]. The assimilation is made on the scale of observations, which requires a change of scale (1–> 25 km) before redistributing increments across the model (Figure 11.6). The state vector is composed here of four soil moisture variables at different depths (the 4 of them defined at 1, 5, 40 and 80 cm). The observation vector is of variable height (m) and the matrix K (defined in section 11.4) is therefore of size $4 \times m$. On Figure 11.6, several SMOS observations can contribute if the selected radius of influence is big. In this example, it is established at 30 km (according to Gaspari, described further below and in Figure 11.7).

Matrix K is calculated based on observation matrices H (allowing us to go from the model space to the space of observations, of size m $\times$ 4) and of covariance of errors of the model B (4 $\times$ 4) and observations errors R (m $\times$ m). Since the assimilation is not performed with a generation of ensembles, the matrices of covariance error B and R are B manually set for optimal interpolation. The elements of matrix B are calculated by

$B_{i,j} = \text{cov}(\text{SM}_i, \text{SM}_j)$ using open-loop simulations for each pixel of the model. This matrix B allows updating the soil moisture of all soil layers; although SMOS soil moisture estimates are only assimilated at 5 cm depth (...) deep (it is not diagonal for this reason). Matrix R is diagonal since it is assumed that errors are uncorrelated between observations. In this study, the covariances of errors were very simplified, since B and R were calculated from the variances of the open loop simulations and observations. The average variance of SMOS soil moisture is 0.017 $(m^3/m^3)^2$. The average of matrix B is as follows:

$$B = \begin{bmatrix} 0.022 & 0.015 & 0.010 & 0.003 \\ 0.015 & 0.019 & 0.011 & 0.003 \\ 0.010 & 0.011 & 0.012 & 0.005 \\ 0.003 & 0.003 & 0.005 & 0.006 \end{bmatrix} \quad [11.8]$$

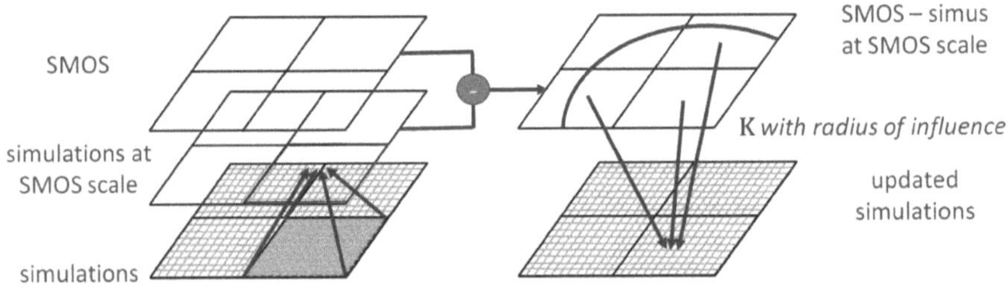

Figure 11.6. *Graphical representation of the changing of scale required for the assimilation of space observations (adapted from [DEL 10, SAH 13])*

We note here that the average coefficient of $B_{2,2}$ and the average value of R are very close (0.019 and 0.017 $m^3/m^3)^2$, which corresponds to a weight roughly equal between the model and the observations. *On the contrary*, the average coefficient of $B_{4,4}$ is very small compared to R, which implies that the observations will have less influence on the fourth layer of soil.

SMOS soil moisture products are assimilated in the second layer. However, SMOS observations do not all have the same impact on the adjustment to be made, particularly depending on the distance between the observation and the model pixel. The weight function used depends on the distance between the observation and the pixel model, and the radius of influence, and is presented in the form of a 5th degree polynomial ([4.10] in Gaspari and Cohn [GAS 99], also used in De Lannoy *et al.* [DEL 10]). This polynomial is also a very good approximation of the weight function used in the SMOS algorithm to represent the antenna pattern (diagram indicating where the power of the received signal is coming from, see Figure 11.7). Column "1" of matrix H (2nd column in this case) is replaced by the weight assigned to different SMOS observations. This method allows to obtain regular adjustments and to avoid side effect on the pixels edges. (as shown in Figure 11.3).

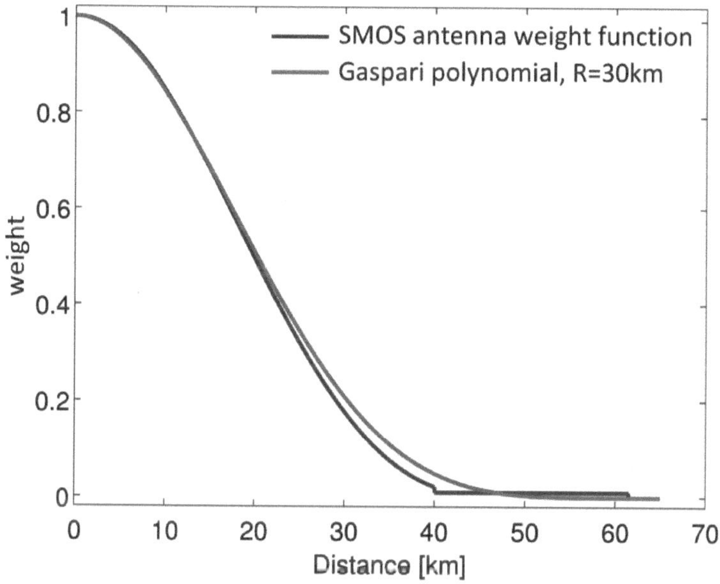

Figure 11.7. *Weight assigned to the observation based on its distance to the model pixel of the considered model. Gaspari's polynomial is actually very close to the modeling function of the antenna pattern of SMOS*

Figure 11.8 shows an example of the impact of assimilation on a simulation of soil moisture at 5 cm depth. Image on the left shows the state of the simulations of the watershed when the model is stopped since SMOS observations are available (second image). Matrix K is calculated and the soil moisture is then updated (third image). In this example, the assimilation of SMOS adds water in the southwestern basin and removes water from the ground to the northeast.

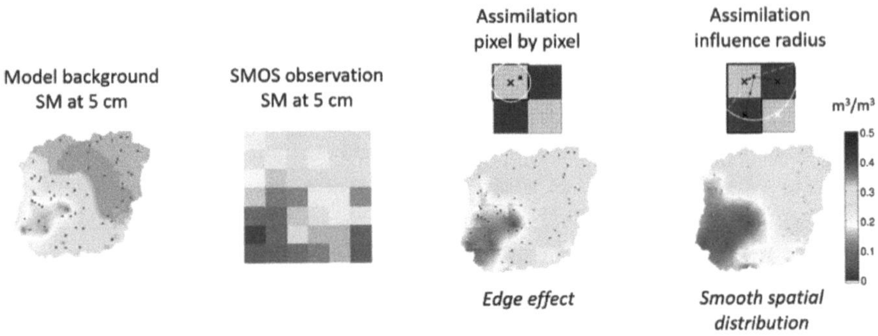

Figure 11.8. *Soil moisture(SM) at 5 cm depth for (i) the model, (ii) SMOS, (iii) the model after assimilation. The two images on the right represent the effect of considering (or not) a smoothing approach based on an influence radius*

By using matrix B, the assimilation in the second layer also affects the water column soil moisture as a whole. The most adjusted layer is obviously the one where SMOS is assimilated at 5 cm, but due to its strong correlation with the surface layer at 1 cm, it strongly impacts the first layer of soil. The layers at 40 cm and 80 cm of depth are slightly less correlated with those at 5 cm, so they are less impacted. Therefore, this method allows correction of the rainfalls that bring a huge amount of water, especially during the dry period.

Located northwest of the watershed, the Bira station measures soil moisture continuously at several depths between 5 cm and 80 cm. Figure 11.9 compares soil moisture observations at 5 cm and 80 cm with the simulations performed in different configurations in 2010. The associated statistical scores are presented in Table 11.2.

Using *in situ* precipitation, soil moisture at 5 cm depth is well represented (R=0.81, RMSE=0.084 m^3/m^3, bias=0.029 m^3/m^3). In the example of the pre-monsoon period (March–April, at the top left of Figure 11.9), we can see the change in soil moisture after a rainfall event that rapidly increases and then gradually declines until returning to the wilting point. *In situ* interpolated rainfalls cause a slight overestimation of soil moisture; especially with PERSIANN, especially with PERSIANN products. The assimilation of SMOS soil moisture at 5 cm can adjust these values too highly, which causes a reduction of the water volume throughout the soil column by adjusting the soil moisture of the different soil layers. In the example of the period from June to July during the rainy season (top right of Figure 11.9), an adjustment is also made with, more rarely, an adjustment where water is added (as in mid-June).

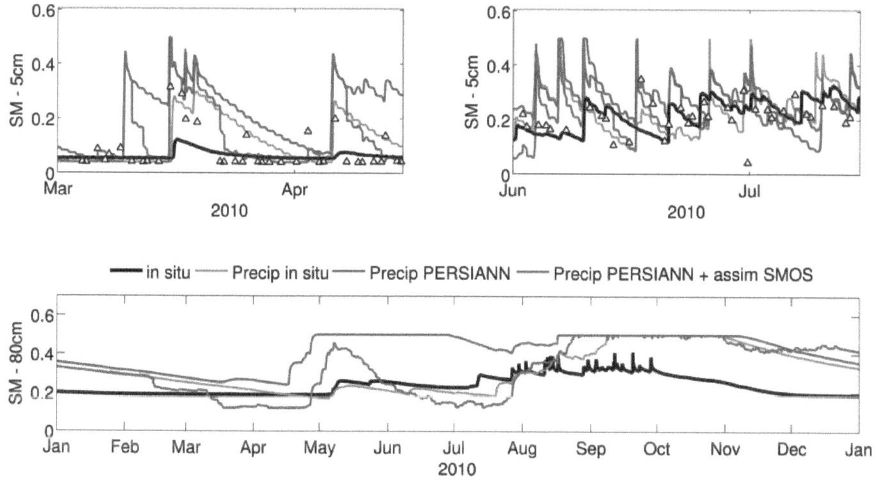

Figure 11.9. *Time series of simulated soil moisture (color) and measured (black) at the Bira station to a depth of 5 cm (upper graph in two periods, for clarity) and at 80 cm (bottom graph). SMOS soil moisture products are represented by triangles. Corresponding statistics are provided in Table 11.2. For a color version of this figure, see www.iste.co.uk/baghdadi/4.zip*

At a depth of 80 cm, assimilation has less effect than at 5 cm since the observations are assimilated with a much lower weight (based on the coefficients of matrix *B*). However, this deeper layer ultimately benefits from a double impact of assimilation: directly during the first-time step of assimilation and with a temporal delay for the water volume coming from the upper layers, which have already been corrected before. Therefore, the time series of the soil moisture at 80 cm is impacted. At the bottom of Figure 11.9, the simulated soil moisture from PERSIANN rainfall saturates starting in May. Assimilation allows ample mitigation of this phenomenon that saturates the end of August even if a sharp increase in May is still visible. This saturation phenomena is not well transcribed in the statistical scores of Table 11.2, however, a net decrease of bias is observed after assimilation (from 0.180 m^3/m^3 to 0.067 m^3/m^3). The assimilation of SMOS allows better constraint of the distribution of water in all soil layers of the model.

Depth	Statistics for the entire year of 2010	Open loop *in situ* Rainfall (green curve)	Open loop PERSIANN Rainfall (blue curve)	SMOS Assimilation PERSIANN Rainfall (red curve)
5 cm	R	0.81	0.69	0.75
	RMSE (m^3/m^3)	0.084	0.127	0.099
	bias (m^3/m^3)	0.029	0.076	0.102
	SEE (m^3/m^3)	0.079	0.102	0.088
80 cm	R	0.50	0.69	0.49
	RMSE (m^3/m^3)	0.123	0.192	0.146
	bias (m^3/m^3)	0.079	0.180	0.067
	SEE (m^3/m^3)	0.095	0.067	0.120

Table 11.2. *Soil moisture statistics at several depths in open loop with precipitations* in situ *and PERSIANN, as well as after SMOS surface soil moisture assimilation. The improved statistics by assimilation are indicated in bold*

The state vector is composed of soil moisture at different depths, but the assimilation has a strong impact on other variables in the model. Indeed, at the time step of assimilation, only soil moistures are updated. However, they are defined by physical equations; the model adjusts other variables in the following time steps, such as surface and subsurface runoff. If less water is contained in the soil, the water volume transported laterally is automatically affected and it ends up in the flow simulations. Figure 11.10 shows simulations of flows at the outlet of the Ouémé basin in 2010 using PERSIANN rainfalls with and without assimilation of SMOS soil moisture (red and blue curves, respectively). Temporal correlation is improved, the bias is decreased and the coefficient of Nash increases from 0.05 to 0.62. These statistics clearly show the positive impact of the spatial soil moisture assimilation of SMOS on the estimation of flows.

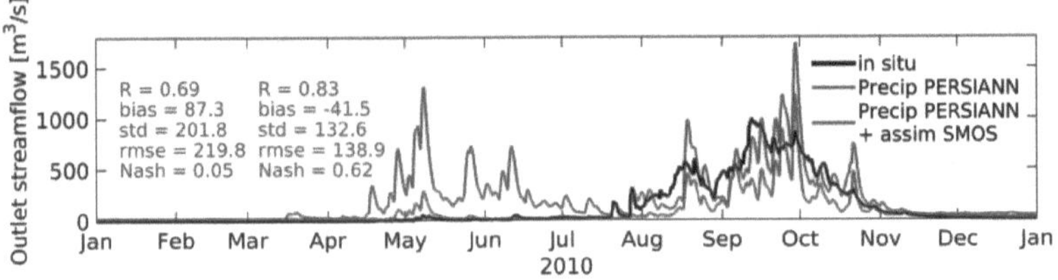

Figure 11.10. *Time series of flows at the outlet: Observed (black) simulated by DHSVM using PERSIANN rainfalls (blue) and using PERSIANN rainfalls with SMOS soil moisture assimilation (red). For a color version of this figure, see Zwww.iste.co.uk/baghdadi/4.zip*

11.7. A favorable future to assimilation in hydrology

In order to better understand, simulate and predict the hydrological cycle of a watershed, it is essential to have field measurements. It is not always easy to establish durable field measurement campaigns over time, making the observation from space an undeniable asset. However, using these observations from space is not trivial when they require incorporation into the models, particularly due to their spatial resolution, or due to their uncertainty. Data assimilation combines model simulations with observations according to their relative accuracies, and thus to better constrain the models.

The Sentinel missions of ESA offer unprecedented spatial and temporal resolution, which is very interesting for catchment hydrology. Satellites Sentinel 1A and 1B, equipped with a synthetic aperture radar, allow us to obtain images at 10 m of resolution which are sensitive to soil moisture. Each of the two Sentinel-2 satellites observe the entire land surface every 10 days, with a resolution of 10 to 60 m in 13 spectral bands ranging from the visible to the infrared range, which allow observation of vegetation dynamics, land occupation, snow or even the evolution of the extent of water surfaces.

Other future space missions have a particular interest for catchment hydrology. The SWOT mission joint mission between CNES and JPL/NASA, will be launched around 2020 and will measure the water depth over land with an accuracy of 10 cm and slopes with a precision of 1 cm/km for rivers more than 100 m wide and lakes with a minimum area of 250 m^2. With two SAR antennas in Ka band, at 870 km altitude, the swath will be of 120 km with an overpass every 22 days. More details on the SWOT Mission can be found on the CNES website (*https://swot.cnes.fr/fr/*) or on the JPL website (http://swot.jpl.nasa.gov).

After SMOS (2009–), Aquarius (2010–2015) and SMAP (2015–), the continuity of Earth observation in L-band microwave radiometry is not yet clearly defined. The needs in spatial resolution in terms of meteorological patterns will rapidly move towards the kilometer and it is important that space observations are able to follow this trend. There are various projects (SMOS-next, for example) that aim to achieve a resolution of

5–10 km directly without disaggregation of the signal. These studies are still in their infancy and many years are necessary for the realization of a space mission.

11.8. Key points

– Several variables observed from space have been assimilated into models (surface soil moisture, water depth, etc.), making it possible to adjust variables which are directly related to other elements of the water cycle such as the flow rates of rivers, snowpack or root-zone soil moisture.

– Assimilation can refine the prediction of extreme events such as droughts or floods, which are relevant social issues. To date, there is no operational structure, since works still relate to past events.

– Spatial hydrological models can simulate the distribution of different hydrological processes within a watershed. The spatial variables of these models can be mapped in correspondence with satellite observations.

– Satellite data assimilation needs to take into account the difference in scale between the model and the field of view of the instrument.

11.9. Bibliography

[AND 07] ANDREADIS K.M., CLARK E.A., LETTENMAIER D.P. *et al.*, "Prospects for river discharge and depth estimation through assimilation of swath-altimetry into a raster-based hydrodynamics model", *Geophysical Research Letters*, vol. 34, p. L10403, 2007.

[ARN 05] ARNOLD J.G., FOHRER N., "SWAT2000: current capabilities and research opportunities in applied watershed modeling", *Hydrological Processes*, vol. 19, no. 3, pp. 563–572, 2005.

[AUB 03] AUBERT D., LOUMAGNE C., OUDIN L., "Sequential assimilation of soil moisture and streamflow data in a conceptual rainfall-runoff model", *Journal of Hydrology*, vol. 280, pp. 145–161, 2003.

[BAT 00] BATES P.D., DE ROO A.P.J., "A simple raster-based model for flood inundation simulation", *Journal of Hydrology*, vol. 236, pp. 54–77, 2000.

[BIA 11] BIANCAMARIA S., DURAND M., ANDREADIS K.M. *et al.*, "Assimilation of virtual wide swath altimetry to improve Arctic river modeling", *Remote Sensing of Environment*, vol. 115, pp. 373–381, 2011.

[BIT 11] BITEW M.M., GEBREMICHAEL M., "Assessment of satellite rainfall products for streamflow simulation in medium watersheds of the Ethiopian highlands", *Hydrology and Earth System Sciences*, vol. 15, pp. 1147–1155, 2011.

[BRO 10] BROCCA L., MELONE F., MORAMARCO T. *et al.*, "Improving runoff prediction through the assimilation of ASCAT soil moisture product", *Hydrology and Earth System Sciences*, vol. 14, pp. 1881–1893, 2010.

[BRO 12] BROCCA L., MORAMARCO T., MELONE F. *et al.*, "Assimilation of surface and root-zone ASCAT soil moisture products into rainfall-runoff modeling", *IEEE Transactions on Geoscience and Remote Sensing*, vol. 50, pp. 2542–2555, 2012.

[CHA 14] CHAKRABARTI S., BONGIOVANNI T., JUDGE J. *et al.*, "Assimilation of SMOS soil moisture for quantifying drought impacts on crop yield in agricultural regions", *IEEE Journal of Selected Topics in Applied Earth Observations and Remote Sensing*, vol. 7, pp. 3867–3879, 2014.

[CHE 11] CHEN F., CROW W.T., STARKS P.J. *et al.*, "Improving hydrologic predictions of a catchment model via assimilation of surface soil moisture", *Advances in Water Research*, vol. 24, pp. 526–536, 2011.

[DAI 03] DAI Y., ZENG X., DICKINSON R.E. *et al.*, "The common land model (CLM)", *Bulletin of the American Meteorological Society*, vol. 84, pp. 1013–1023, 2003.

[DEL 10] DE LANNOY G.J.M., REICHLE R.H., HOUSER P.R. *et al.*, "Satellite-scale snow water equivalent assimilation into a high-resolution land surface model", *Journal of Hydrometeorology*, vol. 11, pp. 352–369, 2010.

[DRA 12] DRAPER C.S., REICHLE R.H., DE LANNOY G.J.M. *et al.*, "Assimilation of passive and active microwave soil moisture retrievals", *Geophysical Research Letters*, vol. 39, p. L04401, 2012.

[DUR 08] DURAND M., ABDREADIS K.M., ALSDORF D.E. *et al.*, "Estimation of bathymetric depth and slope from data assimilation of swath altimetry into a hydrodynamic model", *Geophysical Research Letters*, vol. 35, p. L20401, 2008.

[EVE 94] EVENSEN G., "Sequential data assimilation with a nonlinear quasi-geostrophic model using Monte-Carlo methods to forecast error statistics", *Journal of Geophysical Research*, vol. 99, pp. 10143–10162, 1994.

[FOR 12] FORMAN B.A., REICHLE R.H., RODELL M., "Assimilation of terrestrial water storage from GRACE in a snow-dominated basin", *Water Resources Research*, vol. 48, p. W01507, 2012.

[FRA 03] FRANCOIS C., QUESNEY A., OTTLÉ C., "Sequential assimilation of ERS-1 SAR data into a coupled land surface-hydrological model using an extended Kalman filter", *Journal of Hydrometeorology*, vol. 4, pp. 473–487, 2003.

[GAL 99] GALANTOWICZ J.F., ENTEKHABI D., NJOKU E., "Tests of sequential data assimilation for retrieveing profile soil moisture and temperature from observed L-band radio brightness", *IEEE Transactions on Geoscience and Remote Sensing*, vol. 37, pp. 1860–1870, 1999.

[GAS 99] GASPARI G., COHN S.E., "Construction of correlation functions in two and three dimensions", *Quarterly Journal of the Royal Meteorological Society*, vol. 125, pp. 723–757, 1999.

[GIU 11] GIUSTARINI L., MATGEN P., HOSTACHE R. *et al.*, "Assimilating SAR-derived water level data into a hydraulic model: a case study", *Hydrology and Earth System Sciences*, vol. 15, pp. 2349–2365, 2011.

[HAN 12] HAN E., MERWADE V., HEATHMAN G.C., "Implementation of surface soil moisture data assimilation with watershed scale distributed hydrological model", *Journal of Hydrology*, vol. 416, pp. 98–117, 2012.

[KAL 60] KALMAN R.E., "A new approach to linear filtering and prediction problems", *Transactions of the AMSE – Journal of Basic Engineering*, vol. 82 (Series D), pp. 35–45, 1960.

[KID 11] KIDD C., LEVIZZANI V., "Status of satellite precipitation retrievals", *Hydrological Earth System Sciences*, vol. 15, pp. 1109–1116, 2011.

[KUM 14] KUMAR S.V., PETERS-LIDARD C.D., MOCKO D. *et al.*, "Assimilation of remotely sensed soil moisture and snow depth retrievals for drought estimation", *Journal of Hydrometeorology*, vol. 15, pp. 2446–2469, 2014.

[LAH 10] LAHOZ W., KHATTATOV B., MENARD R., *Data Assimilation – Making Sense of Observations*, Springer–Verlag, Berlin Heidelberg, p. 718, 2010.

[LAH 14] LAHOZ W.A., DE LANNOY G.J.M., "Closing the gaps in our knowledge of the hydrological cycle over land: conceptual problems", *Surveys in Geophysics*, vol. 35, pp. 623–660, 2014.

[LEB 09] LEBEL T., CAPPELAERE B., GALLE S. *et al.*, "AMMA-CATCH studies in the Sahelian region of West-Africa: an overview", *Journal of Hydrology*, vol. 375, pp. 3–13, 2009.

[LI 12] LI B., RODELL M., ZAITCHIK B.F. *et al.*, "Assimilation of GRACE terrestrial water storage into a land surface model: evaluation and potential value for drought monitoring in western and central Europe", *Journal of Hydrology*, vol. 446, pp. 103–115, 2012.

[LIA 94] LIANG X., LETTENMAIER D.P., WOOD E.F. *et al*, "A simple hydrologically based model of land-surface water and energy fluxes for general-circulation models", *Journal of Geophysical Research*, vol. 99, pp. 14415–14428, 1994.

[LO 10] LO M.H., FAMIGLIETTI J.S., YEH P.J.F. *et al.*, "Improving parameter estimation and water table depth simulation in a land surface model using GRACE water storage and estimated base flow data", *Water Resources Research*, vol. 46, p. W05517, 2010.

[MAT 10] MATGEN P., MONTANARI M., HOSTACHE R. *et al.*, "Towards the sequential assimilation of SAR-derived water stages into hydraulic models using the particle filter: proof of concept", *Hydrology and Earth System Sciences*, vol. 14, pp. 1773–1785, 2010.

[MON 09] MONTANARI M., HOSTACHE R., MATGEN P. *et al.*, "Calibration and sequential updating of a coupled hydrologic-hydraulic model using remote sensing-derived water stages", *Hydrology and Earth System Sciences*, vol. 13, pp. 367–380, 2009.

[NOI 89] NOILHAN J., PLANTON S., "A simple parameterization of land surface processes for meteorological models", *Monthly Weather Review*, vol. 177, pp. 536–549, 1989.

[OKI 98] OKI T., SUD Y.C., "Design of Total Runoff Integrating Pathways (TRIP) – a global river channel network", *Earth Interaction*, vol. 2, pp. 1–37, 1998.

[OTT 94] OTLLE C., VIDAL-MADJAR D., "Assimilation of soil moisture inferred from infrared remote sensing in a hydrological model over the HAPEX-MOBILHY region", *Journal of Hydrology*, vol. 158, pp. 241–264, 1994.

[PAU 01] PAUWELS V.R.N., HOEBEN R., VERHOEST N.E.C. *et al.*, "The importance of the spatial patterns of remotely sensed soil moisture in the improvement of discharge predictions for small-scale basins through data assimilation", *Journal of Hydrology*, vol. 251, pp. 88–102, 2001.

[PAU 02] PAUWELS V.R.N., HOEBEN R., VERHOEST N.E.C. *et al.*, "Improvement of TOPLATS-based discharge predictions through assimilation of ERS-based remotely sensed soil moisture values", *Hydrological Processes*, vol. 16, pp. 995–1013, 2002.

[PED 14] PEDINOTTI V., BOONE A., RICCI S. *et al.*, "Assimilation of satellite data to optimize large-scale hydrological model parameters: a case study for the SWOT mission", *Hydrology and Earth System Sciences*, vol. 18, pp. 4485–4507, 2014.

[PER 03] PERRIN C., MICHEL C., ANDRÉASSIAN V., "Improvement of parsimonious model for streamflow simulation", *Journal of Hydrology*, vol. 279, pp. 275–289, 2003.

[RED 06] REDELSPERGER J.L., THORNCROFT C.D., DIEDHIOU A. *et al.*, "African Monsoon Multidisciplinary Analysis: an international research project and field campaign", *Bulletin of the American Meteorological Society*, vol. 87, pp. 1739–1746, 2006.

[REI 02] REICHLE R.H., MCLAUGHLIN D.B., ENTEKHABI D., "Hydrologic data assimilation with the ensemble Kalman filter", *Monthly Weather Review*, vol. 130, pp. 103–114, 2002.

[REI 08] REICHLE R.H., CROW W.T., KEPPENNE C.L., "An adaptative ensemble Kalman filter for soil moisture data assimilation", *Water Resources Research*, vol. 44, p. W03423, 2008.

[REN 14] RENZULLO L.J., VAN DIJK A.I.J.M., PERRAUD J.M. *et al.*, "Continental satellite soil moisture data assimilation improves root-zone moisture analysis for water resources assessment", *Journal of Hydrology*, vol. 519, pp. 2747–2762, 2014.

[RID 14] RIDLER M.E., MADSEN H., STISEN S. *et al.*, "Assimilation of SMOS-derived soil moisture in a fully integrated hydrological and soil-vegetation-atmosphere transfer model in Western Denmark", *Water Resources Research*, vol. 50, pp. 8962–8981, 2014.

[ROW 05] ROWLANDS D.D., LUTHCKE S.B., KLOSKO S.M. *et al.*, "Resolving mass flux at high patial and temporal resolution using GRACE intersatellite measurements", *Geophysical Research Letters*, vol. 32, p. L04310, 2005.

[SAH 13] SAHOO A.K., DE LANNOY G.J.M., REUCHLE R.H. *et al.*, "Assimilation and downscaling of satellite observed soil moisture over the Little River experimental watershed in Georgia, USA", *Advances in Water Resources*, vol. 52, pp. 19–33, 2013.

[SCH 07] SCHUMANN G., MATGEN P., HOFFMANN L. *et al.*, "Deriving distributed roughness values from satellite radar data for flood inundation modelling", *Journal of Hydrology*, vol. 344, pp. 96–111, 2007.

[SU 10] SU H., YANG Z.L., DICKINSON R.E. *et al.*, "Multisensor snow data assimilation at the continental scale: the value of Gravity Recovery and Climate Experiment terrestrial water storage information", *Journal of Geophysical Research*, vol. 115, p. D10104, 2010.

[VIS 11] VISCHEL T., QUANTIN G., LEBEL T. *et al.*, "Generation of high-resolution rain fields in West Africa: evaluation of dynamics interpolation methods", *Journal of Hydrometeorology*, vol. 12, pp. 1465–1482, 2011.

[WIG 94] WIGMOSTA M.S., VAIL L., LETTEMAIER D.P., "A distributed hydrology-vegetation model for complex terrain", *Water Resources Research*, vol. 30, pp. 1665–1679, 1994.

[WOO 91] WOOD E.F., "Global scale hydrology: advances in land surface modeling", *Reviews of Geophysics*, vol. 29, pp. 193–201, 1991.

[ZAI 08] ZAITCHIK B.F., RODELL M., REICHLE R.H., "Assimilation of GRACE Terrestrial Water Storage data into a land surface model: results for the Mississippi river basin", *Journal of Hydrometeorology*, vol. 9, pp. 535–548, 2008.

12

Satellite Data Assimilation: Application to the Water and Carbon Cycles

12.1. Assimilation: what is the purpose?

Land surface processes control the water and energy fluxes between the continental surfaces and the atmosphere, as well as the interaction of the water and carbon cycles. They are complex and characterized by strong heterogeneities both spatially (orography, land use, soil types, etc.) and temporally (variability of diurnal and seasonal cycles). Due to the complexity and the large range of spatial and temporal scales involved in these processes, land surface data assimilation has not yet reached levels of maturity, which are found in the areas of atmospheric (notably, in weather forecast) and oceanographic data assimilation. Nevertheless, considerable progress has been made in recent years in the development of the data assimilation on continental surfaces, in conjunction with the expansion of spatial observation of the Earth.

The observation of continental surfaces by means of remote sensing allows us to monitor biophysical variables of the land and vegetation on a global level, which lead to more and more sophisticated spatial and temporal resolutions. It is based on a variety of sensors, which use different spectral bands and different measurement methods. Certain sensors are used for

Chapter written by Jean-Christophe CALVET, Patricia DE ROSNAY, Alina L. BARBU and Souhail BOUSSETTA.

research activities. For example, the measurements in the thermal infrared field are relevant to estimate surface temperature, which is notably related to the interactions between soil moisture and the evapotranspiration fluxes. The study of the potential of the assimilation of surface temperatures for applications in hydrology is a fully-fledged research topic [VAN 01]. Currently, surface temperatures are not used for applications in operational hydrology. Other sensors supply the information used for the operational applications for monitoring continental surfaces, meteorological forecasting, or hydrology. Information about land use is received through sensors operating in the visible and near infrared spectrum (see [ING 16]). This information is crucial and it is used for mapping the parameters of models simulating the hydrological cycle. Measurements received from these sensors are also highly relevant for monitoring radiative variables (for example, the surface albedo) and for daily monitoring of vegetation (for example, leaf area index). Finally, measurements of sensors with low-frequency active and passive microwaves are sensitive to soil and vegetation water content (see [WIG 16] and Chapters 2 and 3 in this book). They represent a useful source of information for data assimilation applied in the study of the water and carbon cycles.

Satellite data should be used carefully on continental surfaces. On one hand, the physics of measurements is complex and the measured signal is the result of multiple interactions between electromagnetic waves and the heterogeneous media that they pass through (various components of the soil, vegetation, and atmosphere). Disruptive factors can interrupt the observations or reduce their quality. These include atmospheric factors (for example, clouds and dust in the spectral bands influenced by the atmosphere) or anthropogenic factors (for example, electromagnetic interferences in the microwave spectrum). On the other hand, the satellite data represent radiative measurements (for example, brightness temperature, backscatter coefficient, luminance), on the basis of which the variables of thematic interest (for example, soil moisture, leaf area index, etc.) should be estimated (inversion) or analyzed (assimilation). Furthermore, the information that they supply is relatively limited. For example, the satellite data does not provide access to certain key variables, such as surface fluxes (water, CO_2, etc.). For other variables, such as the total water content of the soil, the biomass of vegetation, the snowpack, the information obtained through remote sensing remains very indirect and rarely unambiguous. Moreover, the description of energy, water, and carbon exchanges between land surfaces and the atmosphere requires the representation of the diurnal cycle of radiation variables and a daily monitoring of vegetation. Today, vegetation monitoring is only possible at global scale with a 10-day time step, using satellites currently in orbit. For applications dedicated to monitoring of water and carbon cycles, within the context of weather forecasting and climate monitoring, it is indispensable to use computer programs (models), which reproduce key processes determining surface fluxes and biophysical variables, as well as their temporal and spatial variations. The models of the soil–plant system when coupled with the atmosphere represent the diurnal cycle of sensible and latent heat fluxes and radiation variables, as well as the temporal evolution of slowly developing variables, such as soil moisture of the root zone. The most recent generation of the soil–plant–atmosphere models, now called the land surface models, is capable of coherently simulating several biophysical variables, which can also be observed through remote sensing, such as surface soil moisture or leaf area index (or LAI). These models represent the surface component of the climate models and the numerical weather prediction systems [BAL 07, KRI 05, MAS 13]. Due to the non-linearity of the involved processes and their representation in the models, even a minimal error in the initial value of certain variables of models is likely to cause serious errors in the representation of the evolution (forecasting) of the system as a whole. Data assimilation provides an optimal estimation of the initial conditions of a land surface model on the basis of the estimations of the model on the one hand, the satellite observations on the other hand, and the uncertainties associated with them. The analysis is the result of the data assimilation. The analysis can be obtained for the model variables that are observable (for example, surface soil moisture and leaf area index of vegetation) and for the slowly developing variables, which cannot be directly observed through remote sensing, such as the total water content of the soil.

Figure 12.1 demonstrates the principle of observations assimilation in a land surface model. The assimilation enables the integration of observations in the model. This model is controlled by the atmospheric component of the model, which provides, either by forcing or in coupled (or interactive) mode, the precipitation, the air temperature, the air humidity, the solar and atmospheric radiation, and the wind speed. The surface model takes into consideration the heterogeneity and the diversity of surfaces by using tables or digital maps of parameters (for example, the vegetation type, the depth of soil, parameters of photosynthesis, etc.). In order to ensure an optimal assimilation process, it is important to know to what degree the model and the observations can be trusted. This requires the quantification of errors in modeling and in observations (see [OTT 16]). The *a posteriori* comparison between the observations and the analysis allows us to verify that the errors in observations are represented well and to monitor their development over time (for example, see Barbu *et al.* [BAR 11] in the context of the assimilation of LAI).

It should be noted that satellite observations could also be integrated into the models through the improvement of forcings. In non-coupled mode with the atmosphere, these observations could be atmospheric forcing, such as incident solar radiation or other model parameters [CAR 12, CAR 14].

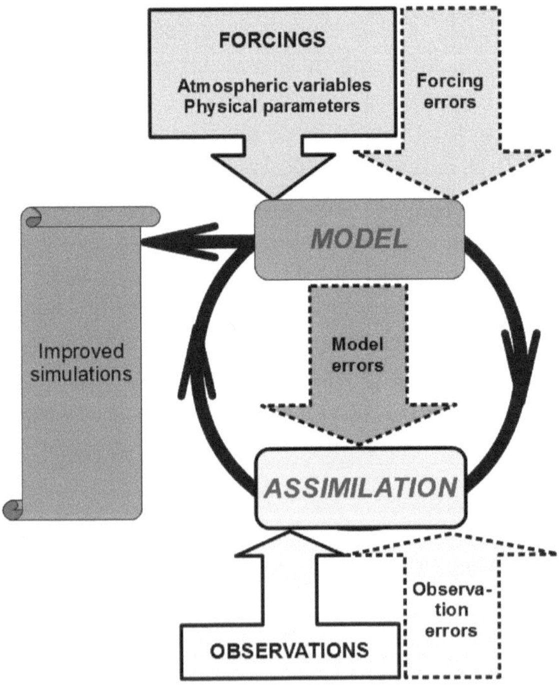

Figure 12.1. *The principle of data assimilation for modeling of the land surface, illustrated here on the example of a surface model forced by the atmosphere*

Assimilation can be implemented differently according to its applications [LIU 12, MON 12]. For climate applications where simulations are realized at the century scale, in the past and in the future, the observations are only available for a relatively short period of time (Earth has only been observed from space for the past three decades). In this case, the main purpose of assimilation is to calibrate model parameters [KAM 12]. There is an assumption that models optimized in this way have a better capability of using forcings in order to recreate a past or present condition, as well as forecasting and better evaluating the impact of climate change on various components of the land system. In those applications where it is important to have the best possible analysis in order to reconstruct the past situations, during which the observations are available, or to have the best initial conditions in order to make a forecast, the assimilation can directly control the temporal trajectory of biophysical variables [BAR 14]. By integrating the observations into models as soon as they become available, it is possible to ensure monitoring in near real time. This is particularly useful in case of extreme events, like agricultural droughts. It is possible to simultaneously assimilate several biophysical variables (for example, LAI and surface soil moisture in [BAR 14]). This last approach is often referred to as the Land Data Assimilation System (LDAS).

In Europe, several services provide access in near real time to biophysical land variables derived from satellite observations on the global or continental level: Copernicus Global Land Service, and EUMETSAT (H-SAF and LSA-SAF).

In the rest of this chapter, examples of space observations assimilation are presented for the following applications:

– numerical weather prediction;

– hydrology;

– monitoring of natural sinks and natural sources of CO_2.

12.2. Analyses of the vegetation and soil moisture for numerical weather prediction

Data assimilation is an essential component of numerical weather prediction systems [KAL 03]. The quality of meteorological forecasts depends both on the model used and on the accuracy of its initial conditions. The initialization of the surface conditions has a considerable impact on the quality of meteorological forecast of the lowest levels of the atmosphere. Surface data assimilation has a strong influence on the forecast variables, such as air temperature and relative humidity or precipitation intensity, at short ranges (a few days) and at monthly and seasonal ranges [BEL 96, KOS 04, DER 14].

The numerical weather prediction systems use data assimilation in order to define the initial conditions of the atmospheric and land variables of the model. In the European Center for Medium-Range Weather Forecasts (ECMWF), the atmospheric assimilation uses a variational approach called 4D-Var [RAB 98] and the assimilation of the continental surface variables is based on a system combining a simplified Extended Kalman filter (EKF) and an optimal interpolation [DER 13]. The analyzed surface variables of the model include soil moisture and soil temperature, as well as snow mass, density and temperature. For snow, the main source of information is provided by *in situ* data and by satellite derived snow cover products. In order to analyze the soil moisture, most operational meteorological centers assimilate *in situ* observations of near-surface air temperature and humidity. These observations are related to the water status of the soil and they allow to indirectly obtain the information about the water content of the soil (see for example [DER 14] for the ECMWF system, [MAH 09] for "Météo-France" system and [BEL 03] for "Environnement Canada"). However, this local meteorological data do not present a uniform global coverage.

Since the 1970s and 1980s, microwave remote sensing has been considered as a method of mapping the soil moisture in a thin soil layer close to the surface (first centimeters of the soil) on various spatial scales [SCH 83]. Since then, several space missions, which provide access to this quantity, have been launched, for example: ERS, ENVISAT, Advanced Scatterometer (ASCAT) on METOP satellites [WAG 07], Advanced Microwave Scanning Radiometer (AMSR-E) Earth Observation System (EOS) [NJO 03], Soil Moisture and Ocean Salinity (SMOS) [KER 01] and Soil Moisture Active Passive (SMAP) [ENT 04].

Several meteorological services assimilate soil moisture data from the ASCAT active sensor either operationally (the ECMWF and the UK Met Office in the United Kingdom [DHA 11]) or for research [MAH 10,

DRA 11]. The covariance matrices of observation and background errors are diagonal and static in most studies applied to the numerical weather prediction and their values are specified from the innovation statistics (meaning the comparison between the observations and the first guess simulation of the model). Thus, the errors in the soil moisture observation of ASCAT of 0.05 m^3m^{-3} and 0.06 m^3m^{-3} are used in the ECMWF and "Météo-France" respectively [MAH 10]. Other studies use the information about the observation error provided with the soil moisture product [DRA 11]. The background error used in the assimilation system of the ECMWF is constant, specified to 0.01 m^3m^{-3}. The experiments carried out at "Météo-France" use different values according to studies either generally constant [MAH 10] or dependent on the soil texture, like in [DRA 11]. A study of Draper *et al.* [DRA 09] compared, in the case of the assimilation of the AMSR-E soil moisture data, the EKF approach (which takes into consideration the temporal evolution, or "cycles", of the covariance matrix of background errors) to a simplified EKF approach, which uses a fixed covariance matrix of background error. The authors demonstrated that even though the two approaches are different, they lead to very similar soil moisture values. Being cheaper in computing time, the simplified EKF is preferred for operational applications.

In addition to the use of soil moisture products, several studies are being performed in France, Canada, the United States, and the United Kingdom to develop the assimilation of passive microwave data of L-band brightness temperatures (1.4 GHz) measured by the missions SMOS (Soil Moisture and Ocean Salinity, see Chapter 2) and SMAP (Soil Moisture Active Passive) for operational meteorological applications.

One of the basic assumptions of data assimilation methods, such as the Kalman filter or variational assimilation, is that the observations and the model are unbiased in relation to each other. In fact, the purpose of the assimilation is to correct random errors of the models but not systematic errors. Therefore, a step prior to the assimilation is necessary in order to eliminate systematic errors. The most classic approach is to use a bias correction, which allows to "correct" the observations in order to ensure that, for a large sample, the mean and the variance of the assimilated data are consistent with those of the model. The so-called CDF-matching (Cumulative Distribution Function matching [REI 04]) is used for matching the distribution functions of the observations and the corresponding variable of the model.

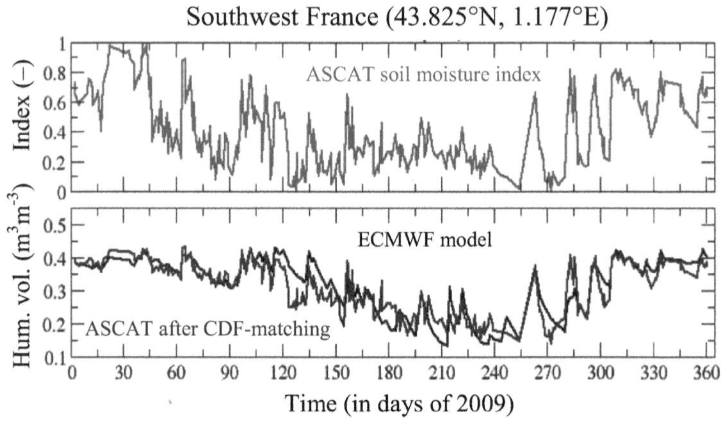

Figure 12.2. *The time series of ASCAT soil moisture index (at the top) and of volumetric moisture obtained by the ECMWF model and ASCAT after CDF-matching (at the bottom) for 2009 for a site in the southwest of France. The top curve shows the ASCAT product before correction of bias. The bottom curve shows that after CDF-matching ASCAT mean and variance are consistent with those of the ECMWF soil moisture. For a color version of this figure, see www.iste.co.uk/baghdadi/4.zip*

Figure 12.2 (top) shows a time series of surface soil moisture derived from the ASCAT sensor for a pixel located in the southwest of France for 2009. The ASCAT soil moisture product is expressed as a relative quantity, in the form of an index, which varies between 0 for dry soil and 1 for saturated soil (see Chapter 3). Figure 12.2 (bottom) shows, for the same site, the volumetric soil moisture simulated by the ECMWF model, as well as the time series of the ASCAT data after bias correction. The bias correction aligns the statistical characteristics of ASCAT and of the model, which guarantees the validity of the data assimilation approach while preserving the temporal dynamics of the observations at the scale of rainfall events. For example, around the 260th day, the ECMWF and ASCAT time series show an increase in the surface soil moisture. This indicates the occurrence of heavy precipitation, of which the effect on the satellite measurement is consistent with the model. Around the 285th day, the model and the observations also indicate an increase in soil moisture, but with an amplitude which is more pronounced in the observations than in the model. With this information (model first guess and observation), the assimilation as it is implemented at the ECMWF calculates an estimation of the soil moisture next to the surface and for the deeper soil layers, which is considered "optimal" according to certain assumptions.

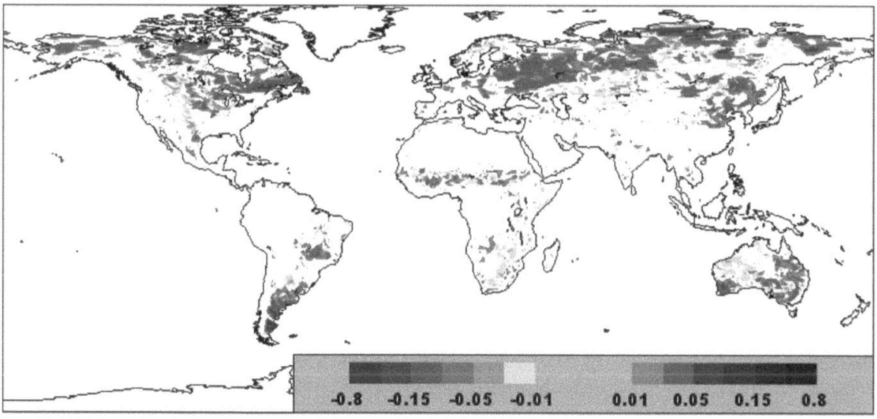

Figure 12.3. *Soil moisture increments expressed in m^3m^{-3} due to the assimilation of the ASCAT soil moisture data collected during the period from July 1st till July 15th, 2013 in the first soil layer (0–7 cm) of the H-TESSEL model of the ECMWF. For a color version of this figure, see www.iste.co.uk/baghdadi/4.zip*

Figure 12.3 shows a global map of increments of the soil surface moisture analysis, in m^3m^{-3}, collected from July 1 to July 15, 2013 generated by the assimilation of the ASCAT surface soil wetness data in the ECMWF numerical weather prediction model. The increment is the difference between the analyzed condition and the first guess of the model, which is shown here for the first layer (0–7 cm) of the ECMWF surface model. Looking at the increments allows for assessing the contribution of the assimilation in the evolution of variables of the model. Since the role of the assimilation is to correct the model random errors, the analysis increments must remain low on average. Figure 12.3 shows that in the ECMWF system the increments are relatively low (almost everywhere lower than 0.05 m^3m^{-3}) and located mainly at high latitudes of the northern hemisphere, as well as in Australia, Argentina, and, to a lesser extent, in Sahel, South African and Brazil. Expressed in mm of water and averaged on the global scale for continental surfaces, the increments are 0.21 mm with a standard deviation of 1.11 mm. Therefore, ASCAT data assimilation mainly corrects surface soil moisture, which makes this data very complementary to the observations of the air temperature and humidity already assimilated by the meteorological models [DEL 15].

Another highly relevant source of information for the soil moisture is derived from low-frequency passive microwaves sensors. The SMOS and SMAP missions have been designed specifically for soil moisture remote sensing (for SMOS, see [WIG 16] and Chapter 2 in this book for SMAP, see [ENT 10]). Their concepts are based on the L-band measurements (1.4 GHz), which represent a good compromise for obtaining a good sensitivity of the signal to the soil moisture, while minimizing the contribution of the vegetation to the signal.

The SMOS products of brightness temperature are available in real time, which makes it potentially possible to use them for operational weather forecast applications. Their assimilation in the forecast system of the ECMWF, as well as in the forecast system of the Canadian Meteorological Center (CMC), is currently being developed [MUN 14]. In order to assimilate the SMOS data, it is necessary to use an observation operator, which allows us to simulate the model equivalent of the observation, meaning the L-band brightness temperature. At ECMWF and CMC, the CMEM (*Community Microwave Emission Model* [DER 09]) modeling platform allows us to simulate L-band brightness temperatures for SMOS and SMAP. Figure 12.4 shows, for 2013, the root mean square difference (RMSD) between the SMOS observations and the ECMWF model first guess, for the XX polarization (in the antenna plane of the SMOS instrument) and for an incidence angle of 40°. Similar to the ASCAT indices, a correction of bias has been applied to the observations of brightness temperature in such a way that the mean and the variance of the observations remain consistent with the model first guess. As such, the RMSD score is a measurement of random differences between the model first guess and the observations, or in other words of the dispersion of innovations. In most regions, the RMSD ranges between 4 K and 8 K. This is the case in Europe, Australia, on a large part of the African and American continents, and on part of Asia. These are random differences of brightness temperatures, which the data assimilation system will try to minimize during the soil moisture analysis. Regions with a weak temporal variability, such as tropical forests, where the signal is dominated by a relatively constant contribution of the vegetation, demonstrate a weaker RMSD, lower than 4 K, which is included in the range of measurement accuracy. Other regions have more significant differences with RMSD values higher than 12 K. Most of the time, these large RMSD values translate a suspicious variability of observations related to a contamination of measurements by electromagnetic interferences, with ground transmitters operating in the same frequency range. Also, they can be related to such processes as snowmelt or ground freezing/thawing, which are difficult to represent in the surface models and in the microwave emission models representing a strong spatial and temporal variability. Quality controls allow for filtering of these situations. The monitoring of the innovation statistics in real time is an important component of assimilation systems, which allows us to identify and to correct the remaining problems in the quality control of the observations.

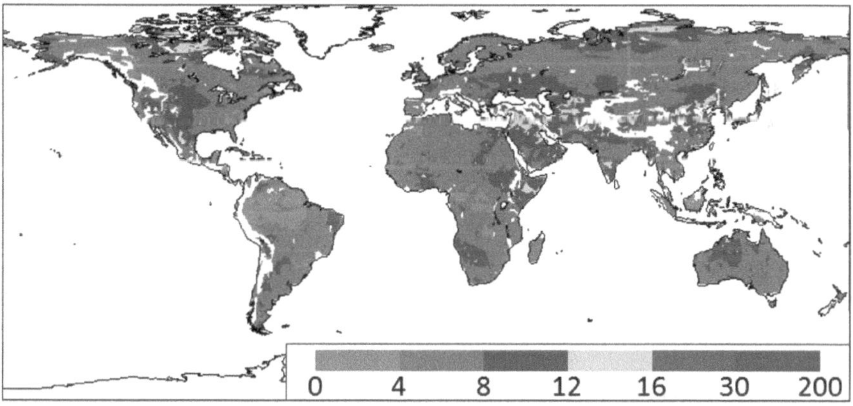

Figure 12.4. *Root mean square error for 2013 of the innovations (observation minus model first guess) of L-band brightness temperatures (1.4 GHz) expressed in K, from the observations of the SMOS satellite and the ECMWF model first guess for XX polarization and for an incidence angle of 40°. For a color version of this figure, see www.iste.co.uk/baghdadi/4.zip*

At ECMWF, the statistics of SMOS innovations are calculated separately for each incidence angle and for each polarization used, meaning in the incidence angles of 30°, 40°, and 50° and in the two XX and YY polarizations of the SMOS antenna plane. For 2013, the RMSD values of the ECMWF SMOS brightness temperatures innovations vary between 7.3 K and 8.5 K depending on incidence angles and polarizations, which indicates a remarkable agreement between temporal dynamics of SMOS observations and simulated brightness temperatures, which is an essential prerequisite for the data assimilation. Current investigations aim to evaluate the impact of the assimilation of SMOS brightness temperatures on soil moisture and numerical weather prediction [MUN 14].

Among the available spatial surface observations, LAI and surface albedo play an equally important role. Boussetta *et al.* [BOU 15] demonstrated that the assimilation of these observations for the numerical weather prediction has a direct impact on the surface atmospheric variables, such as the 2 m air temperature and humidity. In general, the climatological monthly means of LAI and the surface albedo are used in the atmospheric models and the inter-annual variability of these variables is not considered.

Figure 12.5 illustrates the impact, on surface processes, of the assimilation of the LAI and albedo data by optimal interpolation (see [OTT 16]) over Australia in November 2010. In particular, it demonstrates the difference in LAI and albedo (analyzed variables), as well as the difference of sensible heat flux (feedback to the atmosphere) in comparison with a reference experiment, which uses climatological values of these variables. After a drought period, precipitation occurred in the center and the east of Australia in November 2010. This rainy event resulted in a positive anomaly of LAI exceeding the climatological mean for more than 50% and an albedo reduction reaching 20% of its climatological value, especially in the central region of Australia where vegetation cover is relatively scattered. LAI data assimilation allows us to take inter-annual variability into consideration. The LAI positive anomaly in this region in November 2010 (Figure 12.5(c)) influences surface processes, especially in the region characterized by relatively dense vegetation (Figure 12.5(b)). The increase in LAI is reflected in the model through an increase in evapotranspiration associated with an increase in latent heat flux and a decrease in sensible heat flux reaching 20 W m^{-2} (Figure 12.5(a)).

The analyzed albedo shows a negative anomaly in the center of Australia in agreement with the observations. This anomaly increases the energy available at the surface in comparison with the reference experiment, which uses a climatological albedo (Figure 12.5(d)). In these conditions of scattered vegetation cover, this is reflected in an increase in sensible heat flux reaching 20 W m^{-2} (Figure 12.5(a)). This year, the western region of Australia was characterized by a dry anomaly, which induced a slight increase in the albedo associated with a decrease in LAI. While the decrease in LAI reduces latent heat flux in favor of sensible heat flux, the increase in the albedo diminishes the energy available at the surface. In the west of Australia, where vegetation is very sparse (Figure 12.5(b)), the effect of the albedo anomaly mainly dominates the impact on surface processes resulting in a decrease in sensible heat flux (Figure 12.5(a)).

These results demonstrate that combined assimilation of LAI and albedo data allows to take into consideration spatial and temporal variations of these variables in the models with a major impact on surface fluxes.

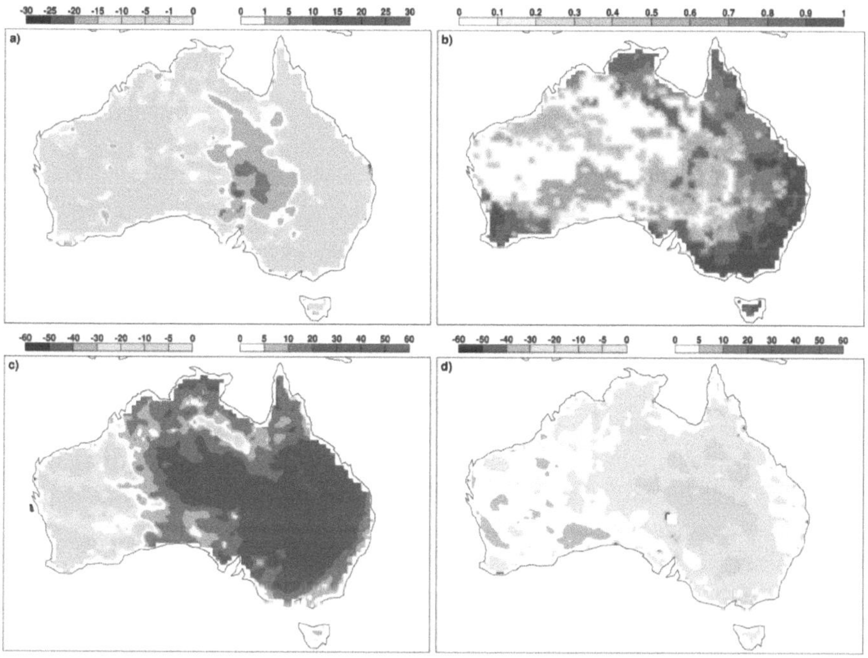

Figure 12.5. *Impact of assimilation in real time of LAI and surface albedo on surface processes in Australia in November 2010 in comparison with the use of the climatological mean of LAI and surface albedo of November for the period of 1999–2013: a) difference of sensible heat flux [W m^{-2}], c) relative anomaly of LAI [%], d) relative anomaly of the albedo [%]. The fraction of vegetation cover is indicated as b). For a color version of this figure, see www.iste.co.uk/baghdadi/4.zip*

12.3. Water: from the soil to the river

The implementation of hydrological models with a physical basis became difficult because of the uncertainty about the value of the parameters of these models and about atmospheric forcings. Satellite data assimilation in these models is a way to facilitate this implementation and remains a very active subject of the research.

Space altimetry provides quantitative information about the height of open water surfaces. The main obstacle in the use of this technique for monitoring the height of rivers and reservoirs is that the current instruments cannot reach a sufficiently fine spatial resolution. The objective of the space mission project Surface Water and Ocean Topography (SWOT) is to meet this observation need. Studies based on synthetic observations demonstrated the potential of this data for hydrology by using assimilation techniques in hydrological models [BIA 11, MUN 15].

Moreover, the analysis of vegetation and soil moisture described above can contribute to the improvement of hydrological models. Data assimilation for modeling land surface offers a framework, which allows to use the observation of surface soil moisture and snow cover from space [HOU 98]. In particular, several studies demonstrated that assimilation of surface soil moisture observations in land surface models improves simulation of soil moisture in the deepest layers of the soil [ENT 94, RAG 95, HOU 98, WAL 01, CRO 03, SAB 07, REI 05, BAL 07] and heat fluxes simulated by the model [PIP 13]. Nevertheless, the relation between the surface soil moisture and soil moisture in the deepest layers is often represented in a simplified way. This has consequences for the results of assimilation and the analyzed value of soil moisture in the root zone is more often far from the reality than when a more detailed model is used [PAR 14].

A difficulty in assimilation of the surface soil moisture (expressed in volumetric moisture) is that this quantity depends on the soil parameters used by the model (notably, the texture). These differences between the model and observations must be corrected or otherwise risk introducing a bias in the water content of the deepest layers [CAL 00]. For this, it is possible to use a simple linear transformation or a more complex adjustment of CDF matching type (see above). It is important to remember that volumetric soil moisture, as opposed to other biophysical variables, is a variable, the value of which may vary depending on a model, without necessarily creating differences in simulation of the water balance and fluxes.

Vegetation intervenes directly into the water cycle by intercepting precipitation (a part of which evaporates before reaching the soil) and returning the water content in the soil to the atmosphere through the action of roots and foliar transpiration. The temporal evolution of the water content in the soil is closely related to foliar transpiration and LAI. Assimilation of LAI observations is likely to improve simulation of soil moisture [BAR 11] and to affect the surface runoff and drainage, especially in winter conditions and during the phase of vegetation growth.

Figures 12.6, 12.7 and 12.8 illustrate the ability of assimilation of observations of the surface soil moisture or LAI to improve simulation of the water content in the deepest layers and river flow.

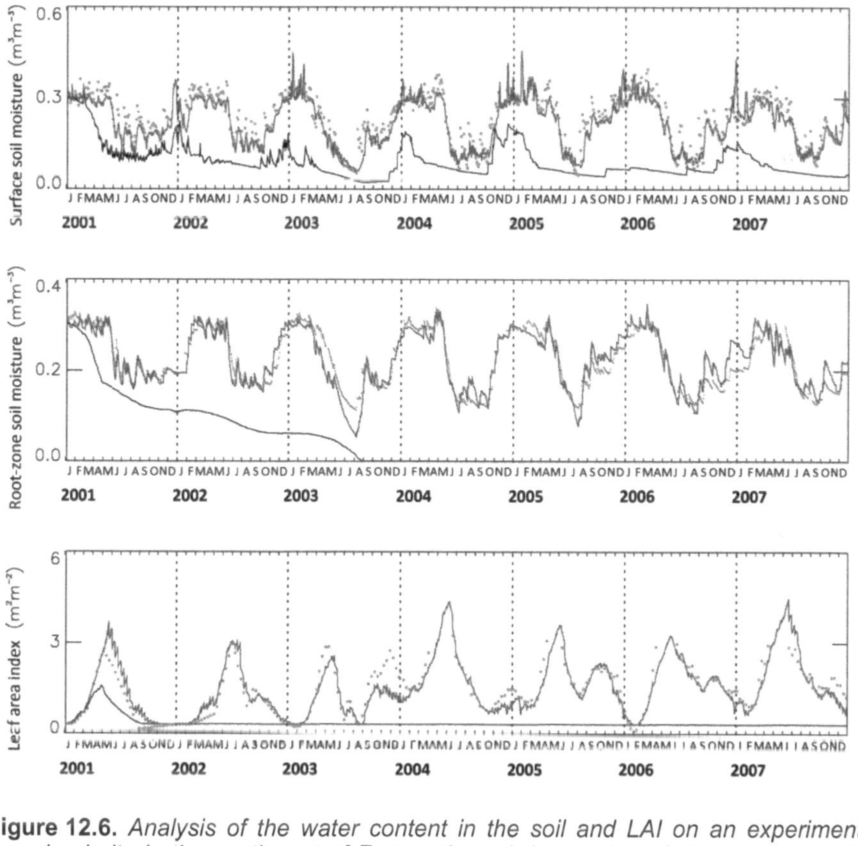

Figure 12.6. *Analysis of the water content in the soil and LAI on an experimental grassland site in the southwest of France through integration of observations of the surface soil moisture (adapted from [ALB 10]). Black line: model with zero precipitation; red line: analysis; green points: in situ observations. From top to bottom: the surface soil moisture (layer of 1 cm thick), the total water content of the soil (layer of 95 cm thick in this study), LAI. For a color version of this figure, see www.iste.co.uk/baghdadi/4.zip*

Figure 12.6, adapted from Albergel *et al.* [ALB 10], demonstrates that the use of observations of the surface soil moisture allows to compensate to a large degree for the absence of information about precipitation. This grassland is located in the southwest of France, where there are local measures of soil moisture and LAI. The authors use a model, which can correctly represent the water balance and vegetation growth if it is provided with a good quality atmospheric forcing, including precipitation. The withdrawal of this atmospheric forcing information (imposing a null value of precipitation) provokes in several months a complete drying of the land and mortality of vegetation (then, simulated LAI moves towards its minimal value). Figure 12.6 shows that integration into the model of daily *in situ* observations of the surface soil moisture through data assimilation allows the analysis to correctly represent the annual vegetation cycle and the water balance.

It is interesting to note that this relation between the information provided by the surface soil moisture and the information provided by precipitation can be exploited in order to estimate cumulative precipitation using time series of satellite observations of the surface soil moisture [CRO 11, BRO 14].

Figure 12.7 illustrates the impact of the vegetation cycle on large-scale hydrology, on more than 300 gauging stations in France covering the range of river basins in France. The function of cumulative distribution of the Nash–Sutcliffe score of simulated daily flow rates is presented for the two SAFRAN–ISBA–MODCOU simulations [HAB 08] using the most recent version of the ISBA surface model of "Météo-France" described in [CAR 13]. In one simulation, LAI is simulated by the model, in the other, LAI is derived from satellite observations and imposed in the model input [FAR 13]. The use of satellite LAI allows to obtain the best scores (closer to 1) for a larger number of gauging stations.

These improvements are explained by the fact that satellite LAI is larger than simulated LAI in winter and in spring. Consequently, this increases evapotranspiration, diminishes the total water content in the soil and reduces flow rates of rivers. This reduction of flow rates allows to reconcile simulations and observations.

This example demonstrates the important role of vegetation on the water cycle. In addition, it should be considered that neither models nor satellite products are perfect. For example, the MODCOU model does not represent the impact of dams and reservoirs, which results in an error in simulation of flow rates. The reservoirs store the water in winter and in spring, which consequently reduces flow rates of rivers. An overestimation of satellite LAI would consequently compensate for this error.

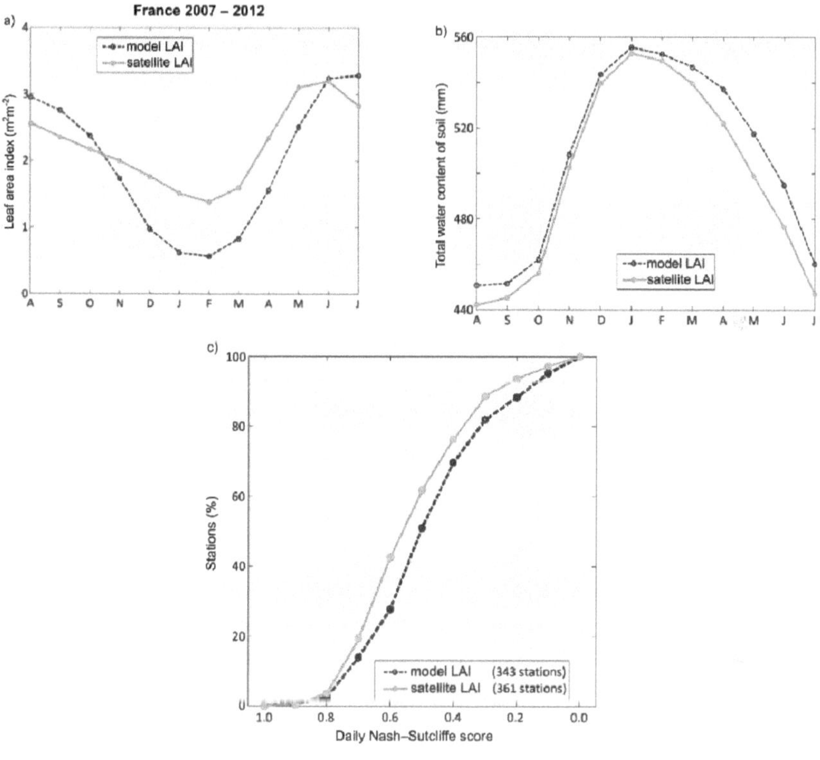

Figure 12.7. *Impact of vegetation on hydrology. Mean monthly values for France, from 2007 to 2012 a) of LAI, b) of the total water content of the land, and c) of the function of cumulative distribution of the Nash–Sutcliffe score, for the two simulations: in blue, LAI simulated by the model, in green, LAI derived from satellite observations [FAR 13]. For a color version of this figure, see www.iste.co.uk/baghdadi/4.zip*

Finally, Figure 12.8, adapted from [PAR 14], illustrates the contribution of the multi-layer soil model in the representation of the link between the surface soil moisture and the water content in the deepest soil layers. In

Parrens *et al.* [PAR 14], an extended Kalman filter is applied to a multi-layer soil version of the ISBA model of "Météo-France", in the case of bare soil in the southwest of France. The central aspect of this assimilation method is the calculation of Jacobian of the observation operator. The observation operator allows to project a condition of the model in the area of observations. The Jacobian provides a linear relation between observations (here, the surface soil moisture) and a state variable of the model (here, soil moisture in the root zone in a given depth). The Jacobian can be calculated through the finite differences method, by imposing disturbances to the analyzed variable at the beginning of the assimilation window and by calculating the ratio of:

– the difference between a simulated observation and its reference condition;

– the disturbance imposed to the state variable of the model.

The Kalman gain, which is calculated from the Jacobian, enables us to convert the difference between the observation and its counterpart issued from the model to the analysis increment (modification of the state variable of the model). Figure 12.8 shows the Jacobian and the Kalman gain values in various depths of the soil during three annual cycles. It illustrates sensitivity of the surface soil moisture simulated by the model to a small disturbance of the water content of the deepest soil layer, on a daily basis. The surface soil moisture is related to the deepest layers (below 20 cm) only in moist conditions (notably, in winter). In summer conditions, the model simulates a separation between surface and deep layers. A simpler model, which would use only one or two soil layers to represent the amount of water in the soil, could not simulate this separation. Another default of simplified approaches to represent water transfers in the soil is the use of a trigger threshold of drainage generally corresponding to the field capacity. With such models, assimilation of the surface soil moisture affects hydrology only when it adds the water beyond the field capacity. While in winter the surface soil moisture is close to field capacity, assimilation tends to systematically add water. Even if this has a little impact on simulated soil moisture (because of the threshold effect), the added water is drained and this increases artificially simulated flow rates. In order to avoid this type of artifact, a model of sufficient complexity should be used, which is capable of representing explicitly diffusion processes.

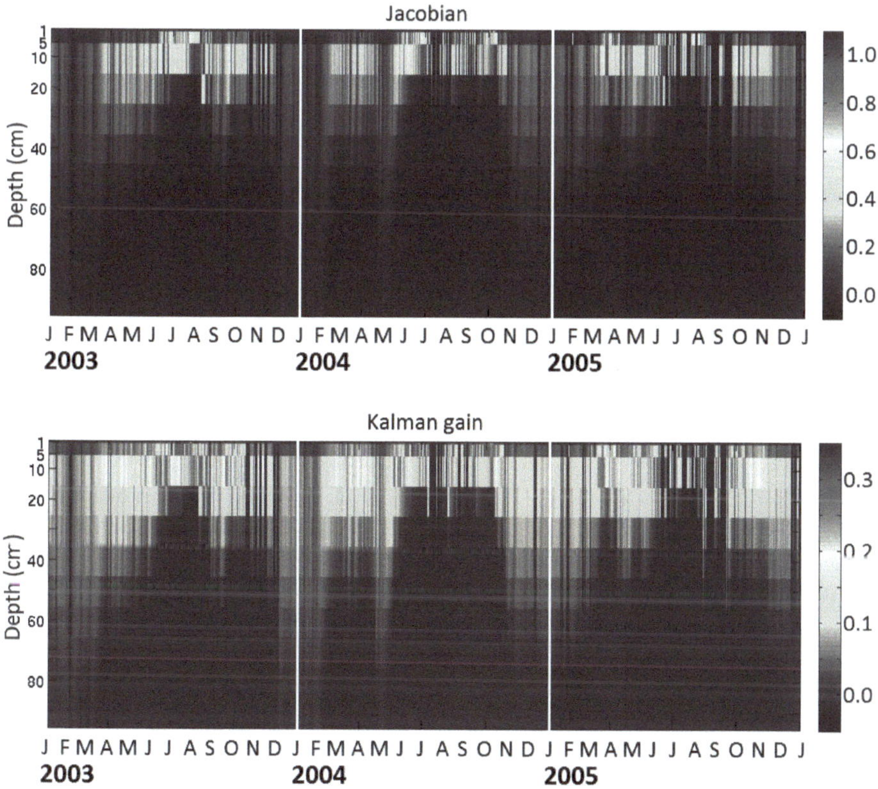

Figure 12.8. *Decoupling between surface and deep soil layers. Jacobian and Kalman gain of bare soil (adapted from [PAR 14]). For a color version of this figure, see www.iste.co.uk/baghdadi/4.zip*

12.4. Natural sinks and sources of CO_2

Land surfaces play an important role in the evolution of atmospheric CO_2, the increase of which is caused by anthropogenic emissions of CO_2. From year to year, vegetation of certain regions of the world can form a sink or a source of CO_2. This inter-annual variability is related to climatic conditions and events, such as droughts or cold waves. The water cycle and the carbon cycle of the land are closely connected. In many regions of the world, vegetation has a strong biological control over hydrological processes, via interception of a part of precipitation by leaves and litter, and stomatal regulation of foliar transpiration. The presence of vegetation also has a considerable impact on direct evaporation of the water in surface soil layers. Inversely, water stress due to a deficit of water in the soil has a direct impact on photosynthesis, autotrophic respiration, mortality of vegetation, and on heterotrophic respiration of the soil. Representing these two aspects (water and carbon) in the data assimilation system ensures a synergy between observations of a different kind. An LDAS based on a land surface model able to simulate the components of the net flux of CO_2 of vegetation, the link between photosynthesis and soil moisture, and the water balance of the soil, allows to implement synergies between satellite observations.

Figure 12.9 demonstrates the example of drought in spring of 2011 in France (adapted from [BAR 14]). This drought was one of the most severe since the 1960s. Very sunny and hot weather during a long period of time, while the soil was still moist at the end of winter, favored photosynthesis and provoked an unusual growth of vegetation. This event can be qualified as a "false spring". Vegetation was more developed than average, which provoked an accelerated evaporation of the water reserve in the soil. In these conditions, a water deficit rapidly developed in May 2011 and reached the values, which are usually observed in July and August. This example shows that the growth of vegetation affects the severity of droughts and that monitoring of vegetation together with monitoring of hydrological variables allows to characterize droughts better.

In particular, the analysis resulting from LAI data assimilation allows to improve other variables, facilitating the evaluation of the impact of extreme climatic events on vegetation, and even agricultural production [ATZ 13]. For example, above-ground biomass of crops or grasslands is not directly measurable by remote sensing, but improving their representation by a model through assimilation of LAI allows to produce more relevant indicators for certain applications, such as monitoring of agricultural droughts.

Assimilation of LAI can be realized on different temporal levels. Long periods (several years) can be considered in order to calibrate the value of key parameters of the soil-plant system, related, for example, to the availability of nutrients or to the maximal water reserve of the soil [KUP 12]. This approach is often called CCDAS (Carbon Cycle Data

Assimilation System [RAY 05]). Also, much shorter periods (from 1 to 10 days) can be considered in order to ensure monitoring of vegetation in near real time. This is the LDAS approach described above. These two approaches are complementary and should be used together in order to obtain the best simulations.

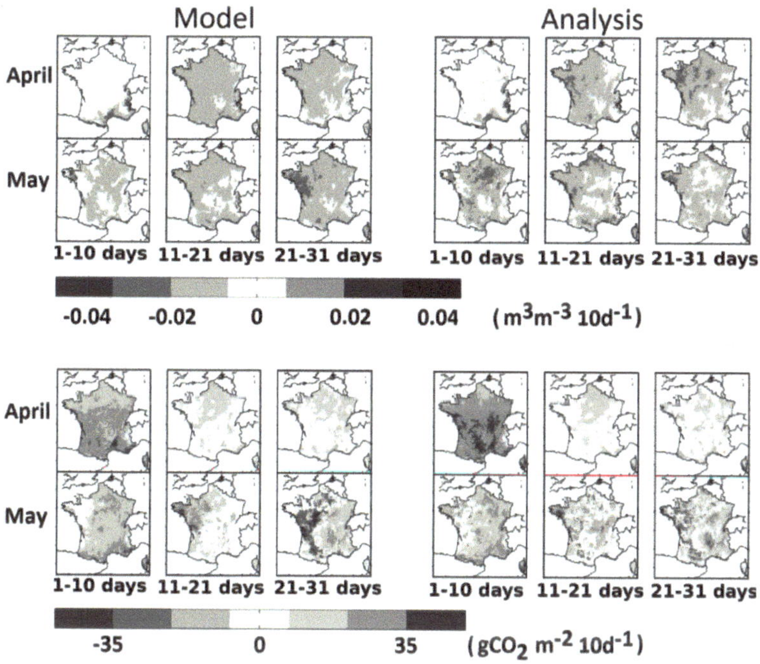

Figure 12.9. *Drought in spring of 2011 in France, in April and in May. Decadal variations of the total water content in the soil (at the top) and of photosynthesis (at the bottom) (adapted from [BAR 14]) simulated by the SURFEX platform (externalized surface) of "Météo-France", without integration of satellite data (on the left) and after assimilation of LAI and surface soil moisture observations (on the right). For a color version of this figure, see www.iste.co.uk/baghdadi/4.zip*

It should be noted that the possibility to simulate the net fluxes of CO_2 allows to use the estimations of these fluxes issued from observations of atmospheric CO_2 [CHE 11] in order to validate the analysis and to detect potential defaults. The synergy between observations is then extended to an atmospheric variable. Similarly, the interaction of LDAS with a hydrological model allows for the validation of surface analysis through *in situ* observations of flow rates of rivers (see above).

12.5. Conclusions and perspectives

Recent studies demonstrated that assimilation of biophysical variables of the soil–plant system derived from satellite data (LAI, soil moisture, surface albedo, snow cover) has a significant impact in applications of different kinds (numerical weather prediction, hydrology, monitoring of droughts and carbon fluxes).

Ensemble methods open the possibility to represent model and forcing errors varying in time. Theoretical studies showed the significant advantage of these methods in comparison with the SEKF, and they will be probably introduced for practical reasons. Ensemble methods are indeed more effective in terms of computing time, especially when a large number of variables have to be analyzed.

Several challenges have to be addressed, notably in the representation of cultivated zones and managed forests, where the impact of anthropogenic effects on the water and carbon balances is crucial. In this field, spatial and temporal high-resolution remote sensing allows to characterize fine-scale agricultural and forestry practices, as well as probably the management of reservoirs used for irrigation or electricity generation. The Copernicus Sentinel missions 1 and 2 start providing an unprecedented data source.

12.6. Key points

– Long series of satellite observations of land surface (30 years and more) that allow reanalyses to be generated are now available, making it possible to improve our understanding of the climate considerably.

– Satellite products are now provided for free in near real time; this enables us to monitor land surfaces, using algorithms that integrate these observations, notably LDAS (Land Data Assimilation Systems).

– Assimilation of satellite data for land surfaces has considerably improved during the last decade, but it has not yet reached the level of maturity found in other fields, such as atmospheric sciences and oceanography; today, the main difficulty is to characterize errors of the models and observations in very heterogeneous conditions.

– Synergies between different types of observations are potentially very beneficial for Land Data Assimilation Systems; in turn, data assimilation is very useful to monitor the quality of satellite products over time and their consistency.

– Research in land surface data assimilation and modeling has to be conducted within a multidisciplinary framework.

12.7. Bibliography

[ALB 10] ALBERGEL C., CALVET J.C., MAHFOUF J.F. *et al.*, "Monitoring of water and carbon fluxes using a land data assimilation system: a case study for southwestern France", *Hydrology and Earth System Sciences*, vol. 14, pp. 1109–1124, 2010.

[ATZ 13] ATZBERGER C., "Advances in remote sensing of agriculture: Context description, existing operational monitoring systems and major information needs", *Remote Sensing*, vol. 5, pp. 949–981, 2013.

[BAL 07] BALSAMO G., MAHFOUF J., BÉLAIR S. *et al.*, "A land data assimilation system for soil moisture and temperature: an information content study", *Journal of Hydrometeorology*, vol. 8, pp. 1225–1242, 2007.

[BAR 11] BARBU A.L., CALVET J.C., MAHFOUF J.-F. *et al.*, "Assimilation of Soil Wetness Index and Leaf Area Index into the ISBA-A-gs land surface model: grassland case study", *Biogeosciences*, vol. 8, pp. 1971–1986, 2011.

[BAR 14] BARBU A. L., CALVET J.C., MAHFOUF J.F. *et al.*, "Integrating ASCAT surface soil moisture and GEOV1 leaf area index into the SURFEX modelling platform: a land data assimilation application over France", *Hydrology and Earth System Sciences*, vol. 18, pp. 173–192, 2014.

[BEL 96] BELJAARS A.C.M., VITERBO P., MILLER M. *et al.*, "Sensitivity to land surface parameterization and soil anomalies", *Monthly Weather Review*, vol. 124, pp. 362–383, 1996

[BEL 03] BÉLAIR S., CREVIER L.P., MAILHOT J. *et al.*, "Operational implementation of the ISBA land surface scheme in the Canadian regional weather forecast model. Part I: warm season results", *Journal of Hydrometeorology*, vol. 4, pp. 352–470, 2003.

[BIA 11] BIANCAMARIA S., DURAND M., ANDREADIS K.M. *et al.*, "Assimilation of virtual wide swath altimetry to improve Arctic river modeling", *Remote Sensing of Environment*, vol. 115, no. 2, pp. 373–381, 2011.

[BOU 15] BOUSSETTA S., BALSAMO G., DUTRA E. *et al.*, "Assimilation of surface albedo and vegetation states from satellite observations and their impact on numerical weather prediction", *Remote Sensing of Environment*, vol. 163, pp. 111–126, 2015.

[BRO 14] BROCCA L., CIABATTA L., MASSARI C. *et al.*, "Soil as a natural rain gauge: Estimating global rainfall from satellite soil moisture data", *Journal of Geophysical Research – Atmosphere*, vol. 119, pp. 5128–5141, 2014.

[CAL 00] CALVET J.C., NOILHAN J., "From near-surface to root-zone soil moisture using year-round data", *Journal of Hydrometeorology*, vol. 1, pp. 393–411, 2000.

[CAR 12] CARRER D., LAFONT S., ROUJEAN J.L. *et al.*, "Incoming solar and infrared radiation derived from METEOSAT: impact on the modelled land water and energy budget over France", *Journal of Hydrometeorology*, vol. 13, no. 2, pp. 504–520, 2012.

[CAR 13] CARRER D., ROUJEAN J.L., LAFONT S. *et al.*, "A canopy radiative transfer scheme with explicit FAPAR for the interactive vegetation model ISBA-A-gs: Impact on carbon fluxes", *Journal of Geophysical Research – Biogeosciences*, vol. 118, pp. 888–903, 2013.

[CAR 14] CARRER D., MEUREY M., CEAMANOS X. *et al.*, "Dynamic mapping of snow-free vegetation and bare soil albedos at global 1 km scale from 10-year analysis of MODIS satellite products", *Remote Sensing of Environment*, vol. 140, pp. 420–432, 2014.

[CHE 11] CHEVALLIER F., DEUTSCHER N., CONWAY T. *et al.*, "Global CO_2 fluxes inferred from surface air-sample measurements and from TCCON retrievals of the CO_2 total column", *Geophysical Research Letters*, vol. 38, p. L24810, 2011.

[CRO 03] CROW W.T., WOOD E., "The assimilation of remotely sensed soil brightness temperature imagery into a land surface model using ensemble Kalman filtering: A case study based on ESTAR measurements during SGP97", *Advances in Water Resources*, vol. 26, pp. 137–149, 2003.

[CRO 11] CROW W.T., VAN DEN BERG M.J., HUFFMAN G.F. *et al.*, "Correcting rainfall using satellite-based surface soil moisture retrievals: The Soil Moisture Analysis Rainfall Tool (SMART)", *Water Resources Research*, vol. 47, p. W08521, 2011.

[DEL 15] DE LANNOY G.J.M., DE ROSNAY P., REICHLE R.H., "Soil moisture data assimilation", In DUAN Q., PAPPENBERGER F., THIELEN J. *et al.* (eds), *Handbook of Hydrometeorological Ensemble Forecasting*, Springer Verlag, New York, 2015.

[DER 09] DE ROSNAY P., DRUSCH M., BOONE A. *et al.*, "The AMMA Land Surface Model Intercomparison Experiment coupled to the Community Microwave Emission Model: ALMIP-MEM", *Journal of Geophysical Research - Atmosphere*, vol. 114, 2009.

[DER 13] DE ROSNAY P., DRUSCH M., VASILJEVIC D. *et al.*, "A simplified extended Kalman filter for the global operational soil moisture analysis at ECMWF", *Quarterly Journal of the Royal Meteorological Society*, vol. 139, no. 674, pp. 1199–1213, 2013.

[DER 14] DE ROSNAY P., BALSAMO, G., ALBERGEL *et al.*, "Initialisation of land surface variables for numerical weather prediction", *Survey in Geophysics*, vol. 35, no. 3, pp. 607–621, 2014.

[DHA 11] DHARSSI I., BOVIS K., MACPHERSON B. *et al.*, "Operational assimilation of ASCAT surface soil wetness at the met office", *Hydrology and Earth System Sciences,* vol. 15, pp. 2729–2746, 2011.

[DRA 09] DRAPER C., MAHFOUF J.F., WALKER J., "An EKF assimilation of AMSR-E soil moisture into the ISBA land surface scheme", *Journal of Geophysical Research - Atmosphere*, vol. 114, p. D20104, 2009.

[DRA 11] DRAPER C., MAHFOUF J.F., CALVET J.C. *et al.*, "Assimilation of ASCAT near-surface soil moisture into the SIM hydrological model over France", *Hydrology and Earth System Sciences*, vol. 15, pp. 3829–3841, 2011.

[ENT 94] ENTEKHABI D., NAKAMURA H., NJOKU E., "Solving the inverse problem for soil moisture and temperature profiles by sequential assimilation of multifrequency remotely sensed observations", *IEEE Transactions on Geoscience and Remote Sensing*, vol. 32, pp. 438–448, 1994.

[ENT 04] ENTEKHABI D., NJOKU E.G., HOUSER P. *et al.*, "The hydrosphere state (Hydros) satellite mission: an earth system pathfinder for global mapping of soil moisture and land freeze/thaw", *IEEE Transactions on Geoscience and Remote Sensing,* vol. 42, pp. 2184–2195, 2004.

[ENT 10] ENTEKHABI D., NJOKU E., O'NEILL P. *et al.*, "The soil moisture active passive (SMAP) mission", *Proceeding of the IEEE*, vol. 98, no. 5, pp. 704–716, 2010.

[FAR 13] FAROUX S., KAPTUÉ TCHUENTÉ A.T., ROUJEAN J.L. *et al.*, "ECOCLIMAP-II/Europe: a twofold database of ecosystems and surface parameters at 1 km resolution based on satellite information for use in land surface, meteorological and climate models", *Geoscientific Model Development*, vol. 6, pp. 563–582, 2013.

[HAB 08] HABETS F., BOONE A., CHAMPEAUX J.L. *et al.*, "The SAFRAN-ISBA-MODCOU hydrometeorological model applied over France", *Journal of Geophysical Research – Atmosphere*, vol. 113, p. D06113, 2008.

[HOU 98] HOUSER P.R., SHUTTLEWORTH W.J., FAMIGLIETTI J.S. *et al.*, "Integration of soil moisture remote sensing and hydrologic modeling using data assimilation", *Water Resources Research*, vol. 34, pp. 3405–3420, 1998.

[ING 16] INGLADA T., "Land-cover mapping from optical images", in BAGHDADI N., ZRIBI M. (eds), *Land Surface Remote Sensing in Agriculture and Forest*, ISTE Ltd, London and Elsevier, Oxford, 2016.

[KAL 03] KALNAY E., *Atmospheric Modelling, Data Assimilation and Predictability*, Cambridge University Press, 2003.

[KAM 12] KAMINSKI T., KNORR W., SCHOLZE M. *et al.*, "Consistent assimilation of MERIS FAPAR and atmospheric CO_2 into a terrestrial vegetation model and interactive mission benefit analysis", *Biogeosciences*, vol. 9, pp. 3173–3184, 2012.

[KER 01] KERR Y., WALDTEUFEL P., WIGNERON J.P. *et al.*, "Soil moisture retrieval from space: The Soil Moisture and Ocean Salinity (SMOS) mission", *IEEE Transactions on Geoscience and Remote Sensing*, vol. 39, pp. 1729–1735, 2001.

[KOS 04] KOSTER R.D., DIRMEYER P., GUO Z. *et al.*, "Regions of strong coupling between soil moisture and precipitation", *Science*, vol. 305, pp. 1138–1140, 2004.

[KRI 05] KRINNER G., VIOVY N., DE NOBLET-DUCOUDRÉ N. *et al.*, "A dynamic global vegetation model for studies of the coupled atmosphere-biosphere system", *Global Biogeochemical Cycles*, vol. 19, p. GB1015, 2005.

[KUP 12] KUPPEL S., PEYLIN P., CHEVALLIER F. *et al.*, "Constraining a global ecosystem model with multi-site eddy-covariance data", *Biogeosciences*, vol. 9, pp. 3757–3776, 2012.

[LIU 12] LIU Y., WEERTS A.H., CLARK M. *et al.*, "Advancing data assimilation in operational hydrologic forecasting: progresses, challenges, and emerging opportunities", *Hydrology and Earth System Sciences*, vol. 16, pp. 3863–3887, 2012.

[MAH 09] MAHFOUF J.F., BERGAOUI K., DRAPER C. *et al.*, "A comparison of two off-line soil analysis schemes for assimilation of screen level observations", *Journal of Geophysical Research – Atmosphere*, vol. 114, p. D08105, 2009.

[MAH 10] MAHFOUF J.F., "Assimilation of satellite-derived soil moisture from ASCAT in a limited-area NWP model", *Quarterly Journal of the Royal Meteorological Society*, vol. 136, pp. 784–798, 2010.

[MAS 13] MASSON V., LE MOIGNE P., MARTIN E. *et al.*, "The SURFEXv7.2 land and ocean surface platform for coupled or offline simulation of earth surface variables and fluxes", *Geoscientific Model Development*, vol. 6, pp. 929–960, 2013.

[MON 12] MONTZKA C., PAUWELS R.N.V., FRANSSEN H.-J.H. *et al.*, "Multivariate and multiscale data assimilation in terrestrial systems: a review", *Sensors*, vol. 12, pp. 16291–16333, 2012.

[MUN 14] MUÑOZ-SABATER J., DE ROSNAY P., JIMÉNEZ C. *et al.*, "SMOS brightness temperatures angular noise: characterization, filtering and validation", *IEEE Transactions on Geoscience and Remote Sensing*, vol. 52, no. 9, pp. 5827–5839, 2014.

[MUN 15] MUNIER S., POLEBISTKI A., BROWN C. *et al.*, "SWOT data assimilation for operational reservoir management on the upper Niger River Basin", *Water Resources Research*, vol. 51, pp. 554–575, 2015.

[NJO 03] NJOKU E., JACKSON T., LAKSHMI V. *et al.*, "Soil moisture retrieval from AMSR-E", *IEEE Transactions on Geoscience and Remote Sensing*, vol. 41, pp. 215–229, 2003.

[OTT 16] OTTLÉ C., MAHFOUF J.-F., "Data assimilation of observations from space", in BAGHDADI N., ZRIBI M. (eds), *Land Surface Remote Sensing in Agriculture and Forest*, ISTE Ltd, London and Elsevier, Oxford, 2016.

[PAR 14] PARRENS M., MAHFOUF J.-F., BARBU A.L. *et al.*, "Assimilation of surface soil moisture into a multilayer soil model: design and evaluation at local scale", *Hydrology and Earth System Sciences*, vol. 18, pp. 673–689, 2014.

[PIP 13] PIPUNIC R., WALKER J., WESTERN A. *et al.*, "Assimilation of multiple data types for improved heat flux prediction: A one dimensional field study", *Remote Sensing of Environment*, vol. 136, pp. 315–329, 2013.

[RAB 98] RABIER F., THÉPAUT J.N., COURTIER P., "Extended assimilation and forecast experiments with a four-dimensional variational data assimilation system", *Quarterly Journal of the Royal Meteorological Society*, vol. 124, pp. 1861–1887, 1998.

[RAG 95] RAGAB R., "Towards a continuous operational system to estimate the root-zone soil moisture from intermittent remotely sensed surface moisture", *Journal of Hydrology*, vol. 173, pp. 1–25, 1995.

[RAY 05] RAYNER P.J., SCHOLZE M., KNORR W. *et al.*, "Two decades of terrestrial carbon fluxes from a carbon cycle data assimilation system (CCDAS)", *Global Biogeochemical Cycles*, vol. 19, pp. 1–23, 2005.

[REI 04] REICHLE R.H., KOSTER R.D., "Bias reduction in short records of satellite soil moisture", *Geophysical Research Letters*, vol. 31, p. L19501, 2004.

[REI 05] REICHLE R.H., KOSTER R.D., "Global assimilation of satellite surface soil moisture retrievals into the NASA Catchment land surface model", *Geophysical Research Letters*, vol. 32, p. L02404, 2005.

[SAB 07] SABATER J.M., JARLAN L., CALVET J.C. *et al.*, "From near-surface to root-zone soil moisture using different assimilation techniques", *Journal of Hydrometeorology*, vol. 8, pp. 194–206, 2007.

[SCH 83] SCHMUGGE T.J., "Remote sensing of soil moisture: Recent advances", *IEEE Transactions on Geoscience and Remote Sensing*, vol. GE-21, pp. 336–344, 1983.

[VAN 01] VAN DEN HURK B.J.J.M., "Energy balance based surface flux estimation from satellite data, and its application for surface moisture assimilation", *Meteorology and Atmospheric Physics,* vol. 76, pp. 43–52, 2001.

[WAG 07] WAGNER W., NAEIMI V., SCIPAL K. *et al.*, "Soil moisture from operational meteorological satellites", *Hydrogeology Journal*, vol. 15, pp. 121–131, 2007.

[WAL 01] WALKER J., HOUSER P., "A methodology for initializing soil moisture in a global climate model: Assimilation of near-surface soil moisture observations", *Journal of Geophysical Research – Atmosphere*, vol. 106, pp. 11761–11774, 2001.

[WIG 16] WIGNERON J.-P., KERR Y., "Passive low frequency microwaves: principles, radiative transfer, physics of measurements", in BAGHDADI N., ZRIBI M. (eds), *Land Surface Remote Sensing in Agriculture and Forest*, ISTE Ltd, London and Elsevier, Oxford, 2016.

Glossary

Aerodynamic temperature: Theoretical temperature of the average source of sensitive heat flux for a complex surface, corresponding for a vegetation cover to the extrapolation within the canopy of the air temperature profile observed above the canopy.

Albedo: Is the fraction of solar energy (shortwave radiation) reflected from the Earth back into space. It is a measure of the reflectivity of the earth's surface. Ice, especially with snow on top of it, has a high albedo. Water is much more absorbent and less reflective.

Altimetric waveform: Histogram of the re-emitted energy by the surface in function of the round-trip time of the altimetric wave, allowing to calculate the height of the observed surface and its main features.

Altimetry: Sensor that measures the round-trip time of a wave transmitted from nadir to the surface. Knowing the speed of propagation of the wave in the different elements of the atmosphere and the position of the satellite relative to the Earth's center of gravity, the height of the observed surface is therefore deduced.

Amazon: With a length of more than 7,000 km, it is the river that drains the largest area in the world (~7 million km^2) and that has the strongest flow at the mouth. The Amazon is sourced in the Andes and flows into the Atlantic Ocean.

Analysis (for assimilation): Vector that contains the state of the system obtained after data assimilation.

Analysis increment: Difference vector between the state analyzed and the model background.

Antenna aperture (altimetry): Aperture at 3 dB of the antenna lobe that controls the shape of the altimetric wave and the observation limit as a function of on the slope of the surface. It depends on the antenna size and the transmission frequency (0.8° for the EnviSat altimeter, 0.305° for that of AltiKa).

Azimuth in radar: Illumination direction along the radar track. The spatial resolution is constant.

Background (in data assimilation): Vector containing the a priori estimate of the state of the system, obtained from simulations or from a climatology allowing to describe seasonal variability.

Background error (data assimilation): The background error measures the difference between the estimated state of the system before analysis (the *prior* state) and the true state.

Backscattering: The reflection of waves back to the direction from which they came (sensor), depending on the surface properties and the sensor parameters.

Bidirectional Reflectance Distribution Function (BRDF): Ratio between the reflected luminance in a given direction and the illumination from an incident direction.

Biomass: In ecology, biomass is the amount of living material in a given habitat. Often, it is estimated in a surface unit or volume rather than absolute mass. The vegetation cover is usually estimated by removing some plants which are weighed fresh and dry (after passing in an oven) in order to obtain dry biomass. In practice, we distinguish aboveground biomass (trunk, branches, leaves) from below-ground biomass (roots).

Bistatic scattering: The electromagnetic scattering phenomenon considered when the radar receiver is distinct of the radar transmitter contrary to backscattering which is considered when using a monostatic radar system where the transmitter and receiver are collocated.

Brightness temperature: Temperature obtained by converting the radiance measured according to Planck's Law at the level of the sensor, weighted by its transfer function; it can be defined for a measurement in a given direction (called directional brightness temperature).

Cartographic reference system: Coordinate system allowing to locate a point in an absolute manner to the surface of the Earth.

Classification (method of): Technique to determine the affiliation of an object (for example, a pixel of an image) to a group (class) using the object's attributes (for example, its spectral signature). It can be supervised or unsupervised.

Conduction flux: Density of conductive flux corresponding to the transfer by diffusion into the ground according to a vertical temperature gradient and thermal characteristics of the medium. SI Unit: Wm^{-2}.

Confusion matrix: Double entry table used to analyze the quality of a classification; the rows correspond to the class of the evaluated objects and the columns correspond to the assigned class; the boxes in the table indicate the number of objects.

Contextual methods (models): Methods that allow estimating a magnitude (for example, evapotranspiration.) depending on the position of n explanatory variable (for example: reflectances, temperatures....) in an n-dimensional space.

Copernicus: A European Earth spatial monitoring program, previously known as Global Monitoring for Environment and Security (GMES).

Correlation length (soil): It is used for the horizontal description of the surface roughness. It is usually defined as the displacement for which the correlation function is equal to 1/e.

Data assimilation: Mathematical methods that serve to optimally correct the previsions of a dynamic model (for example a TSVA model) by using available observations (for example field measurements or measurements issued from remote sensing).

Delay Doppler Map (DDM): Map representing the magnitude of the correlation between direct and reflected GNSS measurements depending on the time delay and the frequency change due to the Doppler Effect.

Delta: Mouth of a river with a triangular shape with several branches, each carrying alluvium toward a sea, an ocean or lake.

Dense medium radiative transfer (DMRT): Modification of the conventional radiative transfer model for dense medium such as snow. This method attempts to take into consideration the dependent diffusion phenomena between closely spaced particles, unlike the conventional radiative transfer which considers each particle as independent broadcasters.

Depth hoar: Large snow grains formed by metamorphism in the presence of a strong thermal gradient. These grains are angular or faceted and can take the form of fragile hollow cups near the bottom of the snowpack.

Dielectric constant: Known physical quantity also known as complex permittivity. It characterizes the electrical properties of the soil in terms of electrical losses due to transportation and absorption of the energy provided by the radar wave. In the case of soil, it depends on moisture content and soil composition.

Differential GPS (Differential Global Positioning System): Instrument that allows measuring geographic positions in a differential manner using a constellation of satellites and a network of fixed ground-based reference stations.

Diffuse reflection: A reflection of waves in different directions and with a distribution of energy following these directions. The distribution between specular and diffuse reflection processes is determined by the roughness and the dielectric properties of that surface.

Direct modeling: Direct modeling aims to predict the return of an electromagnetic wave to the sensor in using a set of physical characteristics representing the observed target (shape, size, dielectric properties).

Directional anisotropy: Difference between the measurements of a same variable of remote sensing according to angular configurations of viewing (zenith and azimuth directions).

Disaggregation: Method that allows improving the spatial resolution (for example, going from a coarse spatial resolution of 25-40 km to a finer spatial resolution of 1 km).

Discharge: Volume of water flowing in a river per unit of time at a given location, it is expressed in m^3/s.

Dry tropospheric correction (altimetry): This correction for the dry gas component of the atmosphere refraction takes into account the path delay in the radar return signal due to the atmosphere. It is far the largest adjustment that must be applied to altimeter measurement since its order of magnitude is about 2.3 m but its temporal variations are low (a few centimetres only).

Electromagnetic radiation: Energy wave that propagates without any material support (in vacuum) and is characterized by a period (cycle length or wavelength – in seconds) and a frequency (number of cycles per second - in hertz). The electromagnetic spectrum is the decomposition of the electromagnetic radiation in function of its wavelength or frequency. The long wavelengths (low frequencies) are radio waves and microwaves while smaller wavelengths (high frequencies) are nuclear or gamma rays.

Emissivity: Ratio between brightness temperature and surface temperature.

Energy balance: Algebraic sum of the vertical components of the energy flux densities made to, or dissipated by, a surface under radiative, conductive and convective forms.

Entropy (Cloude decomposition): Represents the random nature of the mechanisms of scattering present in the pixels whose average was calculated.

Evapotranspiration (ET): Amount of water transferred to the atmosphere by plant covers resulting from plant transpiration and soil evaporation. ET represents the actual water consumption of a crop; it depends on the needs of the crop (themselves dependent on its state of development and climatic demand) and water available in the soil. Its knowledge is essential for monitoring crop water balance.

Extinction (coefficient): Dielectric loss of a wave in a medium by absorption, scattering or internal reflection. It is the inverse of the penetration depth.

Foreshortening: Foreshortening occurs when the radar beam reaches the base of a tall feature tilted towards the radar (e.g. a mountain) before it reaches the top. Because the radar measures distance in slant-range, the slope will appear compressed and the length of the slope will be represented incorrectly at the image plane.

Fractal dimension: The fractal dimension, D, is a quantity that is intended to reflect the manner in that has a fractal set to fill the space, at all scales.

Funicular regime: Metamorphism mechanism of wet snow that occurs when the proportion of liquid water in the cover is more than 7-8% of the total volume. Liquid water forms meniscus around the contact points between the grains and in the concave areas by the capillarity force.

Geographic Information System (GIS): A geographic information system or geographical information system (GIS) is a system designed to capture, store, manipulate, analyze, manage, and present all types of spatial or geographical data, according to Wikipedia.

Geo-referencing: Operation which consists of properly positioning geographical data (like that from a satellite image) in a geographic coordinate system.

Glacier: A glacier is a persistent body of dense ice that is constantly moving under its own weight; it forms where the accumulation of snow exceeds its ablation (melting and sublimation) over many years, often centuries. Glaciers slowly deform and flow due to stresses induced by their weight, creating crevasses, seracs, and other distinguishing features. According to Wikipedia.

Global Navigation Satellite System (GNSS): Generic name used to define all global navigation systems and satisfying all the navigational needs of maritime and aeronautical users.

GNSS Reflectometry (GNSS-R): Set of remote sensing techniques based on the study of GNSS signals reflected by the Earth's surface.

GPS (Global Positioning System): Geolocation system operating at a global scale and based on a measurement of receipt elelectromagnetic signals transmitted by satellites in orbits around the Earth.

GRACE (Gravity Recovery And Climate Experiment): Geodetic mission developed by the NASA (U.S.) and the DLR (Germany) composed of two vehicles in pursuit, it was put into orbit in March of 2002 and placed at low altitude (meaning 400-450 km). It provides a global and temporal mapping of gravity field variations whose finer variations are associated with mass redistributions in the fluid envelopes of the Earth.

Hydrological model: Model allowing the representation of the water cycle processes. It usually incorporates a water balance coupled with an energy balance. We differentiate the empirical and physical models as well as global and distributed models.

Hyperspectral remote sensing: Remote sensing from a sensor allowing the observation in many narrow spectral bands (a few nanometers) in the wavelength range from the visible, near infrared, mid-infrared to the infrared thermal.

Incidence angle (or viewing angle): Angle between the incident radar beam on a surface and the line perpendicular to the surface at the point of incidence.

Independent Component Analysis (ICA): Method of statistical analysis of data, also known as blind source separation method.

Infrared: Electromagnetic radiation of a wavelength between 0.7 μm and 100 μm.

Innovation: Vector of the difference between the observation vector and the state vector of model on the observation points.

Interference Pattern Technique (IPT): GNSS-R single antenna measurement technique based on the study of variations in the signal to noise ratio caused by the interference between the direct signal and the signal reflected by the surface.

Interferometric Complex Field (ICF): In GNSS-R, the ICF represents the time series of the ratio between the correlation peak of the reflected signal and the correlation peak of the direct signal.

Interferometric GNSS-R (iGNSS-R): GNSS-R measurement technique consisting of correlating the signal reflected by the Earth's surface with the direct signal measured by a high gain antenna pointing to the source satellite.

Internal stratification: The snow is a stratified medium whose density and grain size depends on the weather conditions at the time of its fall. Stratification creates internal reflections and dielectric losses. Each layer of snow can be identified by a density or an own texture.

Inverse modeling: The used model allows to obtain the physical characteristics of the medium observed from data extracted from radar, optical or other remote sensing signals.

Inversion: Estimation method of a parameter based on the minimal error between measured and simulated signals.

Ionospheric error (altimetry): The presence of ions in the ionosphere slows down the path of the altimetric radar wave and leads to an error of several centimeters over the distance. It is corrected for the ocean by using measurements at two frequencies, or for land surfaces by modeling.

Kappa Coefficient: A quality estimator. It measures the level of agreement between the classification and the real data. It varies between 0 and 1. The closer it is to 1, the more the agreement is important.

Katabatic wind: Or also Gravity wind, a wind that carries high density air from a higher elevation down a slope under the force of gravity. Katabatic winds can rush down elevated slopes and creates roughness on all a scales.

Latent heat flux: Latent heat is energy released or absorbed, by a body or a thermodynamic system, during a constant-temperature process. SI Unit: Wm^{-2}. It corresponds to evapotranspiration, its mass flux (kg s^{-1} Wm^{-2}) is converted into Wm^{-2} by multiplying it by the latent heat of vaporization.

Layover: It occurs when the radar beam reaches the top of a tall feature before it reaches the base. The return signal from the top of the feature will be received before the signal from the bottom. As a result, the top of the feature is displaced towards the radar from its true position on the ground, and “lays over” the base of the feature.

Leaf area index (LAI): A dimensionless quantity that is defined as the one-sided green leaf area per unit ground surface area (LAI = leaf area/ground area, m^2/m^2); the green LAI is the part of the green vegetation LAI.

Liquid water content (LWC) in snow: Quantity of water at the liquid status in a snow cover. It is expressed as a percentage and can be represented in two different forms, either as a volumetric liquid water content (ratio between the volume of water and the volume of snow) or the mass liquid water content (ratio between the mass liquid water and the mass of the snow).

Local incidence angle: Angle between the incident radar beam on a surface and the line perpendicular to the illuminated surface in case of an inclined surface (local slope).

Mass balance: Is an application of conservation of mass to the analysis of physical systems. For snow, it corresponds to the balance between the inputs in the form of snow and the outputs in the form of cast iron or ice flow. It shows the health status of the glacier or ice cap.

Meteorological forcings: Environmental data allowing to run a model. In general, they are composed of rain fields, temperatures, wind, sunlight, etc. The number and nature of the required forcing data instead of forcings depend on the used model.

Microwaves: A form of electromagnetic radiation with wavelengths ranging from one meter to one millimeter; with frequencies between 300 MHz (100 cm) and 300 GHz (0.1 cm).

Moisture measurement by gravimetry: Consists of taking soil samples which are subsequently dried in an oven at 105°C for 24 hour, to deduce the moisture through the difference between the dry weight and the wet weight of samples.

Multiple scattering (optical): A photon is released outside of the medium after its interaction with several particles.

Multispectral remote sensing: Remote sensing from a sensor allowing to observe several wide enough spectral bands (of the order of several tens of nanometers) in wavelengths ranging from the visible, near infrared, mid-infrared to the thermal infrared.

Multi-view radar data (N views): Radar data where each measurement corresponds to an average (in amplitude or intensity) of “N” statistically independent samples. Radiometric accuracy is better than when the N number of views is bigger.

Net Radiation: Algebraic sum of hemispherical radiation received and transmitted by a surface on the set of wavelengths; net radiations represents the radiative energy balance available (or lost) by the surface. SI Unit**:** Wm^{-2}

Noise Equivalent Delta Temperature (NeDT): It corresponds to the smallest detectable change in temperature of a black body source in the scene.

Normalized Difference Snow Index (NDSI): Based on the difference in reflectance between the visible and the near infrared. This index allows mapping the presence of snow from satellite images.

Normalized Difference Vegetation Index (NDVI): This vegetation index is calculated from the reflectances in the Near Infrared (NIR) and the red (R): NDVI = (PIR–R)/(PIR+R). It measures the intensity of the photosynthetic activity of the surface. It is efficient to distinguish the types of covers provided that they are not too dense as tropical rainforests.

Object-oriented classification: Image classification method used for segmenting it into distinct polygonal objects and classify these segmented objects not only using radiometric properties of the image, but also the geometrical properties of segmented objects, such as their size or form.

Observation error: Difference between the observations and the model counterpart of the true state. The observation error contains the measurement error, the measurement representativeness error and the error due to the observation operator. The observation errors are assumed to be non-biased, non-correlated and with a known variance.

Observation operator: Function that allows the model to simulate the equivalent observations.

Optical reflectance: Proportion of incident light (solar illumination) reflected by the Earth's surface. It is expressed as a percentage.

Optical thickness: The optical thickness of a layer (atmosphere, vegetation) measuring the degree of transparency of the medium. It characterizes the fraction of electromagnetic radiation (or light radiation) scattered or absorbed by the components of the layer traversed.

Outlet glacier: Coastal glaciers that drain most of the continental ice.

Paludification: Accumulation of organic matter in the forest due to an increase in soil moisture.

Panchromatic: A panchromatic image is a comprehensive record of the radiation reflected by the Earth's surface in the range of wavelengths of the visible spectrum without color separation. For example, a SPOT 5 panchromatic image is acquired between 0.48 and 0.71 μm. This data is more spatially resolved that the multispectral image acquired by the same satellite.

Pedotransfer functions: In the field of soil science, pedotransfer functions are tools based on statistical relationships, which allow estimating and predicting the properties or the behavior of the soil which are difficult to measure directly, from other characteristic features of the soil readily observable on the ground.

Pendular regime: Metamorphism mechanism of wet snow that occurs when the proportion of liquid water in the cover is less than 7-8% of the total volume. Liquid water covers completely the snow grains and the capillary force becomes negligible. Snow metamorphism mechanisms are greatly accelerated.

Penetration depth: Depth at which a wave, radar or optical, penetrates in the medium. It corresponds to the depth of a medium which issues 63% of energy. It depends of the medium's properties.

Phenological stages: Development phases of a vegetation cover corresponding to the succession of periodic events during its growth cycle (for example, flowering, heading, budding, etc.); in remote sensing, it can be correlated to the evolution of NDVI.

Pixel: Unitary element constituting an image (deriving from the English phrase "picture element"). Depending on the image type and the encoding used, the pixel "contains" one or more values corresponding to the digital pixel counts in different image channels.

Polar caps: Glaciers forming an area of large ices. Due to their polar position, they play a major role in stabilizing the world's climate.

Polarimetric decomposition: Behavior of scattered signals by a target, measured by polarimetric radar by decomposing it into simpler components.

Polarization: The orientation of the electric field vector in the plane orthogonal to the wave propagation direction.

Potential Evapotranspiration: Theoretical evapotranspiration of a well water-fed surface submitted to real climatic conditions.

Principal Components Analysis (PCA): Is a statistical procedure that uses an orthogonal transformation to convert a set of observations of possibly correlated variables into a set of values of linearly uncorrelated variables called principal components. The number of principal components is less than or equal to the number of original variables, according to Wikipedia.

Profilometer: It's an instrument used to measure the roughness or micro-geometry of a surface.

Quadri-dimensional assimilation: Approach consisting to assimilate series of observations over a time window, called assimilation window (for example, the technique 4DVar).

Radar albedo: Ratio between the intensity of the electromagnetic wave scattered by a surface or a particle and the intensity emitted by the radar antenna.

Radar backscattering coefficient (denoted σ^0): Designates the Radar cross-section (RCS) by observed surface unit, and is expressed in (m^2/m^2). It is preferred instead of the RCS in the case of large surfaces. Due to its strong dynamic, it is usually expressed in decibels: $\sigma^0_{dB} = 10 * \log_{10}(\sigma^0)$.

Radar bands X, C, L, P: Frequency bands used in the microwaves, from 8 to 12.5 GHz for the X-band, 4 to 8 GHz for the C-band, from 1 to 2 GHz for the L-band and from 0.3 to 1 GHz for the P-band.

Radar cross section (RCS): Effective area of a target (in m^2) which reflects the power P_i of the incident radar wave in an isotropical manner throughout the space. For a monostatic radar at a distance R from the target, it is estimated using $RCS = 4\pi R^2 \frac{P_S}{P_i}$, P_S denoting the power received by the receiving antenna.

RADARSAT Constellation: Canadian SAR mission following the success of satellites RADARSAT-1 and RADARSAT-2 (C-band, 3 satellites, compact polarimetry).

Radiative surface temperature: Emission temperature of the surface considered as a gray body; it can be defined for a measurement in a given direction (referred to as directional radiative temperature).

Radiative transfer: Theory based on the conservation of energy flux through an elementary volume. It can describe the behavior of an electromagnetic wave passing through a scattering medium composed of dielectric particles.

Radiometric resolution: It describes the ability of a sensor to distinguish small differences in the intensity of the received energy. The maximum number of levels of intensity depends on the number of bits used. For example, a RapidEye image is encoded in 16 bits or 65536 numeric values available to transcribe the recorded intensity; a SPOT image is encoded in 8 bits or 256 numeric values.

Rayleigh approximation: Rayleigh scattering refers to the scattering of electromagnetic waves on particles sized much smaller than the wavelength (λ).

Real evapotranspiration: Vertical density of water vapor flux corresponding to the sum of evaporation from the soil, the rain intercepted by the foliage and the transpiration.

Reference ellipsoid: In geodesy, a reference ellipsoid is a mathematically defined surface that approximates the geoid, the truer figure of the Earth, or other planetary body. Because of their relative simplicity, reference ellipsoids are used as a preferred surface on which geodetic network computations are performed and point coordinates such as latitude, longitude, and elevation are defined. According to Wikipedia.

Remote sensing: The acquisition of information about an object or a phenomenon without making physical contact with the object.

Reservoir: Body of water closed by an artificial reservoir, which generally consists of a dike or a dam along a river.

Resolution in distance and on the ground: SAR images are recorded in oblique distance (compressed mode) and then they can be projected on the ground (dilated mode) in a given cartographic reference.

Retracking: Altimetry data is done by computing the departure of the waveform's leading edge from the altimeter tracking gate and correcting the satellite range measurement (and surface elevation) accordingly.

Revisit time: The satellite revisit time is the time elapsed between observations of the same point on earth by a satellite. According to Wikipedia:

Rising edge: First part of the altimetric waveform that is controlled by the height of the asperities (ocean waves or snow dunes on the polar caps), the slope of the surface and the penetration of the radar wave in the medium.

Roughness: All irregularities characterizing the geometry of a surface.

Scatterometer: Sideways-looking radar instrument used to perform an accurate measurement of the radar reflectivity of observed surfaces. The improvement on the radiometric precision is thus obtained at the expense of spatial resolution (on the order of tens of kilometers).

Segmentation: Delimitation of a polygon composed of a set of similar pixels from a spectral point of view.

Sensible heat flux: Vertical density of convective heat flux corresponding to the vertical gradient of the fair temperature. SI Unit: Wm^{-2}.

Sensor (satellite): Device embedded on the satellite that collects and measures incident radiation from the observed surface.

Sequential assimilation: The estimate at the previous time (k-1) is used to construct an a priori estimate (prediction) of the state at the present time "k" by using a numerical model. This prediction is coupled with the observation Yk measured at time k in order to produce an estimate of the state at this time.

Simple scattering (optical): A photon is disseminated outside of the medium traversed as a result of its interaction with a single particle.

Snow cover albedo: Fraction of incident solar radiation that is reflected by the snow cover. Its value ranges between 0 and 1.

Snow cover fraction: Fraction of the surface of a given area that is covered by the snow cover (for example, a pixel or a watershed).

Snow density: Mass of a snow sample per unit of volume. The mass includes that of liquid water contained in the sample and the volume includes natural air voids.

Snow depletion curve: This term generally refers to a function that links the snow depth (or its water equivalent) and the snowy surface of an area.

Snow metamorphism: Re-crystallization by mass transfer between ice crystals under the effect of thermodynamic imbalance with consequent modification of the microstructure.

Snow optical radius: The effective radius of the snow grains, i.e. the radius of ice spheres having the same SSA as the snow, is called the optical radius.

Snow specific area: Ratio between the surface of the air-ice interface, and the ice mass of a sample of spheres (m^2/kg).

Snow water equivalent: Total amount of liquid water derived from a snow cover after melting. It is expressed in mm and represents the height of the blade of liquid water per surface unit.

Snow-covered surface: Surface covered by snow.

Soil Vegetation Atmosphere Transfer (SVAT) model: Dynamic model simulating the coupled transfer of mass (water, greenhouse gas effects) and of energy within the soil and between the soil (including the root zone), the plant and the atmosphere (surface boundary layer).

Spatial resolution: The spatial resolution of a geographical document is the real size of the smallest element shown in this document. On a satellite image, the spatial resolution corresponds to the size of a side of a pixel.

Speckle (radar): The speckle is produced by random interference, constructive and destructive, coming from multiple scattering that occurs in every resolution cell of a radar image. This effect manifests itself as a salt and pepper texture on images.

Spectral resolution : Describes the finesse of a spectral band used by a sensor. If the spectral resolution is finer, the windows of the different channels of the sensor are narrower.

Specular reflection: The mirror-like reflection of light (or of other kinds of wave) from a surface, in which light from a single incoming direction (a ray) is reflected into a single outgoing direction.

Standard deviation of heights (soil): Vertical dimension of the roughness of the surface soil, corresponding to the root of the maximum of the correlation function of heights.

Station network: Set of measurement stations installed on a site for calibration and validation purposes.

Stomatal regulation: Regulation of gas exchanges between the inside and the outside of the leaf at the cellular organelles level which are called 'stomata'.

Stream: Long watercourse with many tributaries and significant flow rates that is generally drained into the sea or ocean, sometimes in a closed sea (for example, the Volga in the Caspian Sea).

Stripes (GRACE): Spectral aliasing effects manifested by the north-south strips on GRACE data. They happen due to the inadequate consideration of variations in high-frequency gravitational potential.

Strong Fluctuation Theory – SFT: Method for describing the dielectric properties of a medium in which we observe large variations of permittivity (example: a wet snow cover).

Supervised classification: Automatic learning technique that produces a classification map from a ground truth.

Supervised learning: Automatic learning method which uses reference data, also called assisted learning.

SVM (Support Vector Machines): Name given to the techniques that allow to find a separator (linear or not) between several groups of variables. The separator depends only of a few points called "Support Vectors".

Synthetic Aperture Radar (SAR): A coherent mostly airborne or spaceborne sidelooking radar system which utilizes the flight path of the platform to simulate an extremely large antenna or aperture electronically, and that generates high-resolution remote sensing imagery.

Temporal resolution: Time frequency with which a remote sensing image acquisition system can acquire an image of the same point on the Earth's surface.

Texture (of an image): Spatial arrangement of the image signal levels.

Texture index (image): Local statistic calculated in the neighborhood of a pixel of an image which takes into account spatial organization.

Thermal conductivity: Physical quantity characterizing the behavior of materials during heat transfer by conduction. It represents the energy (quantity of heat) transferred per surface and time unit under a temperature gradient of 1 Kelvin per meter. It is expressed in watts per meter-Kelvin ($W \cdot m^{-1} \cdot K^{-1}$).

Thermal Infrared Emissivity: Average emission spectrum of bodies present on Earth (3-100 μm).

Tibetan Plateau: Vast land area of over 2 million km^2 of high altitude (on average 4000 m), in a semi-arid climate regime, and dotted with hundreds of lakes. It is located at the east of the Himalayan chain.

Trailing edge: Second part of the altimetric waveform which is controlled by the depointing of the satellite, the slope of the surface, the penetration of the radar wave in the snow, the altimeter antenna gain and, in some cases, by the distribution function of roughnesses.

Transboundary basin: River (including its watershed) that traverses several riparian countries, and whose management of resources usually requires international agreements of the countries crossed. There are about 250 transboundary basins on Earth, largely being managed through a basin committee.

Tri-dimensional assimilation: Approach consisting of assimilating the observations by supposing them valid at the same time (for example, using a Kalman filter).

Unsupervised learning: Automatic learning method, sometimes referred to as "clustering", which divides a heterogeneous group of data in subsets of similar data.

Validation: Comparing an estimated parameter and field measurement in order to validate quantitatively the estimation method (inversion) used.

Visible: Part of the electromagnetic spectrum visible to the human eye, between 380 and 780 nm.

Volumetric soil moisture: Volume of water contained in a volume of soil (measured in m^3/m^3). It is often used a soil moisture index ranges between 0 and 1,which represents the saturation rate in water of the soil.

Water stress (index of): Water availability indicator; occurs when the demand for *water* exceeds the available amount during a certain period or when poor quality restricts its use.

Watershed: A basin-like landform defined by highpoints and ridgelines that descend into lower elevations and stream valleys.

Wavelength: Each electromagnetic wave is defined by its wavelength, which represents the spatial periodicity of oscillations. The wavelength is inversely proportional to frequency and is expressed in meters.

Wet troposphere correction (altimetry): The wet troposphere correction is the correction for the path delay in the radar return signal due to liquid water in the atmosphere. It is calculated from radiometer measurements and meteorological models.

List of Authors

Nicolas BAGHDADI
TETIS
IRSTEA
Montpellier
France

Alina L. BARBU
CNRM
Toulouse
France

Monique BERNIER
INRS ETE
University of Québec
Canada

Karen BONIFACE
ISTerre
University Grenoble Alpes
France

Gilles BOULET
CESBIO
IRD
Toulouse
France

Souhail BOUSSETTA
ECMWF
Reading
UK

Stéphane CALMANT
LEGOS-OMP
IRD
Toulouse
France

Jean-Christophe CALVET
CNRM
Météo France
Toulouse
France

Jean-François CRETAUX
LEGOS-OMP
CNES
Toulouse
France

Jean-Pierre DEDIEU
LTHE
CNRS
Grenoble
France

Patricia DE ROSNAY
ECMWF
Reading
UK

Yannick DUGUAY
INRS ETE
University of Québec
Canada

Marie DUMONT
CNRM
Météo France
Toulouse
France

Alejandro EGIDO
NOAA
Washington
USA

Frédéric FRAPPART
GET-GRGS
CNAP
Toulouse
France

Pierre-Louis FRISON
MATIS
University of Paris-Est
Marne-la-Vallée
France

Simon GASCOIN
CESBIO
CNRS
Toulouse
France

Lionel JARLAN
CESBIO
IRD
Toulouse
France

Yann KERR
CESBIO
CNES
Toulouse
France

Jean-Pierre LAGOUARDE
ISPA
INRA
Villenave d'Ornon
France

Delphine LEROUX
CNRM
Toulouse
France

Ramata MAGAGI
CARTEL
University of Sherbrooke
Canada

Erwan MOTTE
CESBIO
Toulouse
France

Eric MOUGIN
GET-OMP
CNRS
Toulouse
France

Thierry PELLARIN
LTHE
CNRS
Grenoble
France

Guillaume RAMILLIEN
GET-GRGS
CNRS
Toulouse
France

Frédérique REMY
LEGOS-OMP
CNRS
Toulouse
France

Nicolas ROUSSEL
GET-OMP
University of Toulouse III
France

Lucía SEOANE
GET-GRGS
CNAP
Toulouse
France

Jean-Pierre WIGNERON
ISPA
INRA
Villenave d'Ornon
France

Mehrez ZRIBI
CESBIO
CNRS
Toulouse
France

Index

E, F

G, H, I

K, L, M

N, O, P

Q, R, S

T, V, W

Scientific Committee

Simonetta PALOSCIA, CNR, Italy

Cécile LOUAMGNE, Irstea, France

Thierry PELLARIN, CNRS, France

Patricia de ROSNAY, ECMWF, UK

Jean-Christophe CALVET, Météo-France, France

Clément ALBERGEL, CNRS, France

Sonia ZINE, University Joseph Fourier University, Grenoble, France

Olivier HAGOLLE, CNES, France

Monique BERNIER, INRS, Québec, Canada

Michel GAY, CNRS, France,

Danielle DE SEVE, Hydro-Québec, Québec, Canada

Philippe MAILLARD, Universidade Federal de Minas Gerais, Brazil

Jerome BENVENISTE, European Space Agency/ESRIN, Italy

Philippe MAISONGRANDE, CNES, France

Monique DECHAMBRE, CNRS, France

Marc LEBLANC, University of Avignon, France

Sylvain BIANCAMARIA, CNRS, France

Jean-Michel LEMOINE, CNES, France

Simon BANVILLE, Natural Resources Canada, Canada

Nicolas BERGEOT, Royal Observatory of Belgium, Belgium

José DARROZES, University of Toulouse, France

Albert OLIOSO, INRA, France

Olivier MERLIN, CNRS, France

Gilles BELAUD, SupAgro Montpellier, France

Simon GASCOIN, CNRS, France

Catherine OTTLE, CNRS, France

Lionel JARLAN, IRD, France

Mehrez ZRIBI, CNRS, France

Nicolas BAGHDADI, Irstea, France

Kolibri 360 GmbH
Ferdinand-Jühlke-Straße 7
99095 Erfurt, Deutschland
produktsicherheit@kolibri360.de